Practical Microwaves

Practical Microwaves

Thomas S. Laverghetta

Purdue University–Fort Wayne

Prentice Hall
Upper Saddle River, New Jersey 07458

Library of Congress Cataloging-in-Publication Data

Laverghetta, Thomas S.
    Practical microwaves / Thomas S. Laverghetta.
        p.   cm.
    Includes bibliographical references and index.
    ISBN 0-13-186875-6
    1. Microwaves.   2. Microwave devices.   I. Title.
TK7876.L384   1996
621.381'3—dc20   95-10971
                                  CIP

Cover art: Tracey B. Ward
Editor: Charles E. Stewart, Jr.
Production Editor: Stephen C. Robb
Design Coordinator: Julia Zonneveld Van Hook
Cover Design: Julia Zonneveld Van Hook
Production Manager: Deidra S. Schwartz
Marketing Manager: Debbie Yarnell
Production Supervision: Susan Geraghty

This book was set in Times Roman and Gill Sans by The Clarinda Company and was printed and bound by R.R. Donnelley & Sons Company. The cover was printed by Phoenix Color Corp.

© 1996 by Prentice-Hall, Inc.
A Pearson Education Company
Upper Saddle River, NJ 07458

Printed in the United States of America

10 9 8 7 6 5 4 3 2 1

ISBN: 0-13-186875-6

Prentice-Hall International (UK) Limited, London
Prentice-Hall of Australia Pty. Limited, Sydney
Prentice-Hall Canada Inc., Toronto
Prentice-Hall Hispanoamericana, S.A., Mexico
Prentice-Hall of India Private Limited, New Delhi
Prentice-Hall of Japan, Inc., Tokyo
Pearson Education Asia Pte. Ltd., Singapore
Editoria Prentice-Hall do Brasil, Ltda., Rio De Janeiro

To Pat, My Love Forever . . .
Plus one day.

Contents

4   Microwave Components   117

8   S-Parameters   319

Preface

The book is written for technicians, technologists, college students, and engineers who have not taken a course in microwaves. It contains the full spectrum of microwave theory in down-to-earth, practical language. The examples, questions, and problems will assist all readers in mastering the material.

Chapter 1 is an introduction that defines microwaves and stresses the importance of dimensions in microwaves, the concept of ground planes rather than points, and the theory of skin effect.

Chapter 2 describes laminates that are used for microwave circuits. The chapter is put together in as much a generic way as possible so that it will not be outdated by new and improved materials. The chapter reflects the wide variety of microwave materials that are available.

Chapter 3 covers transmission lines. Theory and applications of coaxial lines, stripline, microstrip, and waveguide are presented with emphasis on practical applications, and with many examples to illustrate the procedures for selecting transmission lines.

Chapter 4 covers the workhorses of microwaves—the microwave components. Those components covered are directional couplers, quadrature hybrids, detectors, mixers, filters, attenuators, circulators and isolators, and chip components (capacitors, resistors, and inductors).

Chapter 5 describes the microwave transistor. Both bipolar and GaAs FETs are covered, with the addition of HEMTs (High-Electron Mobility Transistors). Chapter 6 stays with solid-state and covers microwave diodes. The PIN, Schottky, tunnel, Gunn, TRAPATT, and IMPATT diodes are covered.

The microwave engineer's most powerful tool—the Smith Chart—is covered in Chapter 7. The chart is dissected and reconstructed to show where each circle comes from. There are many examples presented since this is not a chart that you can memorize. You have to do a large number of problems to get comfortable with it.

Chapter 8 presents S-parameters. Their theory, measurement, and applications are all covered.

Chapters 9 and 10 cover the testing aspect of microwaves, with Chapter 9 devoted to test equipment (including high-frequency oscilloscopes) and Chapter 10 to representative microwave test setups.

With this lineup of material, all readers should be able to become familiar with the concepts of microwaves and feel comfortable working with microwaves in the field or in the lab.

Acknowledgments

The completion of this edition is one that has been accomplished through the cooperation and help of many people and firms. This section will attempt to identify those people. If I should overlook someone in this process, please forgive me. Your help is greatly appreciated, and the book would not have been possible without you.

I would like to thank all of the people from various companies who supplied me with information and photographs for this text: Joan King of Marconi Instruments, Ted Kidane of Hewlett-Packard, Bill Artz of XL Microwave, John Hamilton of Anritsu Wiltron, Carla Slater of Giga-Tronics, and Charles Meyer of Boonton Electronics. These people were a tremendous help. A special thank-you goes to Joanne Zelle, who did her usual superb job on the new drawings that had to be done for the text. Also, to Charles Stewart of Prentice Hall a thank-you for making it possible to put this book back into print and supply students, engineers, and technicians with much-needed microwave information.

I would also like to thank the most important people in the world to me, my family. These folks make life worth living and make ventures such as this book very rewarding. They are always there for me, even when I may not deserve them. They are truly a blessing.

Introduction to Microwaves

Objective

To introduce the reader to the field of microwaves. Since the concepts of microwaves are so different from those dealing with voltages and currents in a wire, it is the main objective of this chapter to get the reader thinking along the lines of waves, rather than currents flowing in a circuit, and then continue with detailed explanations of microwave concepts in the following chapters.

Key Terms

Distributed Component	Skin Depth
Ground Plane	Skin Effect
GHz	Wavelength
Microwaves	

1.1   Microwave Definitions

The term *microwave* brings to mind different things to different people. To the cook, it indicates a fast way of cooking; to the everyday person, it means a large variety of TV channels because of satellites and microwave dish antennas; and to the traveler, it could mean a radar speed trap on the side of the road. These are but a few of the descriptions you would probably get from a variety of people describing what the term *microwave* means to them.

All of the above meanings would, of course, be correct when who is explaining the term is considered. We must, however, arrive at a much more specific standard definition, which will apply to microwave engineering, if we are to establish a base to be used throughout this text.

The elusiveness of a standard definition for the term *microwave* can be seen if we consider the four following dictionary definitions. Note that two of the dictionaries used are common references frequently found in a home, school, or office. The other two are electronics dictionaries. The four definitions are as follows:

An electromagnetic wave of length between 50 cm (600 MHz) and 1.0 mm (300 GHz).

Oxford American Dictionary

A term applied to radio waves in the frequency range of 1000 MHz (1.0 GHz) and upward. Generally defines operations in the region where distributed constant circuits enclosed by conducting boundaries are used instead of conventional lumped-constant circuit elements.

Modern Dictionary of Electronics

Designating or of that part of the electromagnetic spectrum lying between the far infrared and some lower frequency limit; commonly regarded as extending from 300 GHz to 300 MHz.

Webster's New World Dictionary

A term used to signify radio waves in the frequency range from about 1000 MHz (1.0 GHz) upwards.

IEEE Standard Dictionary

The first observation of these definitions is the general nature of both the Oxford and Webster dictionary examples. The only statements they have in common are what they call the upper frequency limit (300 GHz). The rest of the definitions would probably require further investigation to obtain an understandable definition; however, most dictionary definitions usually are very general.

The two electronics dictionary definitions are very similar in nature and provide a more understandable definition. They will still require some explanation, but are more understandable than the previously mentioned definitions. One area they agree on is the starting point for microwaves (1000 MHz or 1 GHz). They also do not restrict the upper limit as did the general reference dictionaries.

To understand some of the terms and areas referred to in the previous definitions, consider Fig. 1–1. Fig. 1–1 is a diagram that indicates frequency ranges extending from those you can readily hear (audio) to gamma rays, which are emitted from radioactive substances. Note the space referred to as *microwaves* just below the center of the chart. It covers part, or all, of these bands designated *uhf (ultra high frequencies)*, *shf (superhigh frequencies)*, and *ehf (extremely high frequencies)*. Frequencies range from 1.0 GHz to 300 GHz. One exception can be taken with this range. The frequencies above 20 GHz are commonly referred to as *millimeter waves* rather than *microwaves* since they involve different methods of fabrication, transmission, and testing because of their high frequencies and ex-

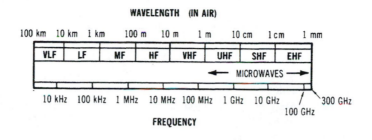

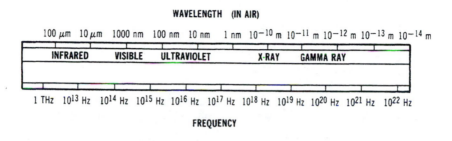

Figure I–I.   Electromagnetic spectrum.

tremely small wavelengths. The full band of 1.0–300 GHz will be used as a reference, however, for our investigation of the term *microwave*. (In practical applications microwaves range from 1–20 GHz.)

We have, thus far, used many terms to proceed toward a definition of *microwave*. Such terms as *frequency, GHz,* and *wavelength* have been generously sprinkled throughout the text and should be clarified.

1.2   Frequency Definition

The term *frequency* simply means how many times the electromagnetic wave repeats itself in one second. One example is the low-frequency end of the microwave spectrum, 1.0 GHz, which says that the wave repeats itself 1000 million times per second or one billion times in one second. This number of repetitions is the *frequency* of that particular wave. The first part of the term is the prefix G. The G stands for *giga* and means that whatever follows it is multiplied by 10^9. A listing of prefixes used in microwaves for frequency description is shown here.

Prefix	Multiply by
Mega (M)	10^6
Giga (G)	10^9
Tera (T)	10^{12}

The second part of the term (Hz) stands for *hertz* and refers to the frequency. The abbreviation (Hz) is capitalized because it comes from the name of a German scientist, Heinrich Rudolf Hertz, who, in 1887, first demonstrated that radio waves did exist and that they were able to travel through space. His name has been used to express cycles per second since the late 1960s. It is, of course, fitting to name the standard for microwave radiation (frequency) after a man who was so instrumental in developing the field of microwaves. In some texts frequency may still be referred to as *Mc (megacycles)* or *Gc (gigacycles)*. Generally, these texts were printed around 1965 or earlier. So, when you see GHz used to express frequency, think "gigahertz" (one billion cycles per second) and remember Mr. Hertz for all his accomplishments.

1.3   Wavelength Definition

The term *wavelength* is a measurement of one cycle of the frequency we have been discussing. The wavelength numbers shown in Fig. 1–1 are referred to as being "in air." That is, they are the wavelengths of those particular frequencies if, and only if, they are traveling through air. If they are traveling through any other medium, the wavelength will be shorter. To emphasize this requirement for specifying the medium that is being used, consider the following formula:

$$\lambda = \frac{C}{f\sqrt{\epsilon}} \qquad (1.1)$$

This formula says that the wavelength of a signal (λ) is equal to the speed of light (C) divided by the product of the frequency (f) and the square root of the dielectric constant of the medium (ϵ). The term *dielectric constant* will be covered thoroughly in Chapter 2 and is the *relative dielectric constant*. Unless specified, we will refer to ϵ as the relative dielectric constant throughout our discussions. For now, we can define ϵ by saying that it causes the energy to be slowed down. The dielectric constant of air is one. This, therefore, makes the formula:

$$\lambda = \frac{C}{f\sqrt{1}} = \frac{C}{f} \qquad (1.2)$$

As an illustration of the effect of dielectric constant of a medium on wavelength, consider the following example:

EXAMPLE 1–1

If we have a microwave signal traveling through air at 3.0 GHz, the wavelength would be

$$\lambda = \frac{C}{f\sqrt{\epsilon}} \qquad (C = 3 \times 10^{10} \text{ cm/sec})$$

$$= \frac{3 \times 10^{10} \text{ cm/sec}}{3 \times 10^9 \text{ Hz}\sqrt{1}}$$

$$= 10 \text{ cm}$$

This corresponds to the wavelength figure shown on the chart in Fig. 1–1 for 3.0 GHz. If we have the same signal traveling through a Teflon® medium instead of air, the wavelength would now be

$$\lambda = \frac{C}{f\sqrt{\epsilon}} \qquad (\epsilon_{Teflon} = 2.1)$$

$$= \frac{3 \times 10^{10} \text{ cm/sec}}{3 \times 10^{9} \text{ Hz } (\sqrt{2.1})}$$

$$= \frac{3 \times 10^{10}}{3 \times 10^{9} \, (1.449)}$$

$$= \frac{3 \times 10^{10}}{4.347 \times 10^{9}}$$

$$= 6.9 \text{ cm} \qquad\qquad\qquad ◆$$

The wavelength has decreased 31% simply by passing the signal through a different medium. As the dielectric constant continues to increase, the wavelength decreases even more. This pattern is how some of the lower frequency microwave circuits are able to be made smaller and smaller. The advantages of using a material for a 1.0-GHz application, for example, with a dielectric constant of 10.2 as opposed to air are readily apparent. A drastic reduction in the length of the lines used would occur, and the circuit would fit on a smaller circuit board. Conversely, as higher frequencies are used, we would want a lower dielectric constant to keep the dimensions to a practical size.

EXAMPLE 1–2

Consider the wavelength of a 20-GHz signal in high-dielectric Teflon® ($\epsilon = 10.2$) and air ($\epsilon = 1.0$):

$$\lambda_{Teflon} = \frac{C}{f\sqrt{\epsilon}}$$

$$= \frac{3 \times 10^{10}}{20 \times 10^{9} \, (\sqrt{10.2})}$$

$$= \frac{3 \times 10^{10}}{20 \times 10^{9} \, (3.19)}$$

$$= \frac{3 \times 10^{10}}{6.380 \times 10^{10}}$$

$$= 0.470 \text{ cm}$$
$$= 0.185 \text{ inch}$$

This number is rather small considering that most of the line lengths used are quarter-wavelengths (0.046 inch), eighth-wavelengths (0.023 inch), or less.

The wavelength for 20 GHz in air is much better as shown:

$$\lambda_{air} = \frac{C}{f\sqrt{\epsilon}}$$

$$= \frac{3 \times 10^{10}}{20 \times 10^{9}\,(\sqrt{1})}$$

$$= \frac{3 \times 10^{10}}{2 \times 10^{10}}$$

$$= 1.500 \text{ cm}$$
$$= 0.590 \text{ inch} \qquad\qquad \blacklozenge$$

Even though the quarter-wavelengths (0.147 inch) and eighth-wavelengths (0.073 inch) in air do not seem that much larger, they can be more easily fabricated in an air dielectric since waveguide components would be used. Also, a construction type that suspends a transmission line in air to take advantage of the low-dielectric constant without going to waveguide is available. Regardless of what type of construction would be used, the air dielectric offers many advantages over a solid dielectric for higher frequency applications.

1.4   Microwave and Skin Effect Definitions

Probably, the best definition of *microwave* for our needs is the one in the *Modern Dictionary of Electronics*:

> A term applied to radio waves in the frequency range of 1000 MHz (1.0 GHz) and upward. Generally defines operations in the region where distributed constant circuits enclosed by conducting boundaries are used instead of conventional lumped-constant circuit elements.

We have already explored the first portion of that definition regarding frequency range and bands. What makes this definition accurate and unique is that following the frequency range, it says that we cannot use conventional lumped-constant circuit elements in microwaves. One reason this statement is true is because of a phenomenon called *skin effect*.

All conductors used in dc, rf, and microwaves are of a metallic construction. At microwave frequencies, however, current does not penetrate deeply into the metal. It travels only on the outer surface, or skin, of the conductor because of the inductance set up in the conductor by the high microwave frequencies. This increased inductance causes the current to be forced to the outside edge or outer surface rather than remain in the center of the conductor.

Because of the skin effect, stripline, microstrip, and flat ribbon leads on resistors or capacitors are very efficient at microwave frequencies. You can see how the radial leads on carbon resistors or tantalum capacitors would not do the job at microwave frequencies.

The current would be only on the outside edge of the leads, and the center would be accomplishing absolutely nothing.

To understand how the skin effect works, consider the following example. Suppose you have a hollow metal tube enclosed at one end. Inside this tube a rubber ball is placed, and the open end of the tube is attached to a rod that will allow the tube to rotate on the rod (Fig. 1–2). When the tube is held stationary or rotated very slowly, the ball will remain down at the rod end of the tube. When the tube is rotated faster, the ball will begin to move toward the enclosed end of the tube. As a certain speed is reached, the ball will be at the end of the tube and will remain there. If we relate the speed of the tube to frequency and say that the faster we spin the tube the higher the frequency, we can begin to relate this action to the skin effect in conductors.

When the tube is at rest or spinning slowly, we will call this condition dc. When the tube is spinning faster and moves the ball toward the enclosed end, that is an rf frequency (vhf, uhf, etc.). When the ball is at the end of the tube, that is the microwave region.

The force that drives the ball to the end of the tube is an old friend from high-school physics, *centrifugal force* (Fig. 1–3). The force that drives the current to the outer edge of a conductor is a *microwave centrifugal force* called *inductance*. This inductance is increased by the high microwave frequencies and causes the current to flow on the outer edges of the conductor as previously described.

When you think about the skin effect as described here, one question may arise in your mind. Obviously, the current cannot just glide across the top of the conductor; it must penetrate some distance into the conductor. How deep is this current penetration into the conductor?

The current density decreases exponentially with the distance beneath the surface. At a depth δ, which is called the *skin depth,* the current decreases to $1/e$ times its surface value; that is, $1/2.718$ or approximately 36.7% of its surface value. (The term δ is also re-

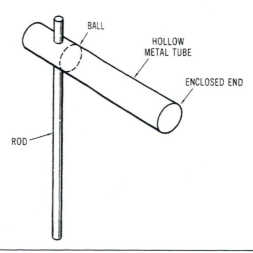

Figure 1–2. Skin effect example.

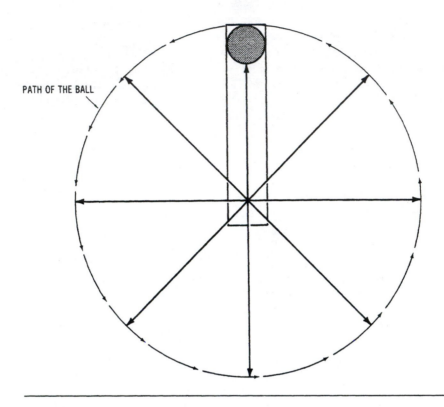

PATH OF THE BALL

Figure 1-3.   Centrifugal forces on the ball.

ferred to as *depth of penetration.*) The distance, δ, can be determined by the following relationship:

$$\delta = \sqrt{\frac{2}{\omega\mu\sigma}} \qquad\qquad (1.3)$$

where, ω is 2π times the frequency of operation in hertz,
    μ is the permeability (or rate of absorption) of the conductor in henrys/meter,
    σ is the conductance (or ability of the conductor to pass current) in mhos/meter.

If we use this formula, we can show just how small δ is for various types of conducting material. Keep in mind that the thickness of copper used in most microwave circuits for conductors is usually 1 ounce (0.0014 inch thick) or ½ ounce (0.0007 inch thick). Table 1-1 shows the depth of penetration for copper, aluminum, silver, and gold at 10 GHz.

    You can see from Table 1-1 that even when using ½-ounce copper, the current penetrates only 0.00002 inch of the total 0.0007 available (or about 2.85%). If 1-ounce copper were used (0.0014 inch), the current would penetrate only the top 1.42% of the conductor. So you can see that the *skin depth* (or *depth of penetration*), δ, is not very deep. To emphasize how the skin effect is prevalent only at microwave frequencies, consider that the

Table 1–1. Skin Depth of Four Elements

Material	σ (mhos/m)	*μ(H/m)	δ (mm)	δ (inches)
Copper	58.00×10^6	$4\pi \times 10^{-7}$	0.00066	0.000025
Aluminum	38.10×10^6	$4\pi \times 10^{-7}$	0.00080	0.000031
Silver	61.70×10^6	$4\pi \times 10^{-7}$	0.00060	0.000023
Gold	40.98×10^6	$4\pi \times 10^{-7}$	0.00050	0.000019

*This is free space permeability.

same conductor, 1-ounce copper, would have a skin depth of 0.0066 mm (0.00025 inch) at 100 MHz. This depth is 28% of the conductor's total thickness. These numbers serve to verify the idea and concept of the skin effect at microwave frequencies.

We can now spell out a definition that can be used throughout the text. Considering everything that has been covered, *microwaves* can be defined as

> Radio waves operating in the frequency range of 1.0 GHz to 300 GHz that require the use of distributed element components for circuit operation due to the skin effect phenomenon. The range of 20 GHz to 300 GHz is generally referred to as the millimeter band although still being considered part of the microwave spectrum.

1.5   Distributed Components

One area in the microwave definition has not been covered, and that is the concept of distributed components.

We are all familiar with lumped components in the field of electronics. They are the little brown resistors with colorful stripes on them and the thin yellow disc capacitors that stand up on a pc board like buildings on a city skyline, to name only two. There are many more, but I am sure you recognize them when you see them. They are termed *lumped* because the resistance, capacitance, or inductance presented by each of the components is concentrated in one confined area. A quarter-watt resistor, for example, has all the resistance concentrated in a body approximately 0.270 inch long and 0.090 inch in diameter.

Distributed components, on the other hand, are spread along an entire length or area of a circuit instead of being concentrated within circuit components. This concept should be more understandable when referring back to the discussion on skin effect. To have the circuit elements spread (or distributed) over a wide area when the only place the current will be flowing is on the outside skin of the conductor makes more sense.

Distributed constant circuits are generally those circuits constructed in microstrip and stripline. These methods of microwave transmission are discussed thoroughly in Chapter 3. Components of the microstrip type are shown in Fig. 1–4.

The components shown actually make up a low-pass filter; but, more importantly, they illustrate how components are fabricated on microwave laminates. The surface area of the C_1 pad, for example, is one plate of a capacitor, the laminate thickness sets the spacing

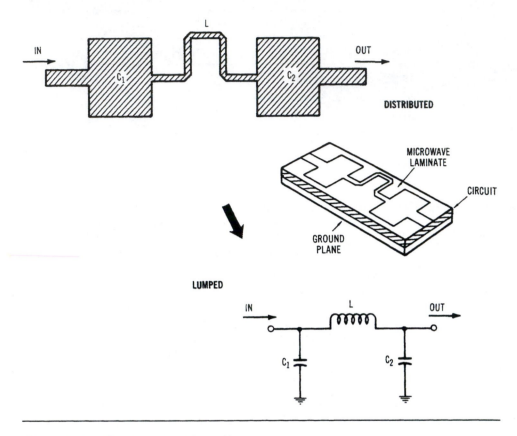

Figure 1–4. Distributed and lumped components.

and is the dielectric of the capacitor, and the ground plane on the bottom of the laminate forms the other capacitor plate. The inductor portion is a very narrow line that is a high impedance. This fact can be understood considering that the resistance of a coil increases as frequency increases. Therefore, at the high microwave frequencies the impedance is high and the line narrow. The lumped-constant equivalent circuit is shown at the bottom of Fig. 1–4.

1.6   Chip Components

Although we usually think of microwave components etched on a circuit board, other methods of obtaining the resistance, capacitance, and inductance needed are available, such as *chip* components. There are chip capacitors, for example, available from 0.010-inch square to 0.5-inch square with values from 0.1 pF to 1800 pF. Chip resistors are obtainable with dimensions as small as 0.100×0.075 inch up to 1.0×1.0 inch for high-power applications. Thin-film manufacturing processes have made microwave inductors available in sizes as small as 0.030×0.030 inch. So a combination of distributive (microstrip or stripline) and chip components gives us a microwave circuit that operates efficiently and is very small.

1.7   Grounds

One last area to be covered before we conclude our introduction to microwaves is the topic of grounds. When grounds are referred to in dc or low-frequency applications, they are considered adequate if they are connected at different points. To see a series of black wires all going to a ground lug on a connector or a ground terminal fastened to a chassis is not unusual. For dc and low frequencies, this type of grounding is perfectly all right. For microwaves, however, using ground points is nothing but trouble. When grounds are referred to in microwaves, *ground planes*, not *points*, are acceptable. Referring back to Fig. 1–4, you may remember that we said the top pad was one plate of a capacitor and the ground *plane* was the other plate. This whole area is a ground and must be in contact with ground over the entire surface if the circuit is to work properly. Consider, for example, what a capacitor would act like if one plate had its full area and the other was simply a point. You would probably agree that the component would not work well at all. The same is true in microwaves. The actual *area* of the ground is the important parameter.

Consider, also, the points that have been emphasized previously about wavelengths, quarter-wavelengths, etc., and how they dictate circuit operation. With two ground *points* a quarter-wavelength apart, you would not have a ground at one point. With a ground *plane* a quarter-wavelength long, you would still have a ground at both points of interest. So whenever you talk grounds in microwaves, talk *planes*, not points.

1.8   Chapter Summary

This chapter provides an introduction to the field of microwaves. In this chapter a definition was formulated, terms defined, and a phenomenon pertaining only to microwaves was presented. This information will be used throughout the following chapters directly and indirectly and will aid greatly in your understanding of the field of microwaves.

1.9   References

1. *Oxford American Dictionary.* New York and Oxford: Oxford University Press, 1980.

2. Graf, R. F. *Modern Dictionary of Electronics.* Indianapolis: Howard W. Sams & Co., 1977.

3. *IEEE Standard Dictionary of Electrical and Electronic Terms,* Second Edition. New York: Wiley Interscience Division, John Wiley & Sons, 1977.

4. *Webster's New World Dictionary.* Cleveland: William Collins Publishing, 1980.

Questions

1.1   Microwave Definitions

    1.   Define *microwaves*.
    2.   What do SHF and EHF stand for?

1.2   Frequency Definitions

    3.   Define *frequency*.
    4.   To convert MHz to GHz, what is the multiplying factor?

1.3   Wavelength Definition

    5.   Define *wavelength*.
    6.   Describe the characteristics of a wave at each half wavelength.
    7.   Describe the characteristics of a wave at each quarter wavelength.
    8.   Why is the value of dielectric constant important when calculating wavelength?

1.4   Microwave and Skin Effect-Definitions

    9.   Define *skin effect*.
    10.   What is δ?
    11.   What is σ?
    12.   Which material has the smallest depth of penetration?

1.5   Distributed Components

    13.   What are *distributed* components?
    14.   Distinguish between lumped and distributed components.
    15.   Why are distributed components used at microwave frequencies?

1.8   Grounds

    16.   What type of grounds are used at microwave frequencies?
    17.   Why don't we use ground points at microwave frequencies?
    18.   What is the most important parameter of a ground plane?

Problems

1.2   Frequency Definition

    1.   A frequency is given as 500 MHz. Express this in GHz.
    2.   An equation to be used to calculate a certain parameter uses frequency in MHz. If we are given a frequency of 5.2 GHz, what is this in MHz?
    3.   Our frequency of operation is 106 MHz. To use this in a specific calculation we need it in GHz. What is this in GHz?

1.3   Wavelength Definition

    4.   What is the wavelength in centimeters in air of a 3 GHz signal?
    5.   We notice that a signal in air repeats every 3.5 cm. What is the frequency of operation?

6. Our frequency of operation is 6.2 GHz, and the medium has a dielectric constant of 2.5. How long is a quarter wave line in this environment?
7. What is the quarter wavelength in Problem 6 in inches?
8. The smallest a wavelength can be is 0.3000. Our frequency of operation is 22 GHz. What should the dielectric constant be to accomplish this?
9. We have a system that uses a material with a dielectric constant of 2.95. If our wavelength is to be no larger than 1.81 cm, what frequency should we operate at?
10. If our frequency of operation is 10 GHz and the dielectric constant used is 2.3, what is the wavelength in meters?

2

Microwave Materials

Objective

To present the concepts of microwave materials and how they differ from conventional printed circuit board materials. Specific types of materials will be presented (woven, nonwoven, high dielectric, and composites) and explained so that the reader can become familiar with the properties of each and be able to make intelligent decisions concerning the *proper* material for their tasks.

Key Terms

Alumina
Anisotropy
Coefficient of Thermal Expansion
Composite
Copper Weight
Dielectric
Dielectric Constant
Dissipation Factor
Dual Dielectric

ED Copper
Microfiber
Peel Strength
PTFE
Quartz
Rolled Copper
Sapphire
Transition Region
Woven

2.1   Introduction

When we speak of microwave material, we are referring to the special circuit board that is used at microwave frequencies. This special board is called a *microwave laminate* as opposed to the epoxy pc (printed circuit) board designation normally given to material for conventional circuits.

If you have studied or worked in digital or rf areas, you will readily recognize the terminology for pc board material. You may also know that components (resistors, capacitors, etc.) may be placed on this board and wired together with very satisfactory circuit operation resulting. The normal type of epoxy board material used for these applications is termed G-10. This material is ideally suited for low-frequency operations, as will be discussed. Our discussions at this point, however, will begin with the higher frequency microwave laminates.

We have stated that *special* circuit board material must be used at microwave frequencies. This statement causes two questions to arise. First, "Why is it necessary to use special materials?" And, second, "What makes them special?"

To answer the first question, we will go back to the standard material, G-10 (also called F-R4). As mentioned before, this material is of glass epoxy composition. With copper bonded to both sides of the epoxy, its primary purpose is to provide mechanical support and separation for circuit interconnections. It is an alternative to putting a large number of standoffs in a chassis and wiring components to them but an alternative that works out very well. Conductors can be etched on both sides of the board, which enables complex circuitry to be realized. The circuitry, however, must be dc or the low-frequency variety. To understand why only low-frequency circuits are realizable, consider the following information taken from a G-10 data sheet:

Dielectric Constant	4.0–4.6
Dissipation Factor	0.018–0.025
Thickness Tolerance	0.031 in ± 0.004 in
	0.062 in ± 0.006 in

(The terms *dielectric constant* and *dissipation factor* will be completely defined in the next sections. For now, it will be sufficient to say that they are the characteristics of the board material, or laminate, that determine how efficiently the energy passes through it.)

The points of interest are the tolerances on the material. There is a 15% change in dielectric constant, the dissipation factor (or energy efficiency of the material) is high enough to be prohibitive for microwave use, and the thickness varies from 10–12%. Remember, we spoke generally about dielectric constant and thickness in Chapter 1. Many statements said that these parameters had to maintain close tolerances. The tolerances shown in the G-10 example are not nearly close enough to be used at microwave frequencies.

Since wavelengths and precise impedance lines are not a concern of dc and low-frequency designers, the G-10 material is an excellent choice. With dc and low frequency, the material does not need to hold close tolerances on dielectric constant or thickness. The value of the dissipation factor also is very acceptable for these applications since losses are not a primary concern as they are at higher frequencies.

To truly understand why special material is needed, compare the same parameters for a woven Teflon® fiberglass laminate in common use today:

Dielectric Constant	2.50 ± 0.05
Dissipation Factor	0.0015
Thickness Tolerance	0.031 in $\pm$ 0.0015 in
	0.062 in $\pm$ 0.002 in

With some quick calculating, you will see that the dielectric constant is held within $\pm 2\%$, the dissipation factor is an order of magnitude better than G-10, and the thickness is held within $\pm 5\%$ in one case and $\pm 3\%$ in the other. You can see how much tighter the tolerances are held in the microwave material than in conventional material. These tighter tolerances guarantee that when calculating a wavelength or line width for a specific material, the calculated value will be exact and remain constant over the frequency range used.

We should now be in a position to answer the two questions posed at the beginning of the section. The answers to both of the questions are very closely related. First, we use special materials for microwave circuits because these circuits depend very heavily on a tightly held value of dielectric constant and material thickness, and these special materials ensure that all lengths and widths will be exactly as designed. Second, the microwave materials are special because they do hold these tolerances (2% for dielectric constant and 3% to 5% for thickness). The dissipation factor is also much lower, which results in less loss through the material over the frequency of operation. Therefore, the two answers rely on one another and result in a closely controlled material that is ideal for microwave circuit use.

Before getting into material terms and describing specific microwave materials, we should mention that the microwave laminates used throughout the industry are controlled by a military specification, MIL-P-13949. Within this specification, a variety of material, or laminate, types is presented. Four types are in common use in microwaves. Each type has a designation assigned to it to characterize the base material of the laminate. These designations are GT, GX, GP, and GR, and they are described as follows:

GT—Glass (woven-fabric) base PTFE resin

GX—Glass (woven-fabric) base PTFE resin for microwave applications

GP—Glass (nonwoven-fiber) base PTFE resin

GR—Glass (nonwoven-fiber) base PTFE resin for microwave applications

(PTFE—Polytetrafluoroethylene—also referred to as *Teflon®*)

Notice that the GT and GX appear similar, and the GP and GR also appear similar. The only difference indicated is that the GX and GR have an added phrase, "for microwave applications." This phrase is only three words, but these are very important words for microwave circuits. These words characterize those laminates as being used specifically for microwave applications, which is not to say that GT and GB base materials cannot be used. They can be and are used, but their applications are somewhat limited. The four laminate types are shown with their applications:

GT and GP—Suitable for looser tolerance designs than conventional circuits or where circuits can be tuned as well as for general-purpose printed circuits requiring the chemical and thermal properties of PTFE material.

GX and GR—Designed for stripline, microstrip, and other applications with close tolerance requirements.

MIL-P-13949 lists ten types of base materials for microwave copper-clad laminates. The four types of laminates were chosen because of their common usage. The types GT and GX characterize a woven Teflon® fiberglass laminate while types GP and GR characterize what is termed a microfiber Teflon® fiberglass laminate. Each of these will be covered individually in the following sections of this chapter.

As was stated previously, MIL-P-13949 has been used for many years to control the quality and testing of microwave materials. Recently, the Institute for Interconnecting and Packaging Electronic Circuits (IPC), along with representatives from the microwave industry, has proposed specification IPC-L-125, specifically designed for microwave materials (called *Specification for Plastic Substrates, Clad or Unclad, for High Speed/High Frequency Interconnections*). In order to use the specification it is necessary to know what the designations in the specification represent. The example below shows these designations.

EXAMPLE 2.1

Given: The following designation for microwave material is present on a sheet of specifications:

IPC-L-125 01 B 2 A 0200 B C1 DH

What material is it?

Solution   The following description results from the above designation:

IPC-L-125	Specification Number
01	Specification Sheet Number
B	Permittivity (Dielectric Constant) This material is 2.50.
2	Permittivity Tolerance ($\pm 1\%$)
A	Pits and Dents Grade (Grade A)
0200	Dielectric Thickness (0.020″)
B	Thickness Tolerance ($\pm 0.003″$)

C1 DH   Metal Cladding Type and Copper
               Weight (ED, ½ ounce)

Thus, the material specified is a woven/glass material with a dielectric constant of 2.5 ±1%, 0.020″ thick, with ½-ounce ED copper.   ♦

We will now continue our discussions of microwave materials by providing much more detail in defining specific parameters used to describe the materials. These parameters include dielectric, dielectric constant, dissipation factor, copper weight, copper type, and anisotropy.

2.2   Dielectric Constant

The term *dielectric constant* has been used many times throughout the earlier sections and was referred to as a *relative* dielectric constant. In the previous section, we said that dielectric constant was one of the characteristics of board material, or laminates, that determine how efficiently the energy passes through the board. This statement is a good, general definition, but what is needed is a more detailed and specific definition. Looking up the definition of the term *dielectric*, we will find that the dictionary says it is a medium or substance that does not conduct electricity. This definition will be used as the basis of an accurate and understandable definition of dielectric and the dielectric constant.

When thinking of a dielectric in a nonmicrowave vein, generally a capacitor comes to mind. You may recall from basic ac theory that a capacitor was two conducting plates with a dielectric (or insulator) between them. This dielectric keeps dc from flowing through the device. The value of capacitance is determined by the following relationship:

$$C = 0.0885 \; \epsilon_r \; \frac{(N - 1)A}{t} \qquad (2.1)$$

where, ϵ_r is the dielectric constant relative to air ($\epsilon_{air} = 1.0$),
   N is the number of plates,
   A is the area of one side of one plate in square centimeters (cm^2),
   t is the thickness of the dielectric in centimeters.

Or, if using inches,

$$C = 0.225 \; \epsilon_r \; \frac{(N - 1)A'}{t'} \qquad (2.2)$$

where, ϵ_r is the dielectric constant relative to air,
   N is the number of plates,
   A′ is the area in square inches (in^2),
   t′ is the thickness in inches.

(These formulas will give the capacitance in picofarads [pF] or 10^{-12} farads. They also neglect any "fringing" at the edges of the plates.)

You will notice in these equations that the capacitance is directly related to the term ϵ_r, or dielectric constant. The higher the dielectric constant, the higher the capacitance; the lower the dielectric constant, the lower the capacitance. The association of dielectric and capacitance is definitely a valid one.

We called a dielectric an insulator in the capacitor example, and the dictionary definition said it was a material or substance that would not conduct electricity. These statements mean that dielectric is a material that causes the path of the electrical energy to be obstructed.

To understand how such a material would work, consider the following illustration. Suppose you were to shoot a hockey puck into an open net, and no one would try to stop the puck from going into the net. If the puck were lifted a foot or so into the air when it was hit, it would go into the net virtually unobstructed by the air at just about the same speed it was going when it was hit. If the puck were slid across a wooden floor, it would be slowed down somewhat by the floor and take more time to get into the net than it would when going through the air. If, finally, you tried to put the puck into the net by shooting it through the grass, it would move considerably slower, and chances are pretty good it would never reach the net. So, obviously, the type of medium an object goes through makes a difference in the speed of that object. Microwave energy passing through a medium or dielectric is exactly the same condition. The velocity of the puck depends upon the makeup of the medium through which it travels. The velocity of the microwave energy is also slowed down depending on the medium. The puck will travel in the neighborhood of 100 miles per hour in air and be slowed to nothing if shot through the grass. Microwave energy travels at the speed of light (186,000 miles/second) in air and is decreased by the ratio $1/\sqrt{\epsilon}$ when traveling through a medium with the dielectric characteristic, ϵ.

We have reached a point where we should now be able to provide an understandable definition of *dielectrics* and *dielectric constant*. A *dielectric* is a material that, due to its internal construction, obstructs the microwave energy as it passes through. This obstruction causes a reduction in speed of the energy and thus changes the characteristics of the circuit being designed.

The term *dielectric constant* comes about from the fact that the material used for microwaves must be uniform in construction, as we have discussed in Section 2.1. Uniform construction means that the dielectric construction, and thus its effect, must remain *constant*. (We compared all of our examples to air in our previous discussions because the *dielectric constant* of air is 1.0. We, therefore, reference everything to this constant value.) The term *dielectric constant* thus indicates how close the material being discussed comes to free-space conditions. The tolerance on the dielectric constant tells how close these conditions are held throughout the material.

The velocity of propagation, wavelength, and time delay through a material are affected by the factor $1/\sqrt{\epsilon}$. It can be seen that the higher the dielectric constant, the shorter the wavelength, the slower the velocity of the energy, and the more time delay there is through the circuit.

EXAMPLE 2.2

Given: A circuit uses a microwave material with $\epsilon_r = 2.20$ and operates at a frequency of 5 GHz. What is the velocity of the signal and wavelength in this material and in air?

Solution   In the material:

$$\lambda = C/(F\sqrt{\epsilon}) \qquad\qquad \text{Velocity} = C/\sqrt{\epsilon}$$
$$\lambda = 3 \times 10^{10}/5 \times 10^9\sqrt{2.2} \qquad \text{Velocity} = 3 \times 10^{10}\ (\sqrt{2.2})$$
$$\lambda = 4.04\ \text{cm} \qquad\qquad \text{Velocity} = 2.022 \times 10^{10}\ \text{cm/sec}$$

In air:

$$\lambda = C/(F\sqrt{\epsilon}) \qquad\qquad \text{Velocity} = C/\sqrt{\epsilon}$$
$$\lambda = 3 \times 10^{10}/5 \times 10^9\sqrt{1} \qquad \text{Velocity} = 3 \times 10^{10}\ (\sqrt{1})$$
$$\lambda = 6.0\ \text{cm} \qquad\qquad \text{Velocity} = 3 \times 10^{10}\ \text{cm/sec} \ \blacklozenge$$

Some common materials and their dielectric constants are shown here:

Material	Dielectric Constant (ϵ)
Air	1.0
Epoxy/glass	4.0–4.6
PTFE*/glass cloth	2.55
Polyphenylene oxide (PPO)	2.55
Irradiated polyolefin	2.32
Ceramic filled PTFE/glass	6–8
Ceramic filled PTFE	10.0
Teflon®	2.10
Polyethylene	2.25

2.3   **Dissipation Factor**

Another term that has been generously sprinkled throughout the previous sections is *dissipation factor*. To understand this term, we must think in terms of losses within a material.

Whenever energy is passed through any medium, there is a loss. The value of this loss depends on the makeup of the medium (or dielectric) under consideration. To illustrate this statement, consider a simple dc circuit consisting of a voltage source and two series resistors (Fig. 2–1). With voltage applied to the circuit, a current (energy) flows through the resistors (the medium). There will be a voltage drop across each of the resistors, creating a power loss through them also. By going through the basic Ohm's law equations, we will see that 5 volts will be dropped across each resistor (E_2, for example). The power at each resistor is the voltage squared divided by the resistance, or 0.25 watt. A loss (or dissipation) in the overall circuit is caused by current (energy) flowing through the resistors (the medium). These resistors have an ability to dissipate a certain amount of energy and to *store* or *pass* a certain amount. By "store," we do not mean retain but rather to pass the energy through the circuit. If you think about it, this is a measure of the dissipation capability of a device, or a *dissipation factor*.

---

*PTFE—polytetrafluoroethylene

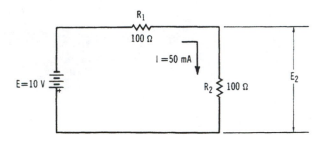

Figure 2–1. DC circuit example.

Each of these microwave laminates also has a capacity for storing and dissipating energy. What we are looking for, ideally, is a material that will dissipate and store energy because the more energy that is stored in the material, the lower loss the material has. These losses are called *dielectric losses* and cannot be compensated for because they are inherent in the material. We, therefore, want a material with a low value of dissipation factor and, thus, low losses.

The term *dissipation factor* is usually represented by the term *tan* δ, which is the tangent of the loss angle. Some representative dissipation factors of material are shown here. (All values are at 10 GHz.)

Material	Dielectric Constant (ϵ)	Dissipation Factor (tan δ)
PTFE glass	2.55	0.0018
Ceramic filled Teflon®	10.20	0.0020
Glass epoxy	4.50	0.0180
Alumina	10.20	<0.0001
Beryllia	6.80	0.0003
Saffire	11.00	<0.0001

A point emphasized in Chapter 1 when we compared G-10 glass epoxy material with microwave laminates was the higher dissipation factor of G-10 (0.018) as opposed to that of the microwave material (<0.002). Why special materials are used with microwaves should be even more clear following our discussion of the dissipation factor.

2.4   Copper Weight

Previously we referred to material and said it had 1-ounce copper on it. What exactly does that mean? This particular case says that if we had one square foot of copper it would weigh one ounce. That is, copper weight is the weight of one square foot of copper. Standard weights of copper are ½ oz., 1 oz., and 2 oz. Table 2.1 relates copper weight to the thickness of that copper.

Copper weights of 1/4 and 1/3 ounce are also available from material manufacturers. They are 0.00350 and 0.00460, respectively. Generally speaking, 1-ounce copper is the

Table 2.1 Copper Thickness and Weight	Weight (ounces)	Thickness (inches)
	0.5	0.0007
	1.0	0.0014
	2.0	0.0028

standard copper weight available from most manufacturers. The other weights can be specified for particular applications. If you are going to have a circuit with narrow lines or very narrow gaps between etched lines, you will not want to etch a lot of copper. Thus, 1/2-ounce or 1/4-ounce copper would be specified. If your circuit is required to operate at higher powers, you should specify 2-ounce copper to get the maximum amount of copper on the etched lines so they can handle this higher power.

2.5   Copper Types

If you were to order a standard piece of copper-clad microwave material, you would get material with 1-ounce copper, as we previously mentioned, and the copper would be specified as being ED (electrodeposited) copper. This is a material produced by a chemical building process in which individual copper particles are electrically joined to form the desired sheet thickness. This is similar to a process of placing small individual pieces of wet clay together to form a flat sheet of some specific thickness. The process of electrodeposition is very precisely controlled by monitoring the current and time of the process. This results in a fairly smooth copper surface that can be used for the majority of microwave applications.

A second type of copper, which must be specified when ordering, is rolled copper. This type of copper is used where copper losses are kept to a minimum by having a very uniform surface. This copper is formed by taking a copper block and running it through sets of rollers with a force pressing down from the top and bottom. This is very similar to the wringer washing machines that were used in the 1940s. The wringers would squeeze the water out of the clothes that had just been washed. The roller action for rolled copper similarly squeezes the copper to a specified thinner thickness. The force used is dependent on what final thickness is required. It can be seen that this roller action can produce a very uniform copper that exhibits low loss and is used for very critical applications.

2.6   Anisotropy

When we specify a relative dielectric constant, ϵ_r, we are concerned with the properties of the material in the X-Y plane. Anisotropy is the difference in the dielectric constant in the X-Y plane, ϵ_{xy}, compared to that in the Z-plane, ϵ_z. This is expressed as ϵ_{xy}/ϵ_z. Generally speaking, the anisotropy for nonwoven (microfiber) Teflon®/fiberglass materials is better than those of woven construction. Also, the anisotropy is lower (approaching 1.0) for low dielectric materials and increases with higher dielectric constant materials. This can be un-

Table 2.2 Anisotropy	**Material**	ϵ_{xy}/ϵ_z
	Non-woven ($\epsilon_r = 2.20$)	1.025
	Non-woven ($\epsilon_r = 2.33$)	1.040
	Woven ($\epsilon_r = 2.17$)	1.090
	Woven ($\epsilon_r = 2.45$)	1.160

derstood by remembering that higher dielectric constant materials have more fiberglass in them. Table 2.2 shows some representative relationships.

2.7   Teflon® Fiberglass Laminates

With the critical parameters of microwave laminates discussed, we can now look at specific materials that are in everyday use in the field of microwaves.

One of the most widely used types of microwave laminates is that made of Teflon® fiberglass, or PTFE/glass (polytetrafluoroethylene/glass), material. Its dielectric constant, dissipation factor, and close tolerance thicknesses make it ideal for microwave use.

We will discuss two types of fiberglass material, woven and nonwoven (microfiber). Neither will be recommended for use as a standard laminate for all designs. Each has its own merits, and both have certain areas where they should be used. It is up to the individual designer to make the proper choice.

2.7.1   Woven Teflon® Fiberglass Material

Woven Teflon® fiberglass material was chosen for discussion first because it has been in existence the longest. It first appeared on the market in the mid-1950s, with a dielectric constant of 2.65 to 2.75. It was originally developed for other uses but was found to have very useful dielectric properties although sometimes unruly. By the early 1960s, the dielectric constant of what is referred to as PTFE/glass cloth laminates had been reduced to the present 2.45 and 2.55 values now part of the GX material referred to in MIL-P-13949 discussed earlier.

This material is constructed in a woven pattern, with the fiberglass being woven into the Teflon® material. If you have ever seen the potholders children weave at school or have looked closely at the fabric on a sofa or chair, you will know what type of construction we are talking about. It is an interweaving of fibers in a very uniform pattern. This uniformity ensures that since the pattern of construction is uniform and constant, the electrical properties (dielectric constant and dissipation factor) will be uniform and constant. Fig. 2–2 shows this uniform construction. Notice the symmetry of the material. The top view, Fig. 2–2A, has the appearance of a perfect checkerboard while Fig. 2–2B, the side view, shows the arrangement of fibers at identical spacings. Fig. 2–3 is an actual photograph of PTFE magnified 25 times. With all of this uniformity in construction, it is no surprise that the pa-

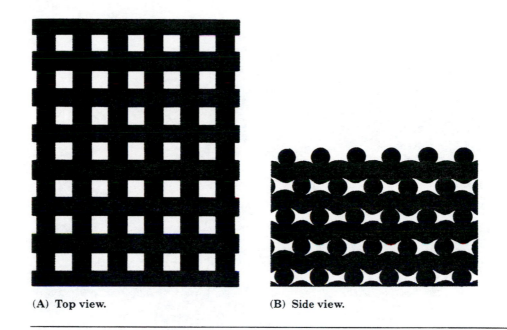

(A) **Top view.**   (B) **Side view.**

Figure 2–2.   Woven-glass laminate construction.

rameters of this material are so good. The material can be cut, sheared, and machined to various shapes to conform to particular physical requirements. The fabric edges should be removed after machining. It can also be drilled, but special care must be taken. Carbide-tip drills should be used with a slow feed and high speed because of the softness of the material.

Woven PTFE/glass cloth material is not only available in dielectric constants of 2.45 and 2.55. By varying the glass content, it is possible to produce other materials with lower dielectric constants. Two such materials have dielectric constants of 2.17 ± 0.02 and 2.33 ± 0.04. Each of these materials has less fiberglass content than the original 2.55 material. The 2.17 material has the minimum amount of fiberglass, which can be seen when its dielectric constant is compared with that of pure Teflon® ($\epsilon = 2.10$), which we used in Chapter 1. This material is as low a content as is possible to get and still call the material PTFE/glass cloth.

We have discussed previously the importance of dielectric thickness for microwave designs. The woven-glass laminates ($\epsilon = 2.17, 2.33, 2.45,$ and 2.55) are available in a wide variety of thicknesses. These thicknesses range from 0.003 inch, 0.005 inch, or 0.010 inch minimum up to 0.100 inch or 0.125 inch maximum. Each of these has specific applications in microwave design. One very big application is quadrature hybrid circuits that use 0.005 inch material and are printed on both sides, using the dielectric thickness as a fixed coupling area. We will cover quadrature hybrids in detail in Chapter 4. When we speak of

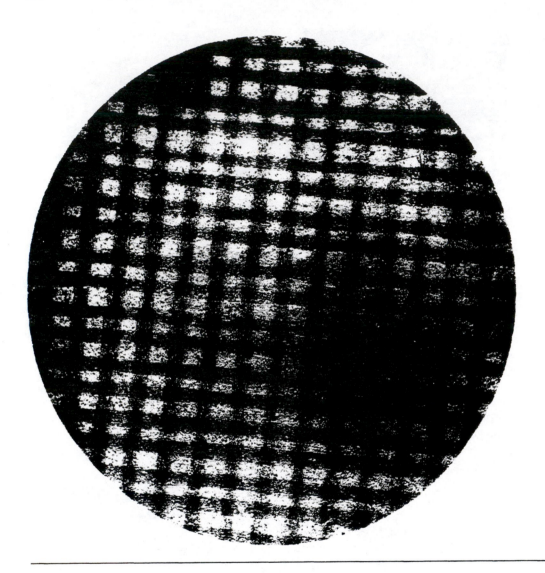

Figure 2–3. Woven PTFE *(Courtesy of Rogers Corp.).*

thickness, we are referring to the thickness of the laminates only. The copper cladding is not included in these dimensions.

   We have previously mentioned how the laminates come in a variety of thicknesses. When this material first came on the scene, there were basically only two thicknesses, 0.030 inch and 0.062 inch. If a coupler, such as the quadrature hybrid mentioned above, was to be built, it was necessary to print the same pattern on both pieces of material as shown in Fig. 2–4. Certain areas were then removed from each piece, as shown in the figure, and a dielectric material (usually thin sheets of Teflon®) of a calculated thickness was

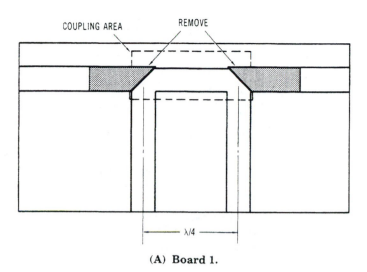

(A) Board 1.

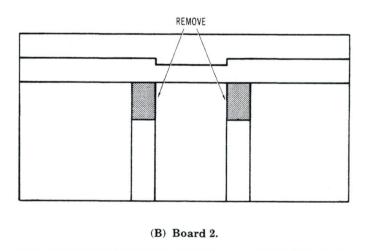

(B) Board 2.

Figure 2–4.   Early coupler construction.

put across the coupling area. The two pieces of laminate were then put together in a milled case and tested. If the coupling value was not right, the case would be opened again and a different Teflon® thickness inserted (either thicker or thinner, depending on the previously run tests). This was a slow but effective means of designing stripline components with the material that was available to the design engineer at that time.

As we have said, there are a variety of thicknesses available today that produce highly reliable and very respectable microwave components. To illustrate this range of thicknesses, consider Table 2–3 for three woven laminates.

Table 2–3. Thicknesses for Three Woven Laminates

Characteristics	Laminate 1	Laminate 2	Laminate 3
Dielectric constant	2.17 ± 0.02	2.33 ± 0.04	2.50 ± 0.04
Dissipation factor	0.0009	0.0015	0.0018
Laminate thickness range	0.010 in to 0.120 in	0.005 in to 0.125 in	0.0035 in to 0.090 in
Peel strength	8 lb/in	8 lb/in	12 lb/in

Notice the wide ranges of thicknesses available in all three laminates. Most of them come in 0.005-inch increments over this range. However, laminate 3 in some instances has less than 0.001 inch between sizes.

One other important area should be noticed in Table 2–3. As the dielectric constant (ϵ) increases, the dissipation factor also increases because of an increased glass content in the material used to increase the dielectric constant. This increased glass content could be considered to be an "impurity" being added to the relatively low loss of pure Teflon® (PTFE) material. There is a point where the quantity of glass added to the Teflon® causes excessive losses in the material. This point is not very useful at microwave frequencies because the losses are too high for practical use, so the glass loading is changed to ceramic loading of the PTFE, which produces a high-dielectric, low-loss laminate. We will cover this material in detail in a later section. (The term *peel strength* refers to the ease with which the copper can be removed from the laminate.)

Our discussions up to this point have been on woven-PTFE/glass cloth laminates. We could probably call this material the grandfather of microwave laminates because even though the 1950s produced styrene-based laminates were not intended for microwave use, the woven laminate is the material that has made microwaves the exacting and sophisticated technology it is today.

2.7.2 Microfiber Teflon® Fiberglass Laminates

The second type of Teflon® fiberglass laminate used in microwaves is of a microfiber or nonwoven construction. This type of laminate first appeared on the market in the 1960s, and when copperclad, conforms to MIL-P-13949 type GP and GR. (You will recall that GP was defined as glass [nonwoven-fiber] base PTFE resin, and GR was glass [nonwoven-fiber] base PTFE resin for microwave applications.)

A question that has probably come to mind is "What is meant by microfiber construction?" This is a structure in which glass microfibers are individually encapsulated with PTFE (Teflon®) and evenly dispersed throughout the material with random orientation in the XY plane of the laminate sheet. The material gives the general appearance of a paper pulp that has been solidified and has a very smooth texture, much smoother than that of the woven type, obviously, because of its basic construction. Fig. 2–5 shows microfiber PTFE magnified 25 times. This type of structure produces a laminate with a dielectric constant

Figure 2–5.   Microfiber PTFE *(Courtesy of Rogers Corp.)*.

variation of less than ±1% and low-dissipation factors. It is also very stable, dimensionally being able to withstand temperatures up to 550°F without warping.

Two microfiber laminates are shown in Table 2–4. Both laminates are commercially available in the thicknesses shown. The microfiber material is easily fabricated, has good dimensional stability, and is resistant to all normally used solvents and etching reagents.

This type of Teflon® fiberglass, like any other reinforced Teflon®, requires special handling and machining procedures. These Teflon® materials are of high-quality plastic, and care should be exercised to prevent contamination. Graphite deposited on the surface from

---

Table 2–4. Thicknesses of Two Microfiber Laminates

Characteristics	Laminate 1	Laminate 2
Dielectric constant	2.33 ± 0.02	2.20 ± 0.02
Dissipation factor	0.0012	0.0009
Thickness range	0.010 in to 0.250 in	0.010 in to 0.250 in
Peel strength	15 lb/in	15 lb/in

---

a pencil mark can significantly change surface electrical properties. Work areas should be kept clean. Material resting on a steel chip can, from its own weight, cause the chip to become embedded. When machining any Teflon®, the area should be well ventilated.

Excessive clamping pressure should be avoided. The use of more clamps and less pressure is recommended. Sharp tools, as stressed earlier, are very important. Sawing can be done easily with skip-toothed or coarse-bladed saws, but fine-toothed saws tend to load with material. The material can be milled and drilled, but drilled holes may become undersized due to the heat generated during drilling and, therefore, should be checked after stabilizing at room temperature.

In general, the processing ease, bond strength, and machinability of the microfiber Teflon® material is the same or better than the woven type. The material is slightly better for punching, and, once again, carbide tools are recommended.

We have discussed the properties and characteristics of both woven and microfiber PTFE/glass cloth copper-clad laminates. Both materials are excellent for microwave applications. There are, however, those who prefer woven and those who prefer microfiber. Those who prefer woven claim that the microfiber material is too soft and that in a stripline application, the copper lines will cause impressions in the opposite ground-plane board and change the characteristics. Conversely, those who prefer microfiber claim that the woven laminate is too "bumpy" because of its interwoven construction and that the copper is not properly bonded to the laminate. As with any product, there are good and bad points. We will present two woven materials and two microfiber materials with matching characteristics and let you make up your own mind. To be sure we are comparing apples and apples, we have chosen materials with similar dielectric constants (ϵ).

You can see from Table 2–5 that there is very little difference between microfiber and woven laminates. For some time, there was a belief that microfiber was the only low-dielectric constant, low-loss PTFE product. Table 2–5, however, indicates that the dielectric constants and dissipation factors of the two materials are very close. As we have previously mentioned, when the glass content is reduced to reduce the dielectric constant, the dissipation and loss decrease accordingly. Thus, the two materials are very similar in dielectric constant and dissipation factor.

Another previously questionable feature was the hardness of microfiber materials. One test of a material's hardness is the amount of warpage it undergoes. Microfiber laminates

Table 2–5. Comparison of Microfiber and Woven Laminates

Characteristics	Microfiber Laminate 1	Woven Laminate 1	Microfiber Laminate 2	Woven Laminate 2
Dielectric constant	2.33 ± 0.02	2.33 ± 0.04	2.20 ± 0.02	2.17 ± 0.02
Dissipation factor	0.0012	0.0015	0.0009	0.0009
Thickness range (inches)	0.010–0.250	0.005–0.125	0.010–0.250	0.010–0.120
Peel strength (lb/in)	15	8	15	8
Water absorption (%)	0.02	0.024	0.02	0.024
Heat capacity (BTU/lb/°F)	0.23	0.24	0.23	0.24
Surface resistance (Megohms)	3×10^8	$>10^4$	3×10^8	$>10^4$

can withstand temperatures up to 550°F with no warpage. With this type of performance, there is no way that a 1-ounce run of copper is going to make an impression in the laminate. So the question of hardness has been resolved.

What the whole bundle of arguments boils down to is that the material used depends on personal application and preference. Sometimes woven Teflon® will give the best performance; sometimes the application will demand microfiber laminates. The main idea is to remain flexible and open-minded so that you are able to evaluate each material and make an objective judgment that will be the best for the job you have to do.

2.7.3 Composite Materials

As the microwave industry advanced over the years and low-cost laminates were more in demand, the need for new material technology became very apparent. To accommodate these needs, composite PTFE materials were introduced. These are ceramic fluoropolymer composites and allow the material manufacturer to obtain a higher dielectric constant, good loss characteristics, and excellent material stability.

One outstanding parameter of composite materials is its temperature stability with regards to dielectric constant. The standard PTFE/glass laminate has a transition region at approximately 19 degrees Celsius (68.2 F) where the dielectric vs. temperature curve has a hitch in it. This can be seen in Figure 2.6 (curve A). It can be seen that the dielectric constant is very consistent at lower and higher temperatures but has a transition point (discontinuity) at the 19 degree point, which can cause problems in some design areas. The composite material is shown in curve B. This is much more constant and shows no indication

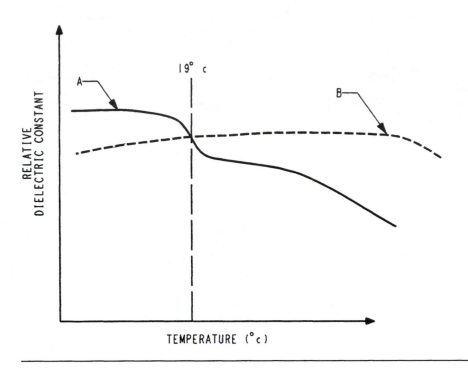

Figure 2–6.   Dielectric constant vs. temperature.

of a transition at any temperature. This is a terrific advantage for circuits that must operate over a wide temperature range.

2.8   High-Dielectric Teflon® (PTFE)

In our earlier discussions, we discussed increasing and decreasing the glass constant of a substrate (laminate) to increase or decrease the dielectric constant. We said that the dissipation factor, and thus the loss, also increased or decreased as the glass content changed. These discussions were based on a fiberglass filler material. If we now use a ceramic material as a filler for the PTFE (Teflon®), we have a whole new ball game.

Ceramic materials have been used in microwaves for some time. The base material for chip resistors and capacitors is ceramic as well as circuit board materials such as alumina, to be covered in the next section. So the idea of using ceramic in one form or another is not new to microwaves. It is new, however, to the fabrication of a material with a high-dielectric constant, low-dissipation factor, and the flexibility of woven or microfiber PTFE/glass.

The material that meets all of the criteria mentioned previously is a ceramic-loaded PTFE or the commonly referred to terminology of *high-dielectric Teflon®*. This copper-clad laminate has been a very popular material since its arrival on the microwave scene in the mid-1970s. It has found many applications where size or form factor was of vital importance to a microwave system. It has both the high-dielectric constant (10.2) and low loss of its long-time predecessor, alumina, with the added advantages of machinability and flexibility.

As we have mentioned previously, the high-dielectric material can be machined; that is, it can be milled, drilled, or routed as desired. Machining is a large advantage since many complex shapes are virtually impossible with alumina. These complex shapes are practically second nature with the ceramic-loaded PTFE.

A typical set of specifications for this high-dielectric material follows:

Characteristics	Specifications
Dielectric constant	10.2 ± 0.5
Dissipation factor	0.002
Thickness range	0.005 in to 0.100 in
Peel strength	7–8 lb/in
Coefficient of thermal expansion	20–25 ppm/°C
Processing	Standard pc methods

Note from these specifications that the material can be processed by standard pc methods. This processing is a big plus since no special etching facilities or deposition units are needed. Only the standard printing and etching capabilities are needed.

Another term that needs to be defined is *peel strength*. Peel strength is the force needed to pull the copper free of the laminate. It is an important parameter since the copper lines must stay adhered to the substrate for proper operation and not break loose at high temperatures, either naturally or during assembly soldering.

The last term that needs an explanation is the *coefficient of thermal expansion*. This term means how much the material changes physically with a change in temperature. To understand how important this term is, consider the fact that each of these laminates must be secured to a ground plane, which is usually the case of the circuit being used. The circuits are usually secured with a conduction epoxy that attaches the substrate firmly and still maintains a good ground. At elevated temperatures, the case and the laminate both expand. The trick is to have both of them expand at approximately the same rate. The coefficient of thermal expansion of the high-dielectric laminate (20–25 ppm/°C) is nearly the same as that of aluminum (28 ppm/°C). (Ppm/°C is parts per million per degree Celsius of temperature.) Since the large majority of cases are milled from aluminum, this is a significant plus for this type of laminate.

The high-dielectric PTFE material is also available with a metal backing already attached to it. (Manufacturers will also put metal backing on low-dielectric materials if requested, but generally it is reserved for higher dielectric material because of the ideal coefficient of thermal expansion and that high-dielectric materials are generally more flexible and need to be more rigid.) Fig. 2.7 shows the board arrangement for metal-clad high-dielectric laminates. It can be seen that the figure consists of a conventional microwave laminate with copper-cladding that is physically attached to a metal ground plane. This metal can be aluminum, copper, or brass, depending on individual requirements. This, as previously stated, results in a microwave laminate that is very dimensionally stable and has the bottom of the final case already built in.

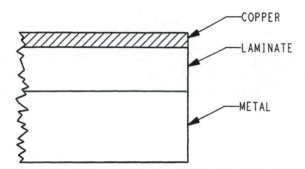

Figure 2–7.   Metal-clad high-dielectric laminates.

An interesting comparison is shown in Table 2–6. It compares the high-dielectric material with the regular PTFE laminate discussed in Section 2.6.

Table 2–6 shows that the high-dielectric material has advantages in some areas and disadvantages in others. It provides a considerable space savings at 3 GHz, for example, over PTFE but should only be used in a microstrip mode and not stripline, primarily because the line widths in stripline are smaller (0.0085 inch for 50 ohms) than those in microstrip (0.022 inch for 50 ohms). It would be much more practical to use PTFE if stripline usage were required.

Table 2–6.   Comparison of High-Dielectric Material with PTFE Laminate

Characteristics	High Dielectric (0.025 in Thick)	PTFE (0.030 in Thick)
Dielectric constant	10.2 ± 0.5	2.50 ± 0.05
Dissipation factor	0.002	0.0015
Peel strength	12 lb/in	10 lb/in
Thickness range	0.005 in to 0.100 in	0.003 in to 0.100 in
Coefficient of thermal expansion	20–25 ppm/°C	130 ppm/°C
λ at 3 GHz	1.233 in	2.491 in
λ at 10 GHz	0.369 in	1.212 in
25-ohm line (stripline)	0.033 in	0.072 in
25-ohm line (microstrip)	0.075 in	0.198 in
50-ohm line (stripline)	0.0085 in	0.046 in
50-ohm line (microstrip)	0.022 in	0.088 in
70-ohm line (stripline)	0.0055 in	0.025 in
70-ohm line (microstrip)	0.009 in	0.050 in

On the other hand, a point we brought up before concerning coefficient of thermal expansion would make the high-dielectric material an ideal choice if the circuits were to operate over a wide temperature range. We noted earlier that the high-dielectric material has a coefficient of thermal expansion similar to aluminum (20–25 ppm/°C for the laminate, 28 ppm/°C for aluminum). These similar coefficients of thermal expansion ensure that the laminate and the case expand and contract at basically the same rate by the same amount. There would be very little danger of the board material being separated from the case ground plane over temperature. The PTFE material, however, has a drastically different coefficient of thermal expansion (130 ppm/°C). This coefficient would undoubtedly cause problems over a wide temperature range and thus should only be used when temperature extremes are small.

So the high-dielectric PTFE laminate has good and bad points. Generally, the good points far outweigh any limitations the laminate may have. This laminate has provided the microwave industry with a material that reduces the size of many existing circuits, provides a low-loss material that can be drilled or cut to any desired pattern, and eliminates problems that arise when standard-sized alumina substrates are strapped together. All in all, it can be considered to be the greatest breakthrough in microwave laminates since PTFE replaced the styrene materials in the mid-1950s.

2.9   Dual-Dielectric Material

Dual dielectric is a microwave material that has a thick metal center (core) with laminates on both sides of it that have two different dielectric constants. The core may be brass, aluminum, copper, invar, or copper-invar-copper. The metal provides dimensional stability to the material and also makes it much more rigid. The choice of dielectrics is dependent on the application. If, for example, you had one circuit operating at 800 MHz and another operating at 15 GHz and needed them to operate in close proximity to each other, one side of the material should have $\epsilon_r = 10.2$ (for the lower frequency) and the other have $\epsilon_r = 2.17$ (for the higher frequency). Figure 2.8 shows the arrangement of a dual-dielectric material.

2.10   Alumina Substrates

Alumina, aluminum oxide, Al_2O_3—these are some of the terms used to describe the material that began the true development of microstrip as a major method of microwave circuit fabrication. This development occurred in the mid-1960s. Prior to this time alumina was available, but the purity of the material (96%) was not high enough. The purity figure arrived at in the mid-1960s was 99.5% and has since moved up to 99.9%. The figure of 99.5% is still the one that generally is used today, however. The idea of impurities is an important one, as you can see from previous discussions. The dielectric constant and the loss tangent (dissipation factor) of the alumina are affected by these impurities. Some of the impurities used are silicon (Si), titanium (Ti), magnesium (Mg), and calcium (Ca). By choosing the proper impurities, the ideal dielectric constant and loss characteristics can be achieved.

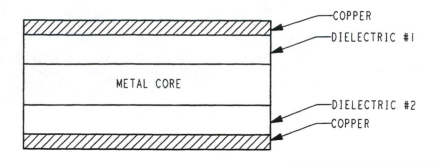

Figure 2–8.   Dual-dielectric material.

A typical data sheet for alumina substrates used for microwave circuitry is shown here:

Material	Al_2O_3 (99.5%)
Outside dimensions	0.5 in $\times$ 0.5 in $\pm$ 0.002 in
Thickness	0.025 in $\pm$ 0.0005 in
Surface finish	2 microinches
Camber	0.001 inch per inch
Dielectric constant	10.1 $\pm$ 0.25
Dissipation factor	0.0001
Metallization	Chrome—gold

You can see right away that, as mentioned previously, we are using 99.5% pure alumina. This particular substrate will be ½ inch $\times$ ½ inch and nominally 0.025 inch thick. This thickness, along with 0.050 inch, has become a standard for the microwave industry since both the alumina and high dielectric laminates are available in those thicknesses, and they can be directly interchanged.

The surface finish on this particular piece is 2 microinches, which is the thickness of the metallization on top of the substrate. Most substrates are metallized on both sides and can be metallized on the edges also if required. Metallized edges are very handy when designing a bandpass microstrip filter that requires that quarter-wave stubs be grounded. These are transmission lines that are one quarter-wavelength at the frequency of operation. You can design the stubs, lay them out so that they come to the edge of the board, and then let the edge metallization provide the ground connection to the bottom of the substrate. When grounding is not needed, such as the input and output of the filter, the metallization can be removed during processing. Such a filter is shown in Fig. 2–9.

The camber is shown as 0.001 inch per inch. *Camber* basically means how flat the material is over its entire surface. You do not want a substrate to bow on the ends; you want it to sit flat on a ground plane. This is why camber is so important.

The dielectric constants of the alumina substrate and the high-dielectric Teflon® laminates are comparable (10.1 $\pm$ 0.25 and 10.2 $\pm$ 0.5); however, the dissipation factors are considerably different. The low value of the dissipation factor results in a very low-loss

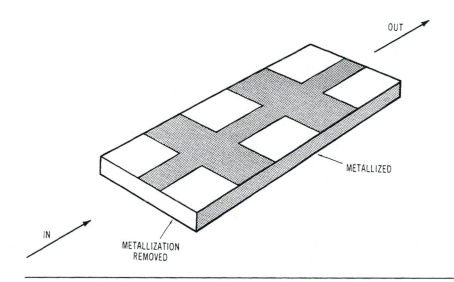

Figure 2–9.   Example of metallized alumina filter.

material. Low loss is the primary characteristic that makes alumina suitable for microwave circuits (although its thermal properties are also very desirable). You might ask why high-dielectric laminates would ever be used instead of alumina if this substrate has such low losses. The answer was covered in the previous section and can be classified as flexibility. Working with alumina can be likened to working with a dinner plate. Without the proper equipment, the plate cannot be drilled or cut. If you try to machine it, you will probably end up with a pile of ceramic chips suitable only for pasting on a canvas to produce modern art. As we stated previously, the high-dielectric laminates are soft and machinable, making nonstandard shapes and sizes very attainable. For standard shapes and sizes, however, certainly use the ½-inch × ½-inch or 1-inch × 1-inch or whatever sized alumina substrate that would fit the need.

The last term on the data sheet is metallization. What you see on a finished substrate is a gold layer on top of a ceramic material. There is, however, more to the substrate than meets the eye. Gold will not bond to ceramic material. It is like trying to use a felt-tipped pen to write on a piece of aluminum to mark one of your circuits. The marks can be seen, but they wipe completely off very easily. The same is true of gold and ceramics. To get the gold to become a part of the ceramic substrate, another material is needed, which usually is chrome. Gold will bond to chrome, and chrome will bond to the ceramic substrate. Chrome, therefore, creates an ideal condition. So there is a hidden aspect of the ceramic substrate that is very important to the operation of that substrate. As a matter of fact, without that tiny piece of chrome, we would end up with a piece of ceramic with no metallization at all.

The alumina substrate has been looked at and found to be an excellent material for microwave circuit fabrication. It has a high-dielectric constant, low-dissipation factor, and an additional advantage of metallization on top, bottom, and edges. It is, however, very fragile and can be drilled or cut only with the appropriate equipment.

All in all, the alumina substrate can be considered to be an asset to the microwave field, which finds many new applications every day.

2.11 Sapphire and Quartz Substrates

The majority of microwave circuits use either an alumina substrate or a PTFE laminate (either low- or high-dielectric constant). Occasionally, there is a need for an alternate substrate because of special requirements for a particular system. Two of these substrates are *sapphire* and *quartz*.

Sapphire is a material with a dielectric constant in the neighborhood of 11.0. We say in the neighborhood because it has been recorded to be ranging from 9.3 to 11.7, depending on its makeup. Its dissipation factor is comparable to that of alumina although not quite as good. It is also very difficult to fabricate sapphire substrates because of their brittleness. As we previously said, the use of sapphire is usually limited to special applications.

Quartz (SiO_2) is a material with an in-between dielectric constant. The majority of the laminates and substrates we have discussed have either a dielectric of 10 or are in the range of 2 to 2.5. The dielectric constant of quartz is 3.78. This value makes it appropriate for some higher-frequency applications where wavelengths are smaller and a lower dielectric constant is needed to keep the dimensions within reason.

The dissipation factor of quartz is a rather unimpressive 0.0015. This dissipation is comparable to PTFE laminates with a dielectric constant of 2.3 to 2.35 and somewhat better than that of high-dielectric Teflon® laminate. Quartz is another material that is difficult to fabricate because of its brittleness, which also limits its use to special applications.

These two substrates have a wide range between their dielectric constant and dissipating factors. One point they have in common is that they are difficult to fabricate. They also are more costly to use than previously discussed materials.

So if you have a special application where a standard sapphire or quartz substrate would fit into a circuit, and their properties are suitable, by all means use these substrates. They are both excellent substrates for microwave applications and should be used where applicable.

2.12 Chapter Summary

We have covered a very important building block in the understanding of microwaves and microwave circuits. Since we are not able to connect components together in microwaves with point to point wiring as in dc and low-frequency applications, we must have materials that do this connecting for us efficiently and accurately. The materials covered in this chapter will meet all of these requirements. They have a variety of dielectric constants, dissipation factors, thicknesses, metallization, and machinability properties. Each has a specific application area in the field of microwaves.

It is not the intention of this chapter to recommend any specific type of material for specific applications. It is, rather, our intention to present the materials available, point out their advantages and disadvantages, and let you make your own decision.

With the material covered in this chapter, you should begin to understand how specialized a material must be to be used in a microwave application. Since we now have a medium to travel in, we can now look forward to the components to be used with this material.

2.13 References

1. Laverghetta, T. S. *Microwave Measurements and Techniques.* Dedham, MA: Artech House, 1976.

2. MIL-P-13949E, "Plastic Sheet, Laminated, Metal-Clad (For Printed Wiring) General Specifications For." 10 Sept., 1976.

3. *Microwave Materials.* 3M Company, St. Paul, MN.

4. Olyphant, M., Jr., Demeny, D.D., and Nowicki, T. E. "Epsilam-10-A High Dielectric Constant Conformable Copper-Clad Laminate." Cu-Tips #6, 3M Company, Electronics Products Division, St. Paul, MN.

5. Olyphant, M., Jr., and Nowicki, T. E. *MIC Substrates-Review.* 3M Company, Electronics Product Division, St. Paul, MN, 1980.

6. *Reference Data for Radio Engineers,* Sixth Edition. Indianapolis: Howard W. Sams & Co., 1975.

7. *RT/Duriod, The Non-Woven, Glass Microfiber-Reinforced PTFE Structure.* Rogers Corp., 1981.

8. Laverghetta, T. S., Microwave Materials & Fabrication Techniques, Second Edition, Norwood, MA., Artech House, 1991.

Questions

  2.1 Introduction

     1. Why do we need special materials for microwaves?

     2. What makes microwave materials special?

     3. What is the designation for PTFE/glass woven material for microwave use when referring to MIL-P-13949?

     4. What is the designation for PTFE/glass nonwoven material for general use?

     5. If the IPC-L-125 designation for a material is IPC-L-125 01 B 2 A 0150 B C1 DH, how thick is the material?

  2.2 Dielectric Constant

     6. Define *dielectric*.

     7. Define *dielectric constant*.

     8. How does the dielectric constant affect the velocity of propagation?

     9. How does the dielectric constant affect the wavelength?

2.3   Dissipation Factor

    10.   Define *dissipation factor*.

    11.   Which dissipation factor is better for microwave applications, 0.0115 or 0.00115? Why?

2.4   Copper Weight

    12.   Define *copper weight*.

    13.   What copper weight is recommended for high-power applications?

    14.   What copper weight is recommended for narrow line applications?

2.5   Copper Type

    15.   What is ED copper?

    16.   What is rolled copper?

    17.   Which copper is smoother, ED or rolled?

2.6   Anisotropy

    18.   Define *anistrophy*.

    19.   Which material has better anisotropy, woven or nonwoven?

2.7   Teflon® Fiberglass Laminates

    20.   Describe the construction of a woven PTFE material.

    21.   How is the dielectric constant of a PTFE/glass laminate increased?

    22.   Describe the construction of a microfiber PTFE material.

    23.   How are composite materials different from conventional PTFE/glass materials?

2.8   High-Dielectric Teflon®

    24.   What is different about high-dielectric constant materials compared to conventional PTFE/glass materials?

    25.   Define *coefficient of thermal expansion*.

    26.   What is the advantage of having metal-backed laminates?

2.9   Dual-Dielectric Materials

    27.   Describe the construction of dual-dielectric materials.

    28.   Why would dual-dielectric materials be necessary?

2.10   Alumina Substrates

    29.   What is another name for *alumina*?

    30.   Define *surface finish*.

    31.   Define *camber*.

    32.   Why is there more than one metal on an alumina substrate?

2.11   Sapphire and Quartz Substrates

    33.   What is the dielectric constant of sapphire?

    34.   Compare sapphire to alumina for loss characteristics.

    35.   Compare quartz to conventional materials in regards to dielectric constant and losses.

Problems

1. A circuit operates at 6.5 GHz and has a dielectric constant of 3.5. What is the length of a quarter wavelength in inches?

2. What is the length of the quarter wavelength in Problem 1 if the circuit is in air?

3. The circuit to be used can have lines whose wavelengths do not exceed 7.25 cm. The frequency of operation is 3.1 GHz. What value of dielectric constant is needed to fulfill this requirement?

4. The dielectric constant of a material is found to vary from 3.0 to 3.3. The frequency of operation is 5 GHz. The length of one wavelength cannot exceed 3.3 cm. Will this condition be acceptable?

5. We have two types of material available. One has a dielectric constant of 2.17, and the other has a dielectric constant of 6.0. If we are operating at 14 GHz, which material should be used? Why?

3 Transmission Lines

Objective

To present basic transmission line theory with general terminology introduced and defined. Coaxial cables (flexible and semirigid), stripline, microstrip, and waveguide are also presented along with their associated terms and theory. Practical applications of each type of transmission line are included in each section.

Key Terms

Attenuation	Reflection Coefficient
Characteristic Impedance	Return Loss
Coaxial Cable	Semirigid Cable
Cut-off Frequency	Short Circuit
Effective Dielectric Constant	TE Mode
Ground-Plane Spacing	TM Mode
Open Circuit	Transmission Line
Pulse Response	VSWR

3.1 General Theory

When working with dc, digital, or low-frequency circuitry, the *transmission line* used to tie two points together is a wire. In the broadest of terms, this wire would be classified as a

43

transmission line since, in reality, it is a device that transfers energy from one point to another. Our applications in microwave, however, require a much more expanded definition than this.

To understand why a more comprehensive definition is needed, an understanding of the makeup of transmission lines that are used at microwave frequencies is necessary. To understand this makeup, we should concentrate on what goes on in these transmission lines between the two points that are being tied together.

To begin with, we could certainly use a plain wire to tie two microwave components together, but neither of them would work right because there would be too much loss in the wire. Energy would be transferred, but not much. So obviously a transmission line involves more than just connecting two components together.

We should probably call a transmission line, at this point, a device that transfers energy from one point to another with a minimum amount of loss. In other words, the transmission line must be *efficient*. Efficiency is the real key to a microwave transmission line whether it is a coaxial line, stripline, microstrip, or waveguide. If the transmission line is not efficient, or is lossy, it cannot be used for microwave applications.

To see how a transmission line differs from a plain wire, refer to Fig. 3–1. Four components we must be concerned with are in this line: capacitance (C), inductance (L), resistance (R), and conductance (G). When an energy source is applied to the line, it "sees" this section in Fig. 3–1. The voltage generated within the line is a *voltage wave*, and the current induced is a *current wave*. The resulting flux linkages per unit of current set up by these waves account for the inductance (L) component shown. The shunt capacitor (C) is the result of an actual capacitor being set up by the transmission line. One plate is the center conductor, and the other plate is the ground plane. Between these plates is a dielectric material. If you recall the definition of a capacitor, this is exactly what one is. If the dielectric medium (discussed in Chapter 2) between the conductors is not perfect, the conductive element (G) results. This result is referred to as conductance per unit length of line. This unit length may be feet, inches, meters, or any other convenient unit. Siemens per meter (S/m) is an example of this definition. The final component is series resistance (R). This term is present because the perfect conductor cannot exist in

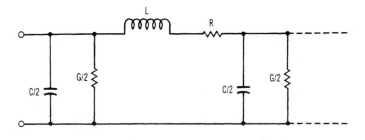

Figure 3–1.   Equivalent circuit of a transmission line.

practice. This resistance depends on the resistivity of the material used, the length and cross section of the conductor, and the distribution of currents in the cross section. (This series resistance and the inductance are shown on one line only but are actually in both lines.)

In most microwave transmission lines, the losses are extremely small as previously mentioned, one criterion for using them at microwave frequencies. If the parameters R = 0 and G = ∞ are adapted, the equivalent circuit would be as shown in Fig. 3.2 and consist of only sections of series inductance and shunt capacitance. In reality, however, this can never occur, but the transmission line designer strives for very low resistance and a high conductance. This results in minimum loss and maximum efficiency.

3.1.1   Reflections and Characteristic Impedance

A phenomenon that occurs in microwave transmission lines and that is not as prominent in low-frequency or dc application is the idea of *reflections*. You will recall from your very first ac and dc courses that there is a maximum transfer of power when the load resistance (R_L) equals the source resistance (R_S). If this condition did not exist, the load would not have a maximum transfer of the power being generated by the source, and there would be power dissipated within the circuit itself. This idea can be carried over for the operation of a microwave transmission line.

Every transmission line has a resistance associated with it that comes about because of its construction. This resistance is a *characteristic impedance*. Most of the microwave systems have a characteristic impedance of 50 ohms. Fig. 3–3 shows why the 50-ohm number is used as a standard. The ideal combination for a transmission line would be to have a high-power handling capability and a very low loss. Unfortunately, both of these conditions do not exist at the same time. Fig. 3–3 shows that the maximum power handling capability occurs at an impedance of 30 ohms while a minimum attenuation is when the impedance is 77 ohms. The 50-ohm figure is sort of a compromise between the two ideal values. At 50 ohms, there is a reasonably low attenuation, and there still is adequate power handling capability. (It is interesting to note that cable TV systems use a 75-ohm characteristic impedance. This 75-ohm

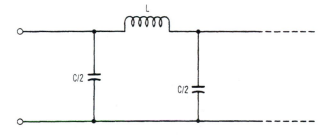

Figure 3–2.   Microwave transmission line with zero loss.

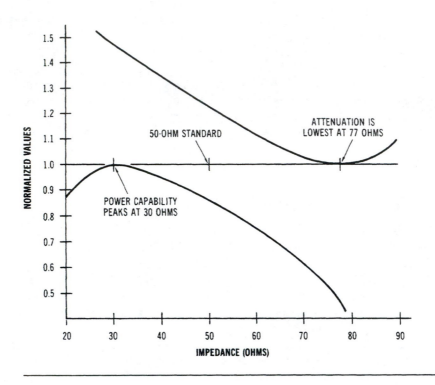

Figure 3–3.   Attenuation and power capability.

characteristic impedance is a good choice since the low loss is beneficial to an industry that relies on many miles of transmission lines for its operation.)

Whenever the resistance at the load is different from the characteristic impedance (50 ohms in the case of most microwave systems), there are reflections set up because all of the energy cannot be absorbed in this noncharacteristic impedance load or mismatch. In other words, the energy that is not absorbed by the load is reflected back toward the generator. This reflection must be kept to a minimum since too large a reflection will degrade the source output level, change the frequency, and even cause damage to the source if the reflections are of sufficient amplitude and proper phase.

To illustrate the idea of mismatches and reflections, we will consider a variety of loads on a transmission line and describe the existing conditions. Keep in mind that all reflections are caused by discontinuities in the line and that we are creating discontinuities at the end of the transmission line in the following examples.

3.1.2   VSWR, Reflection Coefficient, and Return Loss

Before moving into the examples, several terms should be defined; the first is *VSWR (voltage standing-wave ratio)*. The VSWR is officially defined as the ratio of maximum and minimum voltages on a transmission line. It can be expressed mathematically in many ways.

$$\text{VSWR} = \frac{E_{max}}{E_{min}} = \frac{E_i + E_r}{E_i - E_r} = \frac{I_{max}}{I_{min}} = \frac{1 + \rho}{1 - \rho} \tag{3.1}$$

where, E_{max} is the maximum voltage on the line,
E_{min} is the minimum voltage on the line,
E_i is the incident voltage magnitude,
E_r is the reflected voltage magnitude,
I_{max} is the maximum current on the line,
I_{min} is the minimum current on the line,
ρ is the reflection coefficient magnitude.

This equation is all very nice and orderly, but what does it mean? To understand the terminology, refer to Fig. 3–4. A string is tied at one end to a wall. The other end is taken and initially held taut, as shown in Fig. 3–4A. If you now flip your wrist, a wave is created on the string that will move down the line toward the wall, as shown in Figs. 3–4B, C, and D. When the wave hits the end of the line (the wall), it then bounces back down the string toward your hand, as shown in Fig. 3–4E. If you continue to flip your wrist, there will be waves moving down and back on the string. The only difference between the waves going down and those coming back is the amplitude of the one traveling back,

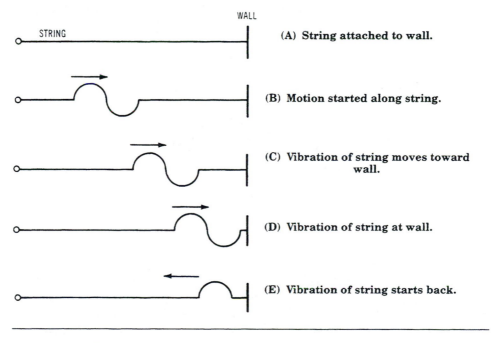

Figure 3–4.   Standing wave example.

which will be lower. It should be obvious by now that if we had a means of absorbing some of this wave at the wall, there would be a much lower amplitude wave coming back than if none of it was absorbed. We can obtain a measure of the absorption of the wall if we measure the amplitude of the wave traveling down the string, measure the amplitude of the wave coming back, and compare the two. In other words, we make a *ratio* of the waves on the string. We are not able to measure a specific wave that moves on the line, so an average reading is taken that makes the waves appear to stand still. We now have a ratio of forward and reverse waves in a standing position; we have a *standing-wave ratio (SWR)*. On this wave, we can measure a maximum voltage (E_{max}), a minimum voltage (E_{min}), a maximum current (I_{max}), a minimum current (I_{min}), an incident amplitude (E_i), or a reflected amplitude (E_r). With these values, we can then use the previously presented equations to find VSWR.

In this example, the SWR is measuring the ability of the wall to absorb the waves of the string. In a microwave circuit, the *standing-wave ratio* is measuring the ability of the circuit to absorb (or allow to pass) microwave energy from an rf source. It basically is measuring how much resistance the circuit offers to the signal and the source that is driving it; that is, it is measuring the *microwave impedance*.

We have, sometimes, referred to this ratio as a *standing-wave ratio*. Most commonly, the parameter is called VSWR *(voltage standing-wave ratio)*, as used at the beginning of the discussion. Oddly enough, the *voltage* standing-wave ratio is many times thought of as the ratio of the amplitude of the *power* transmitted in a forward direction to the *power* reflected back toward the source due to a mismatch at the load. It is, however, the ratio of the incident (or transmitted) *voltage* to the reflected *voltage*. The power term comes into play when using *return loss,* which is an expression for the amount of power (in dB) *returned* from a device back toward the source due to a mismatch. It is defined as follows:

$$\text{return loss} = -20 \log_{10} \rho \qquad (3.2)$$

Before leaving VSWR, we should say that it is always referenced to 1.0. A VSWR reading could, therefore, be 3.2:1, for example.

The term ρ is the *reflection coefficient*, as we have previously referred to it in the VSWR equation. The reflection coefficient is a number between 0 and 1.0, which is a percentage of the energy that is reflected back to the source.

Fig. 3–5 is an illustration of the principle of reflections that should help you understand the reflection coefficient. The flashlight is a microwave source, the venetian blind is the input (or output) circuit of a device, and the wall is the device itself. The object, of course, is to get as much light (microwave energy) to the wall (microwave device) as possible. Fig. 3–5A is the ideal case. The blinds are completely open, and almost all of the light gets through to the wall, which is analogous to a system where the device is perfectly matched to the source. That is, the impedance of the device (Z_L) exactly equals that of the energy source driving it (Z_S). This equality between impedance and source, of course, is the ultimate goal in all cases; however, this goal is not reached. Usually, the case shown in Fig. 3–5B applies. Here it can be seen that the blinds are partially closed so that some of the light is reflected off them and back to the source. Obviously, there is not as much energy

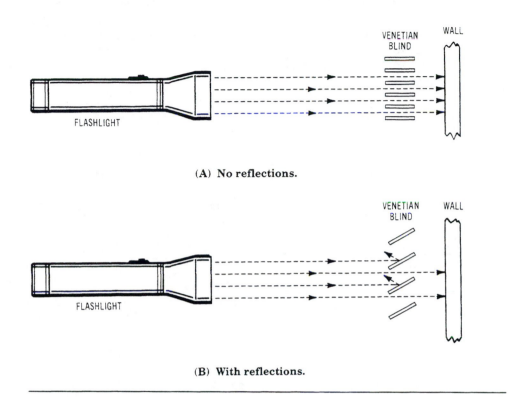

(A) No reflections.

(B) With reflections.

Figure 3–5.   Reflection example.

getting to the device as in Fig. 3–5A. Fig. 3–5B is a very good representation of reflection as a function of impedance. Always a certain amount of energy is reflected back to the source since a perfect device is not possible. The measure of these reflections gives the reflection coefficient and a measure of the value of the microwave impedance of a device.

Examples of reflection coefficient, VSWR, and return loss are shown in Table 3–1. It can be seen from this table how a low value of VSWR can mean a system well matched to the source impedance. As the VSWR increases, the return loss decreases since the ratio of the reflected power to the incident power increases, and the reflection coefficient increases toward 1.0 since a greater percentage is reflected back.

With the ideas of VSWR, reflection, and reflection coefficient fresh in our minds, let us proceed to the transmission line examples to characterize different loading conditions on the line.

3.1.3   Transmission Line Examples

Following are six examples that illustrate the relationships and conditions that exist with various loading conditions on the transmission lines.

Table 3–1. Examples of Reflection Coefficient, VSWR, and Return Loss

VSWR	Return Loss (dB)	ρ
1.0	∞	0.00
1.05	32.25	0.024
1.10	26.44	0.047
1.15	23.12	0.069
1.20	20.82	0.091
1.25	19.08	0.111
1.30	17.69	0.130
1.40	15.56	0.166
1.50	13.98	0.200
1.75	11.20	0.275
1.92	10.03	0.315
3.00	6.02	0.500
5.00	3.52	0.666
10.00	1.74	0.818
∞	0.00	1.000

Matched Load   To establish a frame of reference, we will begin with the ideal condition, $Z_L = Z_0$, a perfect match. Fig. 3–6A shows this matched condition. From this figure, you can see that certain relationships and conditions exist with a matched load on the transmission line. These relationships and conditions are as follows:

- There are no reflections. All of the power transmitted down the line is absorbed by the load. Thus, none of it is reflected back.

- There are no standing waves. Remember that for standing waves to occur there must be an incident wave (E_i) and a reflected wave (E_r). Since there are no reflections, there is no reflected wave and, thus, no standing waves.

- VSWR is equal to 1.0. We said previously that

$$\text{VSWR} = \frac{E_i + E_r}{E_i - E_r}$$

where, E_i is the incident wave,
  E_r is the reflected wave.

Since we have already determined that $E_r = 0$ (no reflections, no reflected wave), the equation now becomes

$$\text{VSWR} = \frac{E_i + 0}{E_i - 0} = \frac{E_i}{E_i}$$

$$= 1.0 \text{ or } 1.0{:}1$$

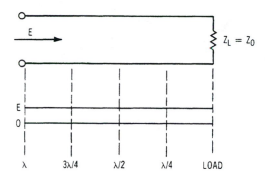

(A) Matched load.

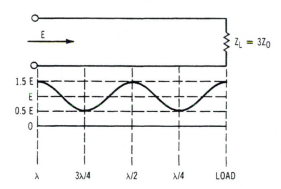

(B) Load resistance greater than impedance ($Z_L > Z_0$).

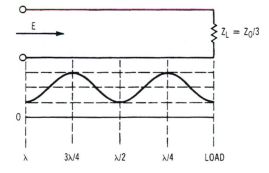

(C) Load resistance less than impedance ($Z_L < Z_0$).

Figure 3–6.   Standing waves for resistive loads.

- The reflection coefficient is zero. If we transpose a previous equation, we can obtain an expression for reflection coefficient. The equation was

$$VSWR = \frac{1 + \rho}{1 - \rho}$$

where, ρ is the reflection coefficient.

If we solve for ρ, we will obtain the expression

$$\rho = \frac{VSWR - 1}{VSWR + 1} \tag{3-3}$$

We have already determined that the VSWR = 1. Therefore,

$$\rho = \frac{VSWR - 1}{VSWR + 1}$$

$$= \frac{1 - 1}{1 + 1}$$

$$= \frac{0}{2}$$

$$= 0$$

- The input impedance of the line does not depend on the length of the line. Since the load is matched to the line, the line appears to be the characteristic impedance anywhere along its entire length.

Load Resistance Greater Than the Characteristic Impedance   The second condition is shown in Fig. 3–6B. In this case the load is a resistance greater than the characteristic impedance (Z_0). In our example, it is $3Z_0$. The conditions that exist on the line with this high resistance termination are as follows:

- There is a voltage maximum at the load and every half-wavelength back. This is understandable since the terminating resistance is higher than the characteristic impedance, so a higher voltage will be present across the higher resistance. This maximum will repeat every half-wavelength, as do all transmission line parameters when the line is not terminated in its characteristic impedance (typically 50 ohms).
- The amplitude of the reflected wave, the magnitude of the reflection coefficient (ρ), and the VSWR will depend on the value of Z_0 and Z_L. The example we use in Fig. 3–6B is a value of $Z_L = 3Z_0$. In other words, in a 50-ohm system, Z_L is 150 ohms. In this example the amplitude goes between 0.5E and 1.5E. This amplitude, of course, would be different if Z_L were two or five times Z_0. (E is the amplitude of the incident, or forward voltage on the line.) The reflection coefficient for $Z_L = 3Z_0$ would be calculated as follows:

$$\rho = \frac{Z_L - Z_0}{Z_L + Z_0} = \frac{150 - 50}{150 + 50} \tag{3.4}$$

$$= \frac{100}{200}$$

$$= 0.5$$

Notice how the reflection coefficient varies if you consider that when $Z_L = 2Z_0$ (Z_L is 100 ohms), ρ is 0.3 and when $Z_L = 4Z_0$ (Z_L is 200 ohms), ρ is 0.6. Obviously, as the resistance increases, the reflection coefficient approaches 1.0 and a VSWR of infinity.

The VSWR of a line with $Z_L > Z_0$ also will vary with the values of Z_L and Z_0. If we go back to our example of $Z_L = 3Z_0$ shown in Fig. 3–6B, we can calculate VSWR as follows:

$$\text{VSWR} = \left| \frac{\rho + 1}{\rho - 1} \right| = \left| \frac{0.5 + 1}{0.5 - 1} \right|$$

$$= \left| \frac{1.5}{0.5} \right|$$

$$= 3.0{:}1$$

To see how the VSWR varies with resistance, consider that when $Z_L = 2Z_0$, the VSWR is 1.85:1, and when $Z_L = 4Z_0$, the VSWR is 4.0:1. Obviously, at this point, the VSWR is approaching the value of infinity, which is present when an open circuit (a very large resistance) is the end of the line. An impedance of 5000 ohms ($Z_L = 100 \, Z_0$) on the output of the transmission line would yield the following parameters:

$$\rho = 0.98$$

$$\text{VSWR} = 99.0{:}1$$

So, you can see the line fast approaching an open circuit condition ($\rho = 1.0$, VSWR = infinity).

- Wavelength locations for voltage and current maxima and minima follow the same pattern as for an open circuit except for the amplitude. After covering the previous section, this statement is very understandable. We have already said that as the resistance increases, it approaches an open circuit. The reflection coefficient (ρ) and VSWR also have the characteristic of an open circuit. It is, therefore, only logical that the voltage and current on the line would behave the same as an open circuit with a lower amplitude.

Load Resistance Less Than the Characteristic Impedance   The third condition is shown in Fig. 3–6C. In this case the line is terminated in a resistance less than the characteristic impedance (Z_0). In this example, it is $Z_L = Z_0/3$. The conditions that exist on the line with a pure resistance less than Z_0 are as follows:

- The incident and reflected voltages are 180° out of phase at the load and at $\lambda/2$ intervals from the load. This phase shift can be understood considering that a low terminating resistance results in a low voltage developed across this termination. To account for the entire wave coming from the generator, a larger voltage must be reflected back, resulting in a maximum voltage reflected at the load where a minimum incident voltage is present (180° out of phase).

- A voltage minimum is present at the termination. The voltage minimum goes back to the previous explanation of a low resistance and, thus, a low voltage at the termination.

- The magnitude of the reflection coefficient (ρ) and VSWR will depend on the values of Z_0 and Z_L. Once again, we will use our examples in Fig. 3–6. The condition under consideration is that of Fig. 3–6C, where $Z_L = Z_0/3$. If we calculate the reflection coefficient for $Z_0 = 50$ ohms, we will get the following number:

$$\rho = \frac{Z_L - Z_0}{Z_L + Z_0} = \left| \frac{16.66 - 50}{16.66 + 50} \right|$$

$$= \frac{33.34}{66.66}$$

$$= 0.5$$

The VSWR for this condition is

$$\text{VSWR} = \left| \frac{\rho + 1}{\rho - 1} \right| = \left| \frac{0.5 + 1}{0.5 - 1} \right|$$

$$= \left| \frac{1.5}{0.5} \right|$$

$$= 3.0{:}1$$

Different values may be used as was done in the $Z_L > Z_0$ case, and various radii of reflection coefficient and VSWR will result. For example, when $Z_L = Z_0/5$, ρ is 0.667 and the VSWR is 4.88. When $Z_L = Z_0/10$, ρ is 0.818 and VSWR is 9.98. As the Z_L decreases, the reflection coefficient approaches 1.0 and the VSWR increases toward infinity. It is important to note that when the load impedance was $3Z_0$ there were values of VSWR = 3.0 and $\rho = 0.5$. Similarly, when $Z_L = Z_0/3$ the VSWR was also 3.0 and the reflection coefficient was also 0.5. So it can be said that these calculations will not tell you if the load impedance is greater or less than the characteristic impedance. They will only tell you there is an impedance that is not the characteristic impedance.

Reactive Load   A fourth condition occurs when a pure reactance is placed at the end of a transmission line, that is, an inductor or capacitor. Fig. 3–7A and B, respectively, show

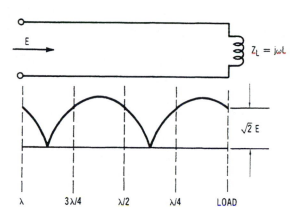

(A) Inductive load.

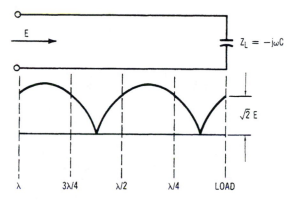

Figure 3–7. Standing waves
for reactive loads.

(B) Capacitive load.

these cases. As before, these are certain relationships that exist with a transmission line terminated in anything but a matched load. For the reactive load they are as follows:

- The incident and reflected voltages are out of phase except at E_{max} and E_{min}, where they can be either in phase or 180° out. This statement can be understood when you consider that a reactance may have either a leading phase angle or a lagging phase angle depending on whether it is an inductance (Fig. 3–7A) or a capacitance (Fig. 3–7B). The amount of reactance will also change the angle of the reflected voltage. There will, therefore, be very few points where a well-defined phase relationship is present. Because there are few well-defined phase relationships, the voltages at the load are not at a maximum or minimum.

- The VSWR is infinite, and the reflection coefficient is 1.0. To understand these statements, let us consider the last section first ($\rho = 1.0$). In this case all of the power coming down the transmission line is reflected back when it reaches the termination. If we go back to basic ac theory, once again you will recall that a reactive component (inductor/capacitor) does *not* dissipate any power. Only the resistive portion of a circuit will dissipate power. Since none of the power can be absorbed in the component to be dissipated, the entire amount must be reflected back. Therefore, the reflection coefficient (ρ) must be maximum, or 1.0. To extend this idea further, we have said that the VSWR and the reflection coefficient are related as follows:

$$\text{VSWR} = \frac{\rho + 1}{\rho - 1}$$

If the reflection coefficient (ρ) is now equal to 1.0, the equation now becomes

$$\text{VSWR} = \frac{1 + 1}{1 - 1} = \frac{2}{0}$$

$$= \infty$$

Short Circuit   The final two conditions for terminating a transmission line are shown in Fig. 3–8. These conditions run the whole range of possible loads on a line—from short circuit to open circuit. The first condition, a short circuit, is shown in Fig. 3–8A and is used in many test setups to perform an initial calibration. This condition is widely used in VSWR and impedance measurements. The relationship on page 57 exists when a transmission line is terminated in a short circuit.

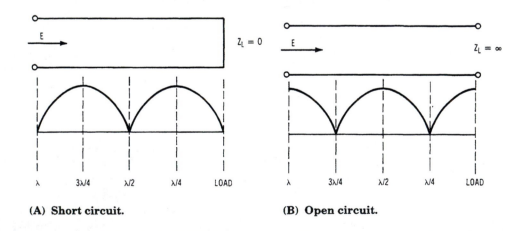

(A) Short circuit.   (B) Open circuit.

Figure 3–8.   Standing waves for a short and open circuit.

- The reflection coefficient is 1.0 at an angle of 180°. This condition is comparable to the previously discussed $Z_L < Z_0$ condition. Remember, as the value of Z_L decreases, the reflection coefficient (ρ) approaches 1.0 and the VSWR increases toward infinity. When the load is a short circuit, it has reached 0 ohms, the lowest point, and thus has a value of reflection coefficient equal to 1.0 and a VSWR equal to infinity.

- A voltage minimum is present at the load and at each half-wavelength along the line. This statement is understandable since a zero resistance load (short circuit) will not develop any voltage across it. We, therefore, have a voltage minimum at the load and, thus, every half-wavelength thereafter. Similarly, a short circuit will cause a current maximum, which also will be repeated every half-wavelength.

- The input impedance of the line is a function of the line length. To understand this statement, refer to Fig. 3–9. This figure shows a transmission line terminated in a short circuit. As you can see, it has a low resistance at the load because it is a short (0 ohm). Back toward the generator, it does not, however, remain a short. It presents an inductive reactance that increases in value until it reaches a quarter-wavelength back and then becomes a very high resistance. (This figure verifies the statement made earlier of how an impedance is opposite every quarter-wavelength.) At this point the impedance is equivalent to an open circuit. As we continue back down the line, the impedance is a high value of capacitive reactance decreasing in value until at the half-wavelength point, we have a low impedance (resistance) that

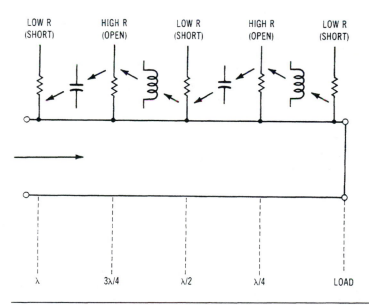

Figure 3–9.   Shorted termination line impedance.

duplicates exactly the load impedance, or a short circuit. The entire process then repeats until one full wavelength (λ) is covered. (This phenomenon of changing impedance will be graphically presented when the Smith chart is covered in Chapter
7.) So you can see that the input impedance truly is a function of the length of the
line.

Open Circuit   The final condition to be covered will be when a transmission line is terminated in an open circuit. (This condition could also be thought of as having no termination at all.) Fig. 3.8B shows the standing wave pattern on such a line. The following relationships exist when the transmission line is terminated in an open circuit:

• The magnitude of the reflection coefficient (r) is 1.0. By looking at Fig. 3–8B,
  you can readily see why the reflection coefficient is equal to 1.0. There is no load
  on the line, so it stands to reason that everything transmitted down the line will be
  reflected back; thus, with a reflection coefficient of 1.0 as in previous cases, the
  VSWR is infinite, which is the result of this relationship:

$$\text{VSWR} = \frac{\rho + 1}{\rho - 1} = \frac{1 + 1}{1 - 1}$$

$$= \frac{2}{0} = \infty$$

• The first voltage minimum is located one-quarter wavelength from the load, and
  the first current minimum is one-half wavelength back. These statements can be
  understood by referring to Fig. 3–8B. The voltage at the output is at a maximum
  because there is an open circuit present. Moving down the line one-quarter wavelength, we would, thus, find a voltage minimum present. Similarly, with an open
  circuit at the load, no current will be at that point or a current minimum. The next
  current minimum back from the load will be one-half wavelength toward the
  generator.

• The input impedance is a function of line length. Fig. 3–10 shows a transmission
  line terminated in an open circuit. Just as in the case of the shorted line, the impedance varies moving down the line. At the load, there is a high resistance since
  there is an open circuit present. Moving back toward the generator, the line appears
  capacitive reactive. This value decreases until, at the quarter-wavelength point,
  a low resistance (short circuit) is seen. The impedance then goes inductive and
  increases in value until, at the half-wavelength point, the high resistance of the
  open circuit is once again present. This sequence repeats itself every half-wavelength for the entire length of the line. Thus, the impedance at the input depends
  on the length of the line and how far from the open the source is placed. Once
  again, this concept will be graphically presented in Chapter 7 when we cover the
  Smith chart.

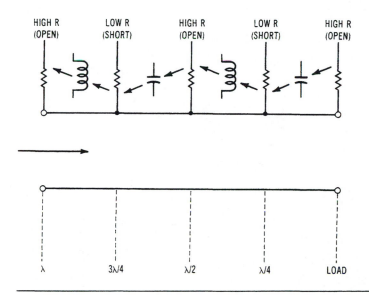

Figure 3–10.   Open circuit termination line impedance.

We have characterized six different conditions for terminations on a transmission line: matched, resistive greater than the characteristic impedance (Z_0), resistive less than the characteristic impedance, reactive (inductive and capacitive), short, and open circuit. All of these have unique relationships that make them useful for particular applications.

It should be pointed out that the relationships previously discussed for transmission line loading will apply to any microwave transmission line. These relationships will also apply equally to coaxial lines, stripline, microstrip, and waveguide because all of these will pass energy from one point to another with a minimum amount of loss—that is, they are all *transmission lines*.

3.2   Coaxial Cables

A *coaxial cable* is defined as a transmission line in which one conductor completely surrounds the other; the two conductors are coaxial and separated by a continuous solid dielectric or by dielectric spacers. Fig. 3–11 shows how this definition holds true.

The first conductor in Fig. 3–11 is the center conductor with the diameter d, which is the outside diameter of the conductor. In most cables, the conductor is usually thought of as being copper. In one sense, this assumption is true. Usually, the material itself is copper but often is tinned or covered with a high-conductance material such as silver. The following list shows material used for both flexible and semirigid coaxial cables. (Both types will be covered in this section.)

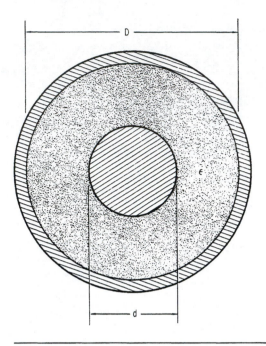

Figure 3–11.   Coaxial cable diagram.

Flexible	*Semirigid*
Aluminum	Silvered copper
Copper	Silvered copper-covered steel
Silver-covered copper	Silvered beryllium copper
Copper-covered steel	Bare copper
Tinned copper	
Silver-covered cadmium bronze	

A wide variety of material is used for the center conductors of coaxial cables. One should be available for your particular application.

The second conductor, the one that completely surrounds the first conductor, is the outside shield in Fig. 3–11. The term D is the inside diameter of the shield. In flexible cable, this conductor is a woven braid and is a solid tube structure in semirigid cable. Materials used for this outer conductor are listed once again for both flexible and semirigid cables:

Flexible	*Semirigid*
Aluminum	Copper
Copper	Gold-plated copper
Silver-covered copper	Silver-plated copper
Copper-covered steel	Cadmium-plated copper
Tinned copper	Tin-plated copper

Silver-covered cadmium bronze	Nickel-plated copper
Galvanized steel	Tin-lead plated copper
	Aluminum
	Stainless steel
	Beryllium copper

The last part of the definition says that the two conductors are separated by a continuous dielectric designated in Fig. 3–11 by ϵ. Some of the material used for both flexible and semirigid cables is shown along with their dielectric constants (ϵ):

Polystyrene ($\epsilon = 2.56$)

Polyethylene ($\epsilon = 2.26$)

Polytetrafluoroethyelene (e 5 2.10) (This material was discussed in detail in Chapter 2. It was called PTFE or Teflon®.)

You can see from these discussions how the definition holds true for both flexible and semirigid coaxial cables.

Before breaking the discussions up into those for flexible and semirigid, it would be beneficial to investigate some of the parameters that must be considered when choosing a coaxial cable. There is much more to choosing a cable than saying that a 50-ohm cable two feet long with SMA connectors on it is needed. Here are listed no fewer than a dozen terms associated with coaxial cables. We do not say that everyone of them need be considered for every application, but usually at least half of them should be. Study the terms and their explanations and make your cable choices accordingly. The terms associated with coaxial cables are as follows:

Characteristic impedance (Z_0)

VSWR

Attenuation

Velocity of propagation

Capacitance and inductance

CW power rating

Cutoff frequency

Pulse response

Shielding

Cable noise

Flexibility

Operating temperature range

We will now undertake a brief explanation of each of these terms and how they relate to coaxial cables.

3.2.1   Characteristic Impedance

We briefly covered characteristic impedance earlier in this chapter and described it as a resistance associated with a cable that comes about as a result of its construction. Fig. 3–12 illustrates how this is possible. Consider the following relationship:

$$Z_0 = \frac{138}{\sqrt{\epsilon}} \log_{10}\left(\frac{D}{d}\right) \tag{3.5}$$

The ratio D/d, which is the ratio of the inner diameter of the outer conductor (D) to the outer diameter of the inner (or center) conductor (d), and the dielectric constant determine the characteristic impedance of the cable.

Most microwave systems use a characteristic impedance (Z_0) of 50 ohms. There are systems, however, that use 75 ohms and 93 ohms for a characteristic impedance. (All of our references to characteristic impedance will be to 50 ohms throughout this text.)

To illustrate the validity of the relationship shown in the equation and to prove how the cable impedance depends on the internal construction and dimension, we will use two examples of cables that are used in microwave applications. One will be a flexible cable (RG-214), and we will jump ahead a little and use 0.085 semirigid cable for our second example.

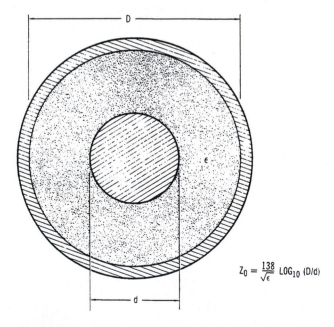

$$Z_0 = \frac{138}{\sqrt{\epsilon}} \text{LOG}_{10} \, (D/d)$$

Figure 3–12.   Cable impedance.

EXAMPLE 3.1 (RG-214)

- Inner diameter of outer conductor (D) 5 0.285 inch.
- Outer diameter of inner conductor (d) = 0.087 inch.
- Dielectric—PTFE (polytetrafluoroethylene)—ϵ = 2.10.
  (Dimensions and dielectric information are from manufacturers' data sheets.)

$$Z_0 = \frac{138}{\sqrt{\epsilon}} \log_{10} \left(\frac{D}{d} \right)$$

$$= \frac{138}{\sqrt{2.10}} \log_{10} \left(\frac{0.285}{0.087} \right)$$

$$= \frac{138}{1.440} \log_{10} (3.275)$$

$$= 95.83 \, (0.5153)$$
$$= 49.38 \text{ ohms} \qquad \blacklozenge$$

You can see that Z_0 comes out pretty close to an ideal 50-ohm cable. Variations in dielectric constant and some dimensional tolerances bring the impedance to a value termed nominally 50 ohms.

EXAMPLE 3.2 (0.085 SEMIRIGID)

Before calculating the impedance of this cable, it would be appropriate to say that the designation of 0.085 semirigid cable comes from the outside diameter of the cable itself. That is, this cable has an outer conductor outside diameter of 0.085 inch. All of the semirigid cables will be discussed in detail in a following section.

- Inner diameter of the outer conductor (D) = 0.066 inch.
- Outer diameter of the inner conductor (d) = 0.0201 inch.
- Dielectric constant (ϵ) = 2.10 (PTFE).
  (Dimensions and dielectric information are from manufacturers' data sheets.)

$$Z_0 = \frac{138}{\sqrt{\epsilon}} \log_{10} \left(\frac{D}{d} \right)$$

$$= \frac{138}{\sqrt{2.10}} \log_{10} \left(\frac{0.066}{0.0201} \right)$$

$$= \frac{138}{1.440} \log_{10} (3.2835)$$

$$= 95.83 \, (0.5163)$$
$$= 49.47 \text{ ohms} \qquad \blacklozenge$$

Once again, tolerances will bring the value to the required 50-ohm impedance.

These two examples clearly illustrate the dependence of the characteristic impedances on the construction of the cable. To sum up the term *characteristic impedance,* we can say that it is the value of resistance, present all along the cable, that would be obtained by an rf measurement at either end as shown in Fig. 3–13.

3.2.2   VSWR

VSWR on a coaxial cable, or any transmission line, is a measure of reflections on that line. These reflections may be due to the cable length, or to breaks or defects in the dielectric or shield, or many times to the connectors attached to the cable. Any condition that upsets the characteristic impedance anywhere on the line or at the end and causes reflections will produce a standing wave. Care should be taken in choosing the cable, the connectors used, and the method for attaching the connectors to the cable.

3.2.3   Attenuation

The second most important cable parameter, next to characteristic impedance, is the attenuation of the cable used for the application. If the requirements are for a receiver, the signal levels are very small, and any excess attenuation in an interconnecting cable can be disastrous. If the requirements are for a transmitter, the levels are high, and any excess loss will mean that the cable will be dissipating more power as well as causing additional requirements on all the circuitry to produce the power lost in the cable. So you can see that attenuation in a cable is a very important parameter no matter what your application is.

Table 3–2 shows various flexible and semirigid cables and their attenuation as a function of frequency. Flexible cables are shown for 50, 75, and 93 ohms. Semirigid cables are shown for 50-ohm systems only with the designations (0.085, 0.141, and 0.250) being the outside diameter.

A couple of points should be discussed before leaving the topic of attenuation in cables. The first is how various cables have widely different values of attenuation. Compare, for example, four 50-ohm flexible cables at 3.0 GHz. We will use RG-8, RG-58A, RG-174, and RG-196. The attenuation figures are 19 dB, 41 dB, 64.3 dB, and 78 dB (all are dB/100 ft). As we have said, the cables are all 50 ohms, so it obviously is not the impedance that determines attenuation. The only variable in these four cables is the size. The outside diameter of each is shown on page 65.

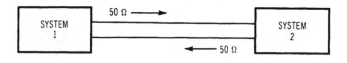

Figure 3–13.   Characteristic impedance.

RG-8A	0.405 inch
RG-58A	0.195 inch
RG-174	0.100 inch
RG-196	0.080 inch

The impedance has remained the same, but the size has decreased in each case. As this size decreases, the attenuation, α, has increased. Since the impedance is the same for all the cables, the inner diameter of the outer conductor (D) and the outer diameter of the inner conductor (d) ratio must have remained the same because the impedance is determined by this ratio. In conclusion, this ratio must have an effect on the attenuation. In reality, it has a great effect on attenuation. This formula illustrates the effect very well:

$$\alpha = 4.34 \frac{R_T}{Z_0} + 2.78\sqrt{\epsilon} \ (f) \tan d \qquad (3.6)$$

Where, α is the attenuation in dB/100 ft,

$R_T = [1/(D + d)]\sqrt{f}$,

f = frequency in MHz,

Tan d = dissipation factor (0.00059 for Teflon® at 10 GHz).

Table 3–2. Cable Attenuation

Cable Type	Z_0	Attenuation (dB/100 ft) Frequency (GHz)						
		0.5	1.0	3.0	5.0	10.0	12.0	18.0
RG-8A	52 Ω	4.6	9	19	28	47	—	—
RG-9A	51 Ω	4.6	9	19	28	47	—	—
RG-58A	52 Ω	11	20	41	—	—	—	—
RG-59A	75 Ω	6.7	11.5	25.5	41	—	—	—
RG-62A	93 Ω	5.2	8.5	18.4	29.5	—	—	—
RG-174	50 Ω	17.5	31	64.3	97	185	—	—
RG-196	50 Ω	27.5	45	78	115	172	—	—
RG-214	50 Ω	4.6	9	19	28	47	—	—
RG-223	50 Ω	8.8	16.5	36	51	90	—	—
0.085— Semirigid	50 Ω	11.6	18.7	34	46	73.2	84	115
0.141— Semirigid	50 Ω	7.2	11.6	21.5	28.5	44.5	51	68
0.250— Semirigid	50 Ω	4.3	7.3	14	18.9	29	33	—

EXAMPLE 3.3

Find the attenuation in dB/100 ft. for RG-214 cable at 10 GHz.

Solution   For RG-214, D = 0.285″, d = 0.0875″, and ϵ_r = 2.1. So,

$$\alpha = 4.34 \ (R_T/Z_0) + 2.78 \ \sqrt{\epsilon} \ (f) \ \text{Tan d}$$

$$R_T = 1/(0.285 + 0.087) \ \sqrt{10{,}000} = 268.8$$

$$\alpha = 4.34(268.8/50) + 2.78\sqrt{2.1} \ (10{,}000)(0.00059)$$

$$\alpha = 47.09 \ \text{dB/100 ft.} \quad \blacklozenge$$

The value of dissipation will decrease slightly at lower frequencies and thus will have some effect on the attenuation calculations. It becomes approximately 0.00055 at 3 GHz and 0.0005 at 1 GHz. These adjustments need to be made in order to achieve accurate attenuation readings.

3.2.4   Cable Characteristics

In addition to characteristic impedance, VSWR, and attenuation, the following cable characteristics also need careful consideration.

Velocity of Propagation   The velocity of propagation of a cable is determined primarily by the dielectric constant (ϵ) of the insulating material between the conductors. It is expressed as a percentage of the velocity of light in free space (3×10^{10} cm/sec). The percentage is calculated as follows:

$$\% = \frac{1}{\sqrt{\epsilon}} \times 100 \tag{3.7}$$

Considering the dielectric listed in the beginning of this section, we would have the following percentages.

Polystyrene (ϵ = 2.56)	62.5%
Polyethylene (ϵ = 2.26)	66.5%
Polytetrafluoroethylene (ϵ = 2.10)	69.0%

These examples say that if we are using polystyrene in a cable, for example, the velocity of the signal traveling through the cable is not 3×10^{10} cm/sec but only 1.875×10^{10} cm/sec. Similarly, polyethylene would result in a velocity of 1.995×10^{10} cm/sec, and polytetrafluoroethylene (PTFE) is 2.07×10^{10} cm/sec. So the higher the dielectric constant, the slower the velocity of propagation in the cable. Conversely, a low-dielectric constant results in a velocity approaching the speed of light. This velocity, of course, occurs when the dielectric constant equals one (air).

Capacitance and Inductance   In Section 3.1 we represented a transmission line in a general way by components of inductance, capacitance, resistance, and conductance. Fig.

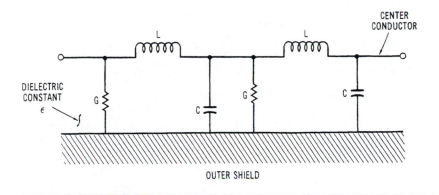

Figure 3–14.   Schematic representation of a coaxial cable.

3–14 is a more practical picture of a coaxial cable. The component C is the natural capacitance set up between the two "conductors" (center conductor and outer plates of the capacitor). The total capacitance of a cable can be calculated by using the following relationship:

$$C = \frac{7.354 \ \epsilon}{\log_{10} D/d} \ \text{pF/ft} \tag{3.8}$$

where, ϵ is the dielectric constant of the insulating material between the center conductor and the shield,

D is the inner diameter of the outer conductor,

d is the outer diameter of the inner conductor.

EXAMPLE 3.4

Calculate the capacitance for RG-214. You will recall the following values for RG-214.

D = 0.285 inch,

d = 0.087 inch,

ϵ = 2.10 (PTFE).

With the cable parameters listed, the following calculation can be made:

$$C = \frac{7.354 \ \epsilon}{\log_{10} D/d}$$

$$= \frac{7.354 \ (2.10)}{\log_{10} 0.285/0.087}$$

$$= \frac{15.443}{0.5153}$$

$$= 29.96 \ \text{pF/ft} \qquad \blacklozenge$$

Notice that the RG-214 cable is representative of 50-ohm cables in regard to capacitance.

The component L in Fig. 3–14 is the inductance set up in the cable by a current flowing through it. It is present in all conductors, and its value can be obtained by the following relationship:

$$L = 0.1404 \log_{10} D/d \ (\mu H/ft) \tag{3.9}$$

If we use the RG-214 cable, once again we can see that there is some value of inductance present, but at higher frequencies. This condition must be considered carefully since the reactance may become significant.

EXAMPLE 3.5

Calculate the inductance for RG-214.

$$\begin{aligned} L &= 0.1404 \log_{10} D/d \\ &= 0.1404 \log_{10} 0.285/0.087 \\ &= 0.1404 \ (0.5153) \\ &= 0.072 \ \mu H/ft \end{aligned}$$ ◆

The component G, shown in Fig. 3–14, is the amount of conductance through the dielectric material. Under ideal conditions where a good dielectric is present, G will be zero since there will be no conduction at all. In reality, it is some small value.

CW Power Rating   The continuous wave (cw) power rating measurement depends on such things as frequency, temperature, altitude, and VSWR; it indicates the amount of power the cable can handle and still perform in the system as originally intended. Basically, this rating indicates how well the cable removes the heat generated by high rf power; its value decreases as the frequency of operation increases. High operating temperatures and high altitude also reduce the rating; both conditions decrease the ability of a cable to dissipate heat. A high VSWR reduces the power rating because there is a "hot spot" at each point of reflection of power. Carefully choosing a cable, taking into consideration proper temperature, altitude, frequency, and system impedances, ensures that it can handle the level of power generated by your system.

Cutoff Frequency   The cutoff frequency is that frequency at which modes of energy other than the normal TEM (transverse electromagnetic) mode can be generated. This cutoff frequency is a function of the cable dimensions (D and d) and the dielectric constant (ϵ).

The higher modes are generated at impedance discontinuities (breaks, bends, etc.), and there are cases where the cable can be operated above the cutoff frequency without substantial VSWR or insertion loss increase. Staying *below* this cutoff frequency is, however, highly recommended.

The coaxial cable cutoff frequency is calculated using the following relationship:

$$f_c = \frac{1.18 \times 10^4 \, (1/\sqrt{\epsilon})}{\pi\left(\dfrac{D + d}{2}\right)} \tag{3.10}$$

where f_c is in MHz.

EXAMPLE 3.6

Calculate f_c for RG-214 cable and for the 0.085-inch semirigid cable. Calculations for RG-214 are first:

$$f_c = \frac{1.18 \times 10^4 \, (1/\sqrt{\epsilon})}{\pi\left(\dfrac{D + d}{2}\right)}$$

$$= \frac{1.18 \times 10^4 \, (1/\sqrt{2.10})}{\pi\left(\dfrac{0.285 + 0.087}{2}\right)}$$

$$= \frac{1.18 \times 10^4 \, (0.69)}{3.14 \, (0.186)}$$

$$= \frac{8142}{0.584}$$

$$= 13.94 \text{ GHz}$$

The following calculations determine f_c for 0.085 semirigid cable:

$$f_c = \frac{1.18 \times 10^4 \, (1/\sqrt{\epsilon})}{\pi\left(\dfrac{D + d}{2}\right)}$$

$$= \frac{1.18 \times 10^4 \, (0.69)}{\pi\left(\dfrac{0.066 + 0.0201}{2}\right)}$$

$$= \frac{1.18 \times 10^4 \, (0.69)}{3.14 \, (0.043)}$$

$$= \frac{8142}{0.1352}$$

$$= 60.22 \text{ GHz} \qquad \blacklozenge$$

The semirigid cable has a much higher cutoff frequency than the RG-214 cable because of its smaller size. In summary, remember to take into consideration the cutoff frequency of the cable and stay *below* it.

Pulse Response   Many microwave applications may require a pulse response instead of a cw signal. In these cases, special parameters (not critical to a cw application) must be considered, such as impedance and reflection, rise time, amplitude, overshoot, and pulse echoes.

The impedance term refers to impedance variation; fluctuations of ±5% are common and tolerable. Greater changes increase reflection and detract from the performance of the cable in a pulse application. The rise time and amplitude are functions of the length of the cable. As the cable grows longer, the rise time increases and the amplitude decreases. The increased capacitance in the longer cable causes exponential (gradual) growth in a pulse and, thus, a longer rise time; that is, it takes more time to attain a specified level. The decrease in amplitude is due to the increased attenuation of the longer cables. An overshoot is when the leading edge of a pulse rises and extends above the desired level, then settles back down. This condition usually arises because of cable construction; techniques using different configurations for the dielectric eliminate the problem. The conditions of rise time and overshoot are shown in Fig. 3–15.

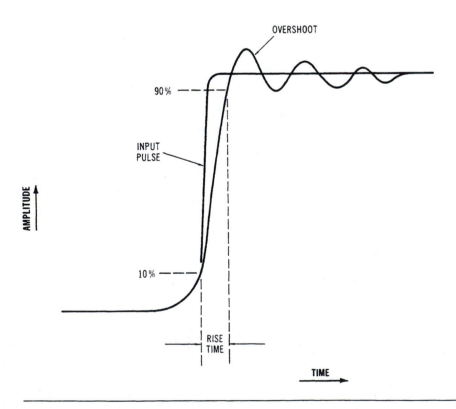

Figure 3–15.   Pulse characteristics.

When a narrow pulse is placed on a cable, a small burst of energy may emerge after the initial one has arrived. That is, two pulses of different levels are present when there should only be one. This pulse echo is caused by finite, periodic reflections within the cable; normally, it is of such a small value that it can be neglected.

Shielding   The shielding efficiency of a coaxial cable depends entirely on the construction of its outer conductor. The most common constructions are

Single Braid	Consisting of bare, tinned, or silver-plated round copper wires (85–90% coverage).
Double Braid	Consisting of two single braids as described earlier with no insulation between them.
Triaxial	Consisting of two single braids as described earlier with a layer of insulation between them.
Strip Braids	Consisting of flat strips of copper rather than round wires (90% coverage).
Solid Sheath	Consisting of aluminum or copper tubing.

This shielding accomplishes two functions: it keeps outside interference from upsetting the rf energy in the cable, and it prevents the internal energy from interfering with outside circuitry.

Cable Noise   A phenomenon noted is that when flexed, a cable generates both acoustical and electrical noise. The acoustical noise is a function of mechanical motion within the cable; the generation of electrical disturbances is attributed to an electrostatic effect. These noise voltages can be minimized by preventing motion between dielectrics and conductors or by using semiconductor layers to dissipate electrostatic charges between the two. Low-noise construction of cables must take into account their life expectancy and the environmental conditions to which they are subjected.

Flexibility   Coaxial cables with stranded center and braided outer conductors are intended for use in applications where the cable must flex repeatedly while in service. Standard braid construction can withstand over 1000 flexes through 180° if bent over a radius equal to 20 times the outside diameter of the cable. Flexible cables may be stored and are normally shipped on reels with a hub radius equal to 10 times their outside diameter. If a flexible cable is to be installed in a fixed, bent configuration, the minimum recommended bend radius is 5 times the outer diameter.

Coaxial cables with a tubular aluminum outer conductor (semirigid cable) cannot withstand more than ten 180° bends over a radius equal to 20 times the outside cable diameter. Semirigid cables are normally shipped on reels with a hub diameter equal to this figure.

Semirigid cables may be bent for installation. The minimum recommended bend radius is 10 times the outside diameter—a bend of only 5 times may cause a cable to exhibit mechanical and electrical degradation. Rebending or reforming of semirigid cables is *not* recommended. Bends and breaks in the center conductor, dielectric, and outer conductor are possible when rebending or reforming is done.

Operating Temperature Range   The operating temperature range of a cable is determined primarily by that of the dielectric and jacketing material.

Operating temperature limits of the most commonly used dielectrics and jacket types are given in Table 3–3. One additional point—only silver-plated conductors are suitable for long-term use at temperatures over 80°C.

3.2.5   Flexible Cable

We stated early in this chapter that a more comprehensive definition of a microwave transmission line (coaxial cable) was needed to understand why microwave cables were special and unique. After covering the twelve parameters associated with coaxial cables, you should now realize how true this statement is. These cables truly are more than devices that transfer energy from one point to another.

Table 3–3.   Operating Temperature Limits

Primary Dielectrics

Polytetrafluoroethylene (PTFE)	−250°C to +250°C
Polyethylene	−65°C to +80°C
Foamed polyethylene	−65°C to +80°C
Foamed or solid ethylene propylene	−40°C to +105°C

Jackets

Fluorinated ethylene propylene (FEP)	−70°C to +200°C
Polyvinylchloride (PVC)	−50°C to +105°C (depending on compound used)
Polyurethane	−100°C to +125°C (depending on compound used)
Nylon	−60°C to +120°C
Ethylene propylene	−40°C to +105°C
High molecular weight polyethylene	−55°C to +85°C
Silicone rubber	−70°C to +200°C
Silicone impregnated fiberglass	−70°C to +250°C
High temperature nylon fiber	−100°C to +250°C

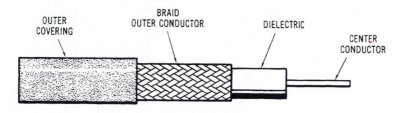

Figure 3–16.   Basic flexible cable construction.

Perhaps the most common type of coaxial cable is the *flexible cable*. The basic construction of a flexible cable is shown in Fig. 3–16. These are the familiar black cables used in so many applications throughout a lab. You have probably seen them many times connecting signal generators to attenuators or detectors to oscilloscopes. Only rarely do labs not use this type of cable.

Note how Fig. 3–16 is very similar to the cable drawings we have presented earlier. There is the center conductor and an outer conductor, with a dielectric separating and insulating the two. The only added item in Fig. 3–16 is the outer covering. This covering provides protection for the cable—mainly environmental protection. It plays no part in the electrical performance of the cable. It simply holds everything together and supplies the protection mentioned previously.

Flexible cable is available in a variety of different types. Most cable catalogs will list cables ranging from RG-4U to RG-309/U (with RG being the military designation for cable). The overall cable diameters vary from approximately $\frac{5}{64}$ inch (0.078 inch) to over one inch. The size used is determined primarily by the particular power requirements of your system. Although there are many types of cables, the basic construction is the same as shown in Fig. 3–16 for all of them. Any variations are usually minor ones. Some have stranded inner conductors, and others have solid-wire conductors; some have single outer conductors, and others have a double braid. Some of these variations can be seen in Fig. 3–17, where nine common cable types have been presented. (Other cable types are presented in Appendix E.) Fig. 3–17A presents electrical data (capacitance, impedance, attenuation, etc.), and Fig. 3–17B shows the mechanical dimensions of each cable. If you refer back to the previous section where we calculated the value of capacitance for the RG-214 cable, you will see that it is close to the 29.5 pF/ft figure used in Fig. 3–17. Also, the dimensions used for D and d can be seen in the mechanical section of Fig. 3–17 as 0.285 inch and 0.087 inch, respectively.

To illustrate how a flexible cable would be chosen for a particular application, consider the following theoretical example:

A cable for a 50-ohm system operating at 3.0 GHz is needed in a pulse operation. The pulse rise time is of a medium duration; its leading edge must be preserved as much as is possible.

RG#	CAP. (pF/ft)	WEIGHT lb/ft	MAX.OPER VOLTAGE (RMS)	Z_0	ATTEN (DB/100ft) FREQ (GHz)					POWER (WATTS) FREQ (GHz)				
					0.4	1.0	3.0	5.0	10.0	0.4	1.0	3.0	5.0	10.0
8A	30.0	0.106	5000	52Ω	4.6	9	19	28	47	350	190	95	65	37
9A	29.5	0.140	4000	51Ω	4.6	9	19	28	47	350	190	95	65	37
58A	26.5	0.029	1900	52Ω	11	20	41			75	44	22	15	
59A	21.0	0.032	2300	75Ω	6.7	11.5	25.5	41		135	77	40	27	15
62A	19.8	0.038	750	93Ω	5.2	8.5	18.4	29.5		138	70	40	27	15
174	30.4	0.008	1500	50Ω	17.5	31	64.3	97	185	25	16			
196	33.5	0.006	1000	50Ω	27.5	45	78	115	172	123	78	41	28	14
214	29.5	0.126	5000	50Ω	4.6	9	19	28	47	350	190	95	65	37
223	31.4	0.034	1900	50Ω	8.8	16.5	36	51	85	90	53	28	20	10

(A) Electrical data for cables.

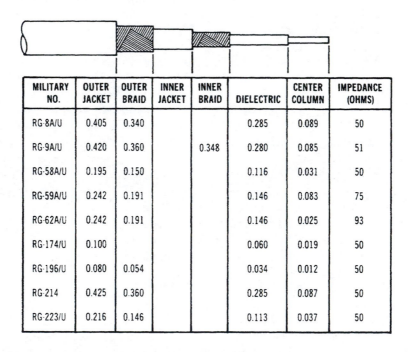

MILITARY NO.	OUTER JACKET	OUTER BRAID	INNER JACKET	INNER BRAID	DIELECTRIC	CENTER COLUMN	IMPEDANCE (OHMS)
RG-8A/U	0.405	0.340			0.285	0.089	50
RG-9A/U	0.420	0.360		0.348	0.280	0.085	51
RG-58A/U	0.195	0.150			0.116	0.031	50
RG-59A/U	0.242	0.191			0.146	0.083	75
RG-62A/U	0.242	0.191			0.146	0.025	93
RG-174/U	0.100				0.060	0.019	50
RG-196/U	0.080	0.054			0.034	0.012	50
RG-214	0.425	0.360			0.285	0.087	50
RG-223/U	0.216	0.146			0.113	0.037	50

(B) Mechanical dimensions for cables.

Figure 3–17. Cable data.

The cable run is to be from a 50-watt transmitter to an antenna 100 feet away. Existing cable runs must be used, and there is a maximum allowable space of ½-inch diameter.

From Fig. 3–17, we should be able to determine an appropriate cable; the process may take a little longer, though, to point out the specific properties of each along the way. We will consider all nine cables and see how each fits our set of requirements.

We can eliminate two cable types immediately—the RG-59A and RG-62A—because each has a characteristic impedance greater than 50 ohms. (The Z_0 for the first is 75 ohms; for the second it is 93 ohms. These cables have applications in 75- and 93-ohm systems, respectively.)

The next specification, pulse response, can be met by choosing a low value of cable capacitance per foot. A low value of capacitance ensures that the pulse shape is preserved. Obviously, it would be a toss-up if this prerequisite were the only one to be satisfied. Values range from 28.5 pF to 33.5 pF, not too broad a spectrum; any of them would be suitable.

The number of available cables dwindles when the 50 watts of power is taken into consideration. Suddenly, only three have this type of power handling capability—8A, 9A, and 214.

Considering the three possible choices, their outer diameters allow each to fit our requirements. We need some additional information to make our final decision. One point is that the RG-214 is a recommended substitute for the RG-9A because the former is a double-shielded cable. That is, instead of having a single layer of metallic braid covering the dielectric and shielding the center conductor, the RG-214 has a double thickness that improves the cable's shielding ability and makes it more rugged. The RG-8A cable also is found to be a single-shielded cable; however, it is recommended for only up to 1 GHz. The RG-214, on the other hand, can be used up to 10 GHz with good results.

Our choice is quite clear at this point. Our friend, the RG-214, has a characteristic impedance of 50 ohms and an adequate cable capacitance/ft figure; it can handle over 50 watts of power easily (it is rated at 95 watts at 3 GHz) and has 29 dB of attenuation in 100 feet; and its outside diameter is 0.425 inch. All these properties indicate that the cable will perform well in our theoretical example.

This hypothetical situation makes it clear that there is more to choosing a cable than calling for a piece of 50-ohm flexible coax. The characteristic impedance is of great importance, but the specific application usually requires much more than that single qualification. As we have seen, capacitance/ft, attenuation/100 ft, power handling capability, and size are all influential; dielectric material, weight/ft, operating voltage, and operating temperature range also may need to be investigated.

One point should be emphasized about flexible cables. The term *flexible* does not mean that they can be flexed in any direction an infinite number of times. Instead, flexible cables have a center conductor (made of some type of metal), a dielectric, and an outer conductor (a metal braid) that can bend a certain number of times at a specified angle (usually 180°) and minimum radius and then return to their original shape with no degradation in operating performance. However, any metal eventually breaks or distorts if it is bent too often. Care should be used regarding flexible cables in your system or test setup. Such components last much longer if they are flat on a bench or in a system instead of kinked up in a tight space.

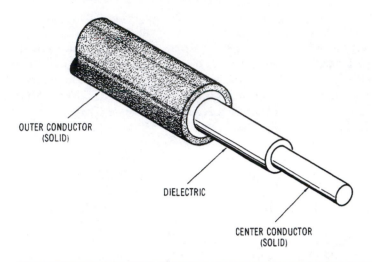

OUTER CONDUCTOR
(SOLID)

DIELECTRIC

CENTER CONDUCTOR
(SOLID)

Figure 3–18.   Semirigid cable.

Thus, the flexible cable becomes out first specific type of transmission line for microwave use.

3.2.6   Semirigid Cable

The name *semirigid* is somewhat misleading. Something called "rigid" is generally associated with items that are solid and will not move. Something called "flexible" brings to mind an object that can be bent and twisted about. To picture a semirigid object as movable is only natural and, to some degree, is true because semirigid cable can be bent. However, it should be bent only *once* because semirigid cable has a solid center conductor and outer conductor. Fig. 3–18 shows the construction of semirigid cable. Note the similarities and differences with all other types we have discussed. The major difference between the other types of cable and the semirigid is that the semirigid has a solid outer conductor. Compared to the flexible-type cable, semirigid offers greater shielding effectiveness because of this solid outer conductor. This increased shielding results in reduced attenuation and increased power rating.

Some of the materials used to construct the semirigid cable are shown here:

Inner Conductor—silver-plated copper, silver-plated copper-clad steel, silver-plated beryllium-copper, silver-plated aluminum, and resistance wire.

Dielectric—polyethylene, polytetrafluoroethylene (TFE Teflon®), polytetrafluoroethylene-hexafluoropropylene (FEP Teflon®), and Kapton®.

Outer Conductor—copper, aluminum, stainless steel, and special copper alloys.

It can be seen how such a cable will, first, be semirigid and, second, offer excellent shielding because its outer conductor may be of copper, aluminum, or stainless steel.

As in the case of the flexible cable, there is more to choosing a semirigid cable than specifying a 50-ohm cable. Such parameters as attenuation, power rating, higher mode cutoff frequency, dielectric strength, and corona extinction voltage must be considered.

Many times throughout this chapter, we have related parameter to the diameter of a cable. Both inside diameter of the outer conductor (D) and outside diameter of the inner conductor (d) have been referred to frequently. The semirigid cable is no exception.

This dimensional relationship is such that the diameter of a cable is chosen so that, for a given space requirement, a cable run has minimum loss and carries maximum power. However, the initial decision also must take into account the fact that a coaxial line (semirigid cable in this case) operated at increasingly high frequencies reaches a point where higher-order modes (other than the desired TEM mode) are supported. The term that describes the frequency where the higher modes are supported is *90% higher mode cutoff frequency*. Care must be taken that the highest frequency at which your system has to operate does not exceed this figure. This mathematical expression shows the dependency of the higher mode cutoff frequency on the dimensions of the cable conductors.

$$90\% \ f_c = \frac{6.75}{\sqrt{\epsilon}\,(d + D)} \ \text{(in GHz)} \tag{3.11}$$

where, d is the inner conductor outside diameter (inches),
  D is the outer conductor inside diameter (inches).

EXAMPLE 3.7

To get an idea of where these cutoff frequencies are for semirigid cable, consider the case of the 0.085 cable we used in a previous example. Remember that for the example, D = 0.066 inch, d = 0.0201 inch, and ϵ = 2.10 (PTFE). With these numbers, let us calculate the 90% higher mode cutoff frequency.

$$90\% \ f_c = \frac{6.75}{\sqrt{\epsilon}\,(d + D)}$$

$$= \frac{6.75}{\sqrt{2.10}\,(0.0201 + 0.066)}$$

$$= \frac{6.75}{1.44\,(0.0201 + 0.066)}$$

$$= \frac{6.75}{1.44\,(0.0861)}$$

$$= \frac{6.75}{0.124}$$

$$= 54.43 \ \text{GHz}$$

So we are saying not to use 0.085 semirigid above 54 GHz.   ♦

3.2.7 Cable Selection

There are times when the selection of a cable involves consideration of a maximum voltage rating. Although it would appear to be safe to operate a cable up to and including the *dielectric strength test potential* (a voltage applied during manufacturing to determine the strength of the cable dielectric), it is better to avoid use above the *corona extinction potential* (voltage where an arcover, or corona, appears and causes noise, electrical damage, or even cable breakdown). There is good evidence that the corona discharge can cause permanent damage to polytetrafluoroethylene, a very common semirigid cable dielectric. (These are two of the terms previously mentioned that were not associated with flexible cables and require definition.)

To meet mechanical or environmental requirements (and thus improve overall electrical performance) when selecting a cable, the following factors should be kept in mind:

1. Inner conductors of copper-clad steel are stiffer and have a higher tensile strength than those of copper. Hence, copper-clad steel wire may be preferable if a cable is to be terminated with an SMA-type connector, the design of which incorporates the cable inner conductor as the connector pin.

2. Copper wire cannot be used in small-diameter cores with relatively heavy insulation; its tensile strength is insufficient to withstand the stresses of the insulation process. Copper-clad steel wire has magnetic properties and hence should not be employed (or else adequate precautions should be taken) near components whose proper operation is disrupted by magnetic interference. Alternatively, beryllium-copper wire is a good choice for a nonmagnetic conductor of high tensile strength and stiffness.

3. Tubular inner conductors reduce the weight of large-diameter cables and enable low VSWR performance over a broad frequency spectrum by using connectors whose pins can be "flush-mounted" by screwing into the cable inner conductor. This latter attribute eliminates the need for a compensation of the discontinuity caused by a contact barrel.

4. When specifying cables, dielectric materials should receive great consideration. As compared to copper, their temperature coefficients of expansion differ by an order of magnitude—construction of a semirigid cable whose dielectric does not "grow" upon heating is impossible. Paste-extruded polytetrafluoroethylene is reported to have a small temperature coefficient of expansion and so would be preferable, but manufacturing limits exist.

FEP dielectric has superior resistance to nuclear radiation effects, but it also exhibits both greater variations in dielectric constant and loss tangent with operating frequency. These problems may be objectionable in broad-band applications.

Polyethylene dielectric, because of its low melting point, is not suitable for use in system designs involving soldering to the cable outer conductor.

Having covered the basic construction and terminology of the semirigid cable, let us see what cables are available for microwave system use.

Types (or sizes) available are 0.020, 0.034, 0.047, 0.056, 0.070, 0.085, 0.141, 0.215, 0.250, and 0.325. (These figures refer to outer conductor outside diameters, expressed in inches.) Of the ten cables listed, three are widely used in microwave, 0.085, 0.141, and 0.250. Table 3–4 shows the characteristics of this trio.

Two important characteristics are not in Table 3–4—cable attenuation and power rating. Because of their great importance, they are listed in Table 3–5 as a function of frequency.

We previously stated that semirigid cables had reduced attenuation and increased power rating. Table 3–5 shows that this statement is indeed true. Table 3–6 is a comparison of the attenuation and power rating of flexible and semirigid cables that illustrates the reduced attenuation and increased power rating of the semirigid cables.

The comparisons made in Table 3–6 were for flexible cables of comparable sizes: 0.085 and RG-196 (0.080 inch), 0.141 and RG-58 (0.195 inch), and 0.250 and RG-214 (0.216 inch). There are some rather interesting numbers in this table. Note, for example, the nearly two-to-one reduction in attenuation for 0.085 and 0.141 cable as compared to their flexible counterparts. Yet when the 0.250 cable is considered, the attenuation follows very closely with the RG-214 until it approaches 10 GHz, where the RG-214 attenuation increases significantly.

Table 3–4. Characteristics of Three Semirigid Cables
(Source: Precision Tube Co., Inc., "Precision Coaxitube," Catalog 3725M)

| | **Part Numbers** | | |
Characteristics	**SRC-085**	**SRC-141**	**SRC-250**
Inner conductor OD	0.0201 ± 0.001 in	0.0359 ± 0.0005 in	0.0641 ± 0.001 in
Dielectric OD	0.066 ± 0.001 in	0.1175 ± 0.001 in	0.210 ± 0.002 in
Outer conductor OD	0.085 ± 0.001 in	0.141 ± 0.001 in	0.250 ± 0.002 in
Safe bend radius	0.125 in	0.250 in	0.357 in
Weight	1.31 lb/100 ft	3.25 lb/100 ft	10 lb/100 ft
Maximum operating temperature	125°C	150°C	175°C
Impedance tolerance	±1 ohm	±1 ohm	±1 ohm
Dielectric constant	2.07	2.07	2.07
Dielectric strength rating	5 kV (rms) 60 Hz	5 kV (rms) 60 Hz	7.5 kV (rms) 60 Hz
Corona extinction voltage	1.5 kV (rms) 60 Hz	1.9 kV (rms) 60 Hz	3.15 kV (rms) 60 Hz
90% higher mode cutoff frequency	54.8 GHz	30.8 GHz	17.8 GHz

Table 3–5. Attenuation and Power Rating Characteristics for Semirigid Cables
(Source: Precision Tube Co., Inc., "Precision Coaxitube," Catalog 3725M)

Frequency (GHz)	0.085 Attenuation (dB/100 ft)	0.085 Power (kW)	0.141 Attenuation (dB/100 ft)	0.141 Power (kW)	0.250 Attenuation (dB/100 ft)	0.250 Power (kW)
0.50	11.6	0.370	7.2	1.000	4.3	2.10
1.0	18.7	0.222	11.6	0.600	7.3	1.20
3.0	34.0	0.115	21.5	0.310	14.0	0.60
5.0	46.0	0.082	28.5	0.230	18.9	0.44
10.0	73.2	0.048	44.5	0.160	29.0	0.28
12.0	84.0	0.040	51.0	0.140	33.0	0.25
18.0	115.0	0.028	68.0	0.100	—	—

Power handling for the semirigid cables far exceeds that of the flexible cables in all cases. The 0.085 cable, for example, will handle 48 watts of power at 10 GHz while RG-196 is capable of handling only 14 watts. The RG-58 is not even specified at 10 GHz, and the RG-214 will handle only 37 watts as compared to 250 watts for 0.250 semirigid cable.

These flexible and semirigid cables all have a solid dielectric with a dielectric constant of 2.10, resulting in a velocity of propagation of approximately 69%. The ideal velocity, of course, is 100% with an air dielectric—an impossibility since the center conductor would have no support. We can, however, come close to this ideal by using a combination of air and dielectric material. If the dielectric is in the shape of a corkscrew (spiral), a much higher volume of air is present in the cable. Although this method accomplishes the initial purposes of lowering the dielectric constant and raising the velocity of propagation, it cre-

Table 3–6. Attenuation and Power Rating Comparisons for Flexible and Semirigid Cables

Cable Type	Attenuation (dB/100 ft) (Frequency—GHz) 0.4	1.0	3.0	5.0	10.0	Power (Watts) (Frequency—GHz) 0.4	1.0	3.0	5.0	10.0
0.085	11.6	18.7	34	46	73.2	370	222	115	82	48
RG-196	27.5	45	78	115	172	123	78	41	28	14
0.141	7.2	11.6	21.5	18.5	44.5	1000	600	310	230	160
RG-58	11	20	41	—	—	75	44	22	15	—
0.250	4.3	7.3	14	18.9	29	2100	1200	600	440	280
RG-214	4.6	9	19	28	47	350	190	95	65	37

ates the problem of periodicity effects because the density of the dielectric inside the cable repeats itself at regular intervals. Periodic VSWR changes also result; thus, any lengths of this sort of cable must be designed very carefully into your particular system. For short runs, though, there is little noticeable effect.

There are other ways to achieve the same initial goal. Cables are available that provide inner conductor support within the outer conductor by using longitudinal splines of paste-extruded polytetrafluoroethylene. These cables are called *air articulated* and are shown in Fig. 3–19. Such construction results in a lightweight component with minimum attenuation, free of any periodicity effect, and having a velocity of propagation of 87%. For equal diameters, air-articulated cables have a lower attenuation figure and a higher maximum operating frequency than do those having solid dielectrics.

As is the case with most components, special types of cables are available upon request—those that are specifically designed to be used at very low temperatures (cryogenic cables), ones with resistance wire as a center conductor to result in a lossy cable, and those that can take excessive vibration or flexing without serious changes occurring in their characteristics.

Since the construction of semirigid cable is different from that of flexible cable, its handling is also different. When working with the former cable, certain procedures should be followed to ensure maintenance of the intended performance. Cutting, measuring, and sol-

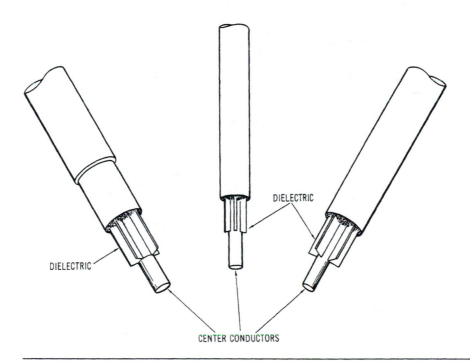

Figure 3–19.   Air-articulated cable.

dering procedures must be observed to result in a usable cable that will operate in your particular system.

3.3   **Stripline**

The terms *tri-plate, sandwich package, strip transmission line,* and *stripline* are all used to describe one of the most useful forms of microwave transmission lines. The source of these names is apparent in Fig. 3–20. Here you see two copper-clad laminates (dielectrics) like the ones explained in Chapter 2, with the circuit "sandwiched" between them.

In its simplest form, the lower laminate would have copper on both sides. One side would have the circuit etched on it, and the other side would be a full ground plane. The upper laminate would have copper on only one side for a ground plane. The two laminates would be put together as shown by means of eyelets, clamping plates, a milled metal case, or by bonding under heat and pressure with a film of resin (lamination). The circuit may also be etched on both laminates as was done for early stripline circuits. We referred to this method in Chapter 2. Another method is to have both laminates either with copper on only one side or no copper at all and etch a circuit on a separate thin laminate to be inserted between the dielectrics. This method is ideal for quadrature hybrid couplers where the coupling spacing is on the order of 0.005 inch. This spacing would be a difficult gap to etch accurately between two lines, and the thin laminate can be controlled and the thickness held over a large-sized piece of laminate.

To further understand stripline, let us show how it evolved to its present form from a now familiar transmission line—the coaxial cable. Fig. 3–21 shows such an evolution. The arrows indicate the electric field between inner and outer conductors at an instant when the inner conductor is positive. When the lines are far apart, the field is weak. There is no field outside a coaxial cable because the outer conductor is continuous.

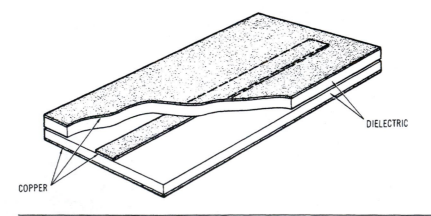

Figure 3–20.   Cutaway view of stripline.

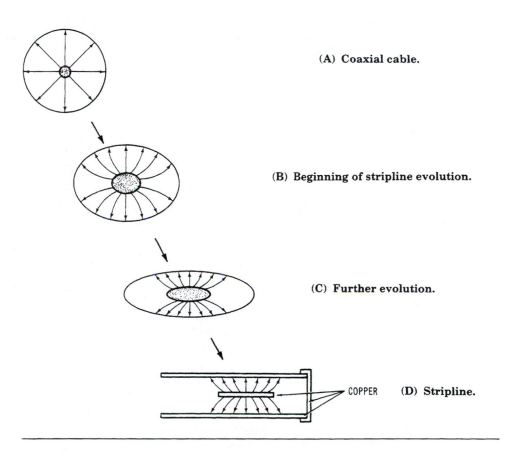

(A) Coaxial cable.

(B) Beginning of stripline evolution.

(C) Further evolution.

COPPER   (D) Stripline.

Figure 3–21. Evolution of stripline from coaxial cable.

Proceeding from step C to step D in Fig. 3–21, the outer conductor is interrupted and it becomes two plates, but the field extends only a short distance on each side of the center strip. If the plates extend well beyond the strip, the field at their edge is essentially zero. Therefore, properly designed stripline, like coaxial cable, is largely self-shielding, though some leakage may occur through the exposed edges of the dielectric. To counteract this potential problem, a conductive tape or strap is folded over the edge of the package to make contact with and to electrically bond together the two ground planes, as shown in Fig. 3–21D. If clamping plates are used, they can be designed to make contact with each other over the edge of the laminate.

3.3.1   Dielectric Constant

With the dielectrics used in stripline, the velocity of propagation and the wavelength are both reduced, as we discussed in Chapter 2.

Table 3–7.   Frequencies and Wavelengths in Air and Two Dielectrics

| Frequency (GHz) | Wavelength | | | | | |
| | λ_0 (Air) | | λ_1 ($\epsilon = 2.55$) | | λ_2 ($\epsilon = 10.2$) | |
	cm	in	cm	in	cm	in
0.50	60.0	23.60	37.70	14.80	18.80	7.40
1.00	30.0	11.80	18.80	7.42	9.40	3.70
3.00	10.0	3.93	6.29	2.47	3.13	1.23
5.00	6.0	2.36	3.77	1.48	1.88	0.74
12.00	2.5	0.98	1.57	0.62	0.78	0.31

The actual stripline wavelength (λ) is equal to the free-space wavelength (λ_0) divided by the square root of the dielectric constant (ϵ):

$$\lambda = \frac{\lambda_0}{\sqrt{\epsilon}} \tag{3.12}$$

To emphasize the importance of the dielectric constant on the physical size of the stripline, Table 3–7 shows five frequencies, the corresponding wavelengths in air, and two widely used dielectrics discussed in Chapter 2 ($\epsilon = 2.55$ and $\epsilon = 10.2$).

Because the dielectric constant controls the wavelengths of signals in a stripline circuit, it is a critical property in all applications; however, the thickness of the dielectric is often of equal importance. The characteristic impedance (Z_0)—a fundamental design parameter for all stripline circuits—depends on the dielectric constant, the width and thickness of the conductors, and the thickness of the dielectric layers.

The importance of the thickness of the dielectric layers can best be shown by indicating all of the areas in which it is used.

3.3.2   Ground-Plane Spacing

First, there is a term that comes up time and time again when stripline circuits are discussed. That term is *ground-plane spacing*. This term is probably the single most important term used in stripline terminology. At the beginning of this section, we described stripline as two laminates on top of one another with the circuit in between. Each of the laminates had copper on one side that was used as a ground plane. Measure from the top ground plane to the bottom ground plane, and you have a distance equal to twice the thickness of the laminate used, which is the ground-plane spacing and is designated by the letter "b" shown in Fig. 3–22. This term is used to calculate every dimension in stripline, with the exception of wavelength.

One dimension that the ground-plane spacing helps to define is the width, w, of a line. Whether the line is 20, 50, or 100 ohms, its width depends on the ground-plane spacing used. What is calculated is the *w/b ratio*, which is width divided by the ground-plane spacing. To aid in these calculations, a chart was developed in 1955. This chart is shown in

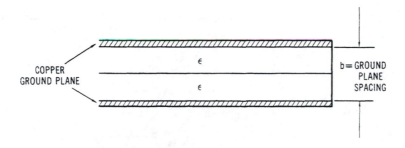

Figure 3–22.   Ground plane spacing.

Fig. 3–23 and will give the width of a line if the dielectric constant (ϵ) and the t/b ratio are known.

3.3.3   T/B Ratio

The *t/b ratio* is the ratio of the thickness of copper on the laminate to the ground-plane spacing. Most copper on microwave laminate is either ½ or 1 ounce. That is, 1 square foot of the copper weighs either ½ or 1 ounce. The thicknesses are 0.0007 and 0.0014 inch, respectively. Copper is available on request in ¼-ounce and 2-ounce sizes (0.0035 and 0.0028 inch). Generally, however, the ½-ounce or 1-ounce copper is used. Considering that the laminate thicknesses usually used are 0.025, 0.030, or 0.062 inch, the thickness of the copper is rather insignificant. If, for example, we were using 0.030-inch material (b = 0.060) and had ½-ounce copper on the laminate, the t/b ratio would be 0.011. In a case like this, we would use the t/b equals 0 curve since the difference between t/b equals 0 and t/b equals 0.011 would be indistinguishable on the curves. If, however, 2-ounce copper (0.0028 inch) on 0.025-inch dielectric material were used, the t/b ratio would then be 0.056, and the t/b equals 0.05 line on the chart in Fig. 3–23 would be used.

3.3.4   Width of Line Calculations

With all of the terms defined, let us now find the width of a line to be used in a stripline circuit.

EXAMPLE 3.8

We would like to find the width of a 70-ohm line that was 0.030-inch material with ½-ounce copper (ϵ = 2.55). To begin with, we calculate t/b.

$$t/b = \frac{0.0007}{0.060}$$

$$= 0.011$$

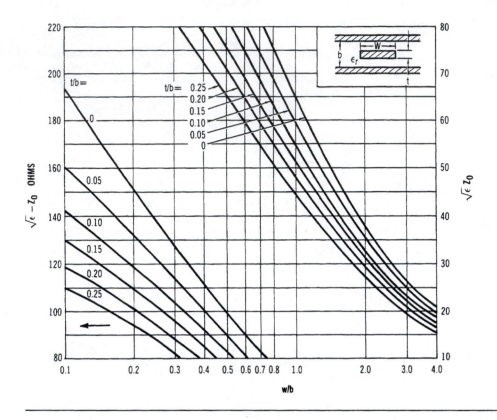

Figure 3–23. Z_0 versus w/b for various values of t/b. *(Reference–S. Cohn, "Problems in Strip Transmission Lines." MTT-3, No. 2, March 1955, Figure 1)*

Therefore, we will use the t/b = 0 curve on the chart. Next, we calculate $\sqrt{\epsilon}\, Z_0$.

$$\begin{aligned} Z &= \sqrt{\epsilon}\, Z_0 \\ &= \sqrt{2.55}\,(70) \\ &= 1.59\,(70) \\ &= 111\ \Omega \end{aligned}$$

We now move up the left-hand vertical axis until 111 ohms is found and move to the right until we intersect the t/b = 0 curve. We then drop straight down to the horizontal axis and read the w/b ratio. In this case, it is 0.45. We now can calculate the value of w for 70 ohms.

$$\begin{aligned} w/b &= 0.45 \\ w/0.060 &= 0.45 \\ w &= 0.45(0.060) \\ w &= 0.027 \text{ in for } 70\ \Omega \end{aligned}$$

♦

Table 3–8. Line Widths	$Z_0(\Omega)$	ϵ	b (inches)	w/b	w (inches)
for Various Impedances	50	2.55	0.060	0.75	0.045
	60	2.55	0.060	0.55	0.060
	70	2.55	0.060	0.45	0.027
	80	2.55	0.060	0.325	0.019
	90	2.55	0.060	0.24	0.014
	100	2.55	0.060	0.18	0.010

A width for any impedance that does not exceed the left-hand vertical axis value can be calculated in the same way as we have calculated the value for 70 ohms. Table 3–8 shows the width of lines for various impedances. Each of these widths was calculated from Fig. 3–23.

3.3.5 Stripline Couplers

The ground-plane spacing also is used to find the spacing between two lines that are used for couplers. *Couplers* are devices that provide a transfer of energy from one circuit to another, and they will be covered in detail in Chapter 4.

Stripline couplers are built in two orientations: *side coupled* and *broadside coupled*. Fig. 3–24 shows these configurations. In both cases, parameters w, s, Z_{0e}, and Z_{0o} are to be considered; w is the line width we have just covered, s is the spacing between the lines, and Z_{0e} and Z_{0o} are the even- and odd-mode characteristic impedances of the line.

The even-mode characteristic impedance (Z_{0e}) is the impedance measured from one strip to ground where the strips (or lines) are at the same potential and have equal current in the same direction.

The odd-mode characteristic impedance (Z_{0o}) is the impedance from each line to ground where the line potentials are opposite and the currents are in opposite directions.

For those who are mathematically inclined, the values of Z_{0e} and Z_{0o} for side-coupled devices are

$$Z_{0e} = \frac{94.15/\sqrt{\epsilon}}{w/b + \dfrac{\ln 2}{\pi} + \dfrac{1}{\pi}\ln\left[1 + \tanh\left(\pi/2 \times s/b\right)\right]} \tag{3.13}$$

and,

$$Z_{0o} = \frac{94.15/\sqrt{\epsilon}}{w/b + \dfrac{\ln 2}{\pi} + \dfrac{1}{\pi}\ln\left[1 + \coth\left(\pi/2 \times s/b\right)\right]} \tag{3.14}$$

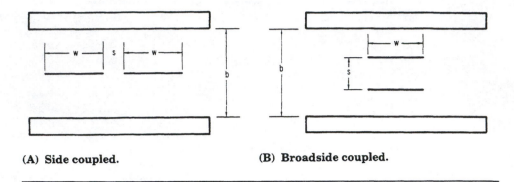

(A) **Side coupled.**   (B) **Broadside coupled.**

Figure 3–24.   Stripline couplers.

For broadside-coupled devices, the expressions are

$$Z_{0e} = \frac{188.3/\sqrt{\epsilon}}{\dfrac{w}{b-s} + \dfrac{\ln 4}{\pi} + \left(\dfrac{1}{\pi} \times \dfrac{c}{\epsilon}\right)} \tag{3.15}$$

and,

$$Z_{0o} = \frac{188.3/\sqrt{\epsilon}\;(s/b)}{\dfrac{w}{b-s} + \left(\dfrac{1}{\pi} \times \dfrac{c}{\epsilon}\right)} \tag{3.16}$$

where,

$$c/\epsilon = \frac{s}{b-s} \ln \frac{b}{s} - \ln(1 - s/b).$$

By going through these calculations, s and w may be determined for a particular coupler design. This determination, however, can be a long and tedious process. Fortunately, charts can be used to make life a little simpler. Fig. 3–25 shows the s/b ratio charts for side coupling, K_0, from -6 to -60 dB. (The K_0 term is the value of coupling you want your device to have in dB.) Fig. 3–26 shows the w/b ratio chart for side-coupled devices.

Similarly, there are charts for broadside-coupled devices. Fig. 3–27 shows s/b ratios for these devices, and Fig. 3–28 shows w/b ratios. These couplers go from -1 dB to -40 dB.

One important point should be brought out about these figures and the characteristics of these devices. The side-coupled devices have a range of -6 to -60 dB, but the broadside devices go from -1 to -40 dB. As mentioned before, there is a limit on how close two lines can be put while still maintaining the constant gap between them. Consider, for example, a case with a -6-dB coupler, where b = 0.060 inch and ϵ = 2.55. The s/b ratio would be 0.025, resulting in s = 0.0015 inch. A 1.5-mil gap is difficult to etch over any distance and maintain a constant spacing. The side-coupled approach is used, therefore, for higher value couplers.

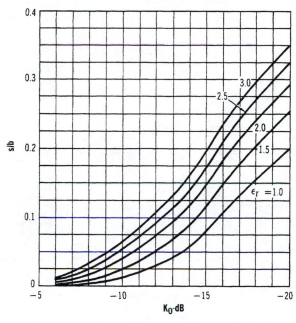

(A) The s/b versus K₀ chart (K₀ from −6 to −20 dB).

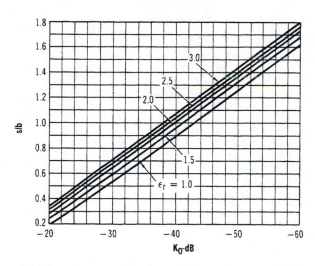

(B) The s/b versus K₀ chart (K₀ from −20 to −60 dB).

Figure 3–25.   The s/b ratio charts for side-coupled devices.

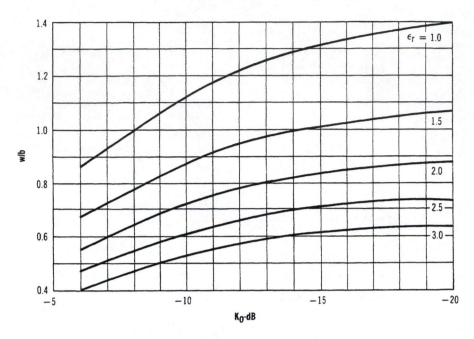

(A) The w/b versus K_0 chart (K_0 from -6 to -20 dB).

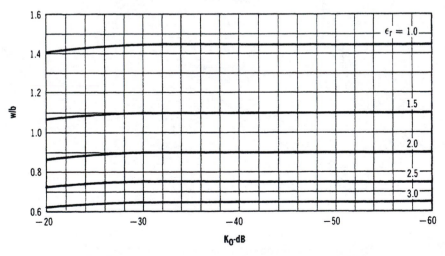

(B) The w/b versus K_0 chart (K_0 from -20 to -60 dB).

Figure 3–26.   The w/b ratio charts for side-coupled devices.

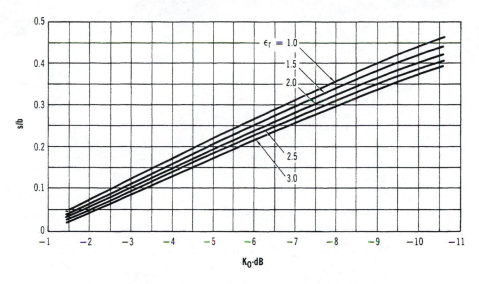

(A) The s/b versus K_0 chart (K_0 from -1 to -11 dB).

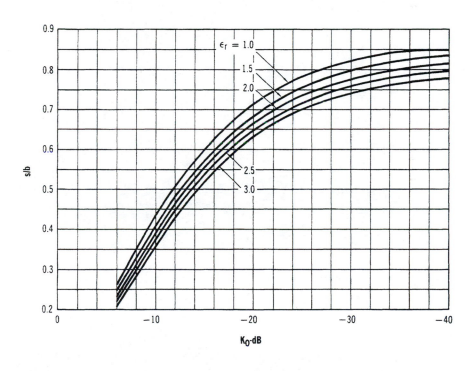

(B) The s/b versus K_0 chart (K_0 from -6 to -40 dB).

Figure 3–27. The s/b ratio charts for broadside-coupled devices.

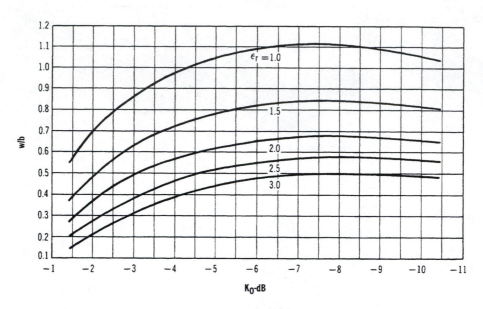

(A) The w/b versus K_0 charts (K_0 from -1 to -11 dB).

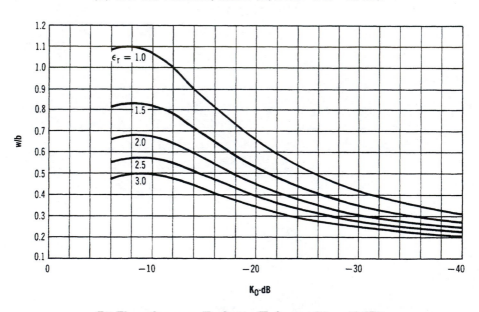

(B) The w/b versus K_0 charts (K_0 from -6 to -40 dB).

Figure 3–28.   The w/b charts for broadside-coupled devices.

Use of the broadside-coupled devices enables the designer to use the thin laminate mentioned in Chapter 2. The typical use is for quadrature hybrids, to be discussed in the next chapter. For now, it is sufficient to say they are −3-dB couplers with a constant 90° phase difference between their output parts. The spacing for these couplers (s) is on the order of 0.005 inch. It is possible to obtain copper-clad laminates of this thickness, etch the circuit on both sides, place this circuit between two 0.030-inch laminates, and have a very repeatable and reliable quadrature hybrid coupler, which would not be possible with side-coupled construction.

The use of stripline—admittedly a great advance in the microwave field—is feasible only within a certain range of frequencies. As seen from Table 3–7, the wavelength at frequencies below 1.0 GHz becomes too large and cumbersome. Similarly, above 18.0 GHz, the wavelength (and therefore the stripline) becomes too small, especially as most components rely on a quarter-wavelength dimension for proper orientation. (A wavelength at 18 GHz = 0.436 inch; λ/4 = 0.109 inch.) Therefore, as a general rule of thumb, we can say that the world of stripline exists between 1.0 GHz and 18.0 GHz.

3.4   Microstrip

The many different types of transmission lines available today vary widely in the majority of their characteristics and parameters; however, they all have only one type of dielectric. The dielectric may be air, Teflon®, or any other appropriate material, but only one variety is used. Conversely, microstrip lines have two types of dielectrics—air and solid.

A microstrip transmission line consists of a metallized strip and a ground plane separated by a solid dielectric as shown in Fig. 3–29. This configuration has become a popular alternative to the classic shielded stripline, especially for miniature microwave circuits.

Microstrip is usually chosen wherever active devices (transistors) are used. Because the transistors can be placed on top of the board in many instances, no cutting of the ground planes is needed. This convenience is also true for any circuit whose components must be mounted to a board. Any resistors, capacitors, or diodes that are to be attached can be sol-

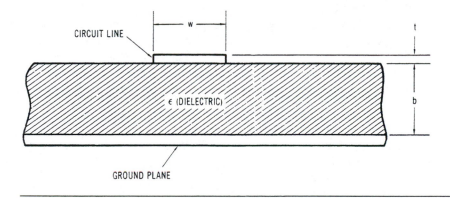

Figure 3–29.   Microstrip transmission line.

dered directly onto the microstrip board. This is also an excellent medium for Surface Mount Technology (SMT).

3.4.1 Effective Dielectric Constant

The dielectric constant and thickness are as important in microstrip as they were in the stripline covered previously. There is one difference, however, when considering microstrip. As we have mentioned, the stripline has the same dielectric above and below the circuit line. This use of the same dielectric results in the use of the dielectric constant published by the laminate manufacturer, which is called a *relative dielectric constant* (ϵ_r). With microstrip construction, the same type of dielectric is used below the line as with stripline, but there is air above the line. The dielectric constant used for design, therefore, must take into consideration the dielectric constant of air ($\epsilon = 1.0$) and that of the material used (ϵ_r). This new value is called the *effective dielectric constant* (ϵ_{eff}).

To find the effective dielectric constant (ϵ_{eff}), the following relationship is used:

$$\epsilon_{eff} = 1 + q\,(\epsilon_r - 1) \tag{3.17}$$

where, q is the filling factor,

    ϵ_r is the relative dielectric constant from the manufacturer's data sheet.

The filling factor (q) is the ratio of the dielectric area to the total area that would be used if the medium were stripline. The chart in Fig. 3–30 shows the relationship of the w/b ratio, the free-space impedance ($Z_0 \sqrt{\epsilon_r}$), and the filling factor (q).

To aid in finding the w/b ratio, first refer to Fig. 3–31, where strip widths in inches versus impedance are shown for four different types of material. The most common type shown is $\frac{1}{32}$-inch Teflon® fiberglass material ($\epsilon = 2.6$), which is the solid line on the chart. To find, for example, the width of a 50-ohm line, locate 50 on the vertical axis and move across until the solid line is intersected. By reading down to the horizontal axis, we find that a 50-ohm line is about 0.090 inch wide. Dividing this width by the $\frac{1}{32}$-inch board material (0.031 inch) gives a w/b ratio of 2.9. Similarly, a 70-ohm line would be 0.050 inch wide, resulting in a w/b ratio of 1.61.

Applying the line width and the w/b ratio to Fig. 3–30 obtains a filling factor and the effective dielectric constant to be used. Consider, for example, the 50-ohm line with a 0.090-inch width and a w/b of 2.9. Locate 2.9 on the left-hand side of the chart and move across until intersecting the q line. Notice that the filling factor for a 50-ohm line in a Teflon® fiberglass material with a relative dielectric constant of 2.55 is about 0.73.

EXAMPLE 3.9

Find the effective dielectric constant of a circuit with $\epsilon_r = 2.55$ and w/b = 2.9.

$$\begin{aligned}
\epsilon_{eff} &= 1 + [q\,(\epsilon_r - 1)] \\
&= 1 + [0.73\,(2.55 - 1)] \\
&= 1 + [0.73\,(1.55)] \\
&= 1 + 1.13 \\
&= 2.13
\end{aligned}$$

                                                                   ◆

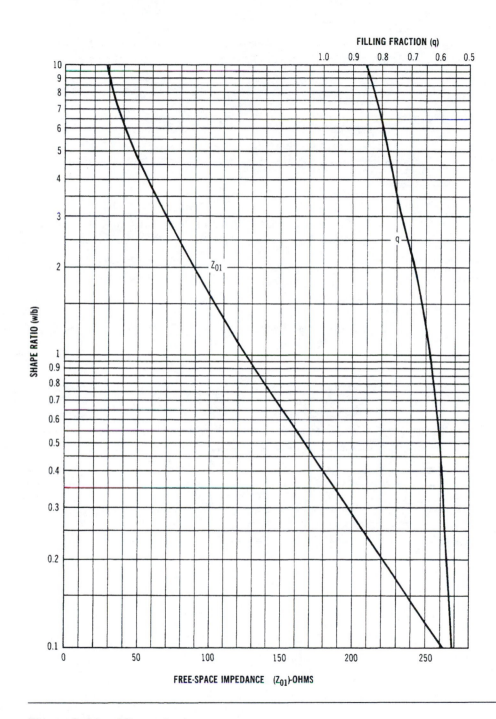

Figure 3–30.   Microstrip chart.

Also, ϵ_{eff} may be found by taking the same w/b ratio of 2.9, moving to the Z_{01} line, dropping down, reading the value of Z_{01} on the bottom axis, and then substituting in the equation

$$\epsilon_{eff} = \left(\frac{Z_{01}}{Z_0}\right)^2 \tag{3.18}$$

To illustrate this process, consider the 50-ohm example again. Once again, w/b is 2.9 and Z_0 is 50 ohms. Moving to the Z_{01} line and down shows that Z_{01} reads approximately 73 ohms. By substituting this value into the previous formula, the following results:

$$\epsilon_{eff} = \left(\frac{Z_{01}}{Z_0}\right)^2$$
$$= \left(\frac{73}{50}\right)^2$$
$$= (1.46)^2$$
$$= 2.13$$

Notice that the same answer as before results for the effective dielectric constant (ϵ_{eff}). Obviously, the effective dielectric constant will be different for every value of imped-

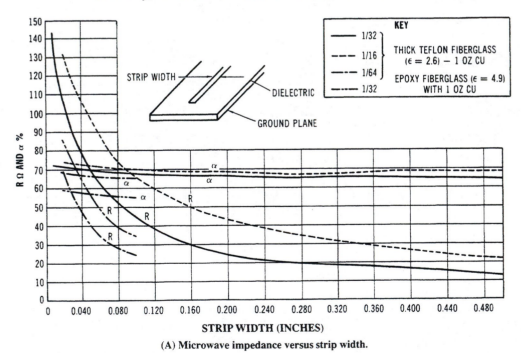

(A) Microwave impedance versus strip width.

Figure 3–31. **Microstrip characteristics.** *(Courtesy Microwave Engineer's Handbook, Vol. 1. Artech House, Inc., 1971)*

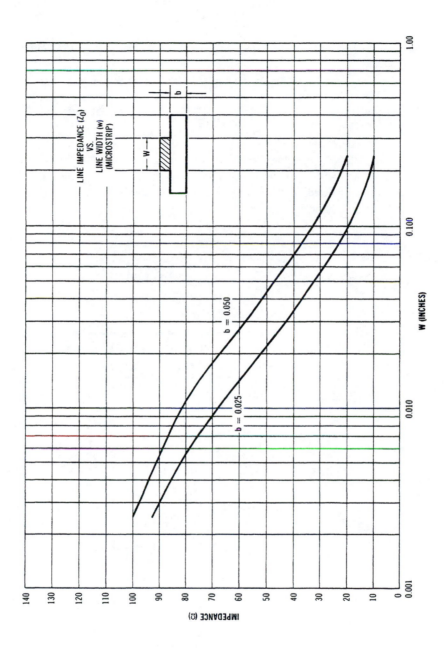

(B) Microstrip width for $\epsilon = 10.2$.

Figure 3–31 (Continued).

ance used. The constant will vary also with the dielectric thickness since the w/b ratio will change.

To aid in finding some of these values, Table 3–9 shows the line impedance, line width, relative dielectric constant (ϵ_r), and effective dielectric constant (ϵ_{eff}) for two common types of microstrip material: 0.030-inch PTFE glass ($\epsilon_r = 2.55$) and ceramic loaded PTFE ($\epsilon_r = 10.2$).

To understand how important it is to have the proper effective dielectric constant, consider the following example. Suppose we have a microstrip system that operates at 4.0 GHz and has a narrow band of operation, say from 3.95 to 4.05 GHz. Using ceramic loaded PTFE material ($\epsilon_r = 10.2$) with a thickness of 0.025 inch, we have a dielectric constant of 10.2 and a wavelength figure for our lines as follows:

$$\lambda = \frac{C}{f\sqrt{\epsilon}}$$

$$= \frac{3 \times 10^{10} \text{ cm/sec}}{4 \times 10^9 \, (2.54)\sqrt{10.2}}$$

Table 3–9. Values for Two Common Microstrip Materials

0.030-inch PTFE Glass Material, $\epsilon_r = 2.55$

Z_0(ohms)	w (inches)	ϵ_r	ϵ_{eff}
100	0.030	2.55	1.97
90	0.035	2.55	1.99
80	0.040	2.55	2.00
70	0.050	2.55	2.02
60	0.060	2.55	2.05
50	0.090	2.55	2.13
40	0.110	2.55	2.16
30	0.150	2.55	2.17
25	0.200	2.55	2.24

0.025-inch Ceramic Loaded PTFE, $\epsilon_r = 10.2$

Z_0(ohms)	w (inches)	ϵ_r	ϵ_{eff}
100	0.0030	10.2	6.17
90	0.0048	10.2	6.252
80	0.0070	10.2	6.339
70	0.0080	10.2	6.397
60	0.0105	10.2	6.451
50	0.0237	10.2	6.789
40	0.0365	10.2	7.050
30	0.0588	10.2	7.424
25	0.0770	10.2	7.661

$$= \frac{3 \times 10^{10}}{4 \times 10^9 \, (2.54) \, (3.19)}$$

$$= \frac{3 \times 10^{10}}{3.244 \times 10^{10}}$$

$$= 0.924 \text{ in}$$
$$= 2.35 \text{ cm}$$

By using the proper dielectric constant for microstrip (eff), we would have the following results (50V transmission line):

$$\lambda = \frac{C}{f\sqrt{\epsilon}}$$

$$= \frac{3 \times 10^{10}}{4 \times 10^9 \, (2.54) \, \sqrt{6.789}}$$

$$= \frac{3 \times 10^{10}}{4 \times 10^9 \, (2.54) \, (2.6)}$$

$$= \frac{3 \times 10^{10}}{2.64 \times 10^{10}}$$

$$= 1.13 \text{ in}$$
$$= 2.88 \text{ cm}$$

A 22% difference in the wavelength results when using the different dielectric constants. In the previous example, if we were 22% off in our calculations, we would be completely out of the band in which we wanted to operate; we would actually be operating at 4.92 GHz. So be sure when using microstrip to use the proper dielectric constant.

If there is one word that describes microstrip, it is simplicity. It is possible to build a microstrip circuit that can perform highly complex operations with the ease more commonly associated with using a carving knife. Some people have looked at a microstrip amplifier and asked "Where are all the parts?"

3.4.2   Grounds

When building microstrip, one precaution should always be taken—be sure that any grounds present in the circuit are established on the underside of the board. One method often used is soldering copper straps to the edges of the board at the top and bottom, along its entire length. If the ground is located at some point other than along the outside edge, a long narrow slot should be cut into the board, through which a copper strap is run and soldered to the board at its top and bottom. Do not use a piece of bus wire as a jumper through a microstrip board. Remember that in microwaves, grounds are not points, but planes. Also, when using metal-backed substrates a good ground connection can be accomplished by using grounding pins, which are driven through the ground point to the metal below and then soldered on top of the board. Many types of pins are used by fabricators. Plated-

Through-Holes, a technology said to be impossible on Teflon®-based material when this book first came out, is an excellent and very reliable method of obtaining a ground on a microstrip circuit board.

3.5   Waveguide

In actuality, every type of transmission line is, in a sense, a waveguide since it is designed primarily to "guide" or conduct energy (an electromagnetic wave) from one point to another. Whether the waves are thought of as electromagnetic field quantities or simple sinusoidal distributions of current and voltage, the meaning is the same. A transmission line is essentially like a track that steers the energy wave along a certain direction. Common usage, however, dictates that the term *waveguide* be restricted to indicate what might more completely be called a hollow-pipe waveguide, or at least a waveguide in which two distinct conductors are not present.

Waveguides are used at microwave frequencies for two reasons. Initially, they are often easier to fabricate than are coaxial lines; secondly, they can often be made to have less attenuation. Coaxial line requires that a center conductor be supported somehow in the center of a cylinder; a waveguide requires no center conductor, and its dielectric is air. Losses in a coaxial line over any distance (usually greater than a couple of feet), along with cable leakage at the higher frequencies, make the waveguide structure even more desirable.

One way of commonly deriving a rectangular waveguide is from a two-wire transmission line, as illustrated in Fig. 3–32. The transmission line is supported by two quarter-wavelength sections that, since the input impedance of each is theoretically infinite, have no effect on the transmission of power. If the number of stubs is increased to infinity, the rectangular waveguide is formed as shown. The "a" dimension of the waveguide cannot be less than one-half wavelength; in fact, it must be slightly greater to completely accommodate the transmission line function while preserving the insulating properties of the quarter-wave sections. Any frequency that makes the "a" dimension less than one-half wavelength causes the circuit to become an inductive shunt that allows no propagation. The frequency at which the "a" dimension is one-half wavelength (free-space wavelength) is called the cutoff frequency and is designated f_c. The free-space wavelength associated with this cutoff frequency is the cutoff wavelength, designated λ_c ($\lambda_c = 2a$). At dimensions greater than 2a some attenuation is experienced. Except for complications that may arise in attempting to avoid higher order modes when $\lambda_c < a$, all shorter wavelengths are passed readily. Thus, a waveguide is very similar to a high-pass filter in this respect.

3.5.1   TE and TM Modes

As mentioned before, microwave energy in a transmission line is, in effect, a wave. This concept is especially appropriate to the propagation of rf energy in a waveguide.

Waves of this type are of great importance, and more than a mere combination of two-plane transverse waves, they take on an identity of their own. They exist in either of two forms, designated as TE or TM modes.

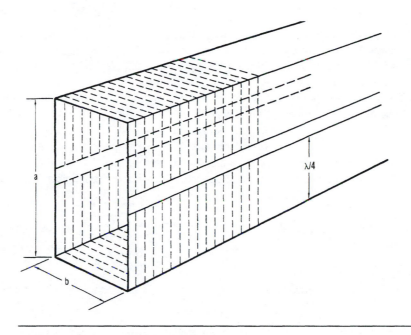

Figure 3–32.   Waveguide derived from quarter-wavelength sections.

In the *TE (transverse-electric) mode*, the electric field is transverse to, or across, the direction of propagation. In the *TM (transverse-magnetic) mode*, the magnetic field is transverse to the direction of propagation.

In addition, subscripts are used to describe the electric and magnetic field configurations. The general symbol is TE_{mn} or TM_{mn}, where the subscript "m" indicates the number of half-wave variations of the electric field intensity along the "a" (wide) dimension of the guide. The second subscript, "n," indicates the number of half-wave variations of the electric field in the "b" (narrow) dimension of the guide. The TE_{10} mode has the longest operating wavelength and is designated as the dominant mode in which the lowest frequency that can be propagated in a waveguide is found. Therefore, knowing the mode of operation and the dimension of the waveguide, the cutoff wavelength can be determined from this equation:

$$\lambda_c = \frac{2}{\sqrt{(m/a)^2 + (n/b)^2}} \tag{3.19}$$

where, a and be are measured in centimeters.

Rectangular waveguides are usually chosen so that only the dominant mode exists over a certain frequency range. The above consideration determines the "a" dimension. The "b" dimension is important because of the following considerations:

1.   The attenuation loss is greater as the "b" dimension is made smaller.
2.   The "b" dimension determines the voltage breakdown characteristics and, therefore, determines the maximum power capacity.

The power capacity depends on the maximum electric field intensity, "a" and "b." Therefore, "a" and "b" should be made as large as possible for higher power capacity. In practice, the dimension "b" is usually chosen to be about one half of "a," forming a ratio of 2:1. Standard waveguide dimensions are shown in Appendix F.

The *characteristic wave impedance* of a waveguide is analogous to the characteristic impedance of two-wire and coaxial lines. The wave impedance represents solely the ratio of the electric and the magnetic fields at a certain point in the waveguide. The actual wave impedance is of little use and is seldom determined. There is no advantage in extending calculations beyond the determination of the per-unit impedance since the actual measurement in a waveguide results in a normalized value. For reference purposes, the wave impedance equations for TE and TM modes are

$$Z_{TE} = \frac{\eta}{\sqrt{1 - (\lambda/\lambda_c)^2}} = \frac{\eta}{\sqrt{1 - (f_c/f)^2}} \qquad (3.20)$$

$$Z_{TM} = \eta\sqrt{1 - (\lambda/\lambda_c)^2} = \eta\sqrt{1 - (f_c/f)^2} \qquad (3.21)$$

where, η is the intrinsic impedance of the transmission line medium (377 ohms for free space).

Notice that the *waveguide impedance* depends on both the frequency for which the waveguide was built and on the medium used (usually free space). Fig. 3–33 shows $Z_{0(TE)}$ and $Z_{0(TM)}$ versus frequency. Note that one is always above the intrinsic impedance of the dielectric, η, the other is below, and they both approach η as the frequency approaches infinity.

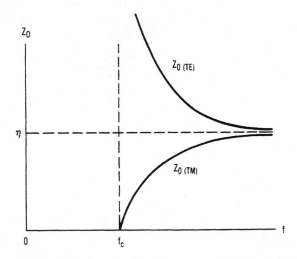

Figure 3–33.   Characteristic wave impedances of TE and TM modes versus frequency.

As is the case with ordinary wired circuits, waveguide impedance usually implies more than just the ratio of two numbers representing the peak, average, or rms values of the sinusoidally varying fields. The impedance may, as a special case, be pure resistance, or it may be reactive.

Table 3–10 shows some interesting similarities between waveguide and transmission lines previously covered.

We have referred to the TE_{10} mode previously. This mode is the primary and dominant one used in waveguide. The field configurations for the TE_{10} mode are shown in Fig. 3–34. Fig. 3–34A is an end view of the guide—that is, the view from looking directly inside the waveguide if it were possible to see the fields. Fig. 3–34B is the top view, and Fig. 3–34C is the side view.

Table 3–10. Transmission Line—Waveguide Analogy

Transmission Line Quantity	Equation or Symbol	Waveguide Quantity	Equation or Symbol
Voltage	V	Transverse electric field	E_t
Current	I	Transverse magnetic field	H_t
Characteristic impedance	$Z_0 = \sqrt{\dfrac{L}{C}}$	Characteristic wave impedance	$Z_{0(TE)} = \eta \dfrac{1}{\sqrt{1 - \left(\dfrac{\omega_c}{\omega}\right)^2}}$ or, $Z_{0(TM)} = \eta \sqrt{1 - \left(\dfrac{\omega_c}{\omega}\right)^2}$
Phase-shift constant	$\beta = \omega\sqrt{LC}$	Phase shift constant	$\beta = \omega\sqrt{\mu\epsilon}\sqrt{1 - \left(\dfrac{\omega_c}{\omega}\right)^2}$
Phase velocity	$\upsilon = \dfrac{1}{\sqrt{LC}}$	Phase velocity	$\upsilon_p = \dfrac{\upsilon}{\sqrt{1 - \left(\dfrac{\omega_c}{\omega}\right)^2}}$
Wavelength	$\lambda = \dfrac{\upsilon}{f}$	Wavelength (in waveguide)	$\lambda_p = \lambda \dfrac{1}{\sqrt{1 - \left(\dfrac{\omega_c}{\omega}\right)^2}}$

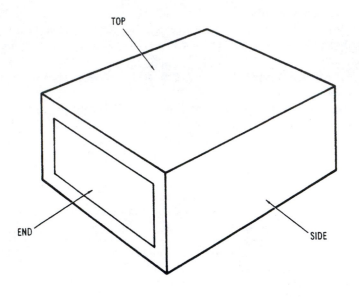

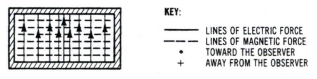

KEY:

———— LINES OF ELECTRIC FORCE
– – – LINES OF MAGNETIC FORCE
•   TOWARD THE OBSERVER
+   AWAY FROM THE OBSERVER

(A) End view of waveguide.

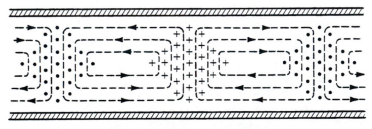

(B) Top view of waveguide.

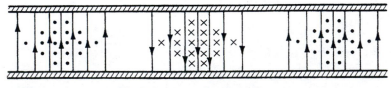

(C) Side view of waveguide.

Figure 3–34.   Field configurations of the TE_{10} mode in the rectangular waveguide.

Although higher order modes within a waveguide are possible, practical use of waveguides is centered around the dominant mode. The higher order modes are discussed only for awareness of their existence. In general, these modes are more highly attenuated and are considerably more difficult to recover from a waveguide system than the dominant mode (TE_{10}). Fig. 3–35 shows the field configuration for five higher order modes, the TE_{20}, TE_{11}, TM_{11}, TE_{21}, and TM_{21}.

The dominant mode (TE_{10}) without the presence of higher order modes is the desirable condition. With this condition there is lower power dissipation, and smaller, lighter, and cheaper waveguide structures are used.

Waveguide is not always found in the familiar rectangular form—sometimes it is circular. The circular variety is used in many special applications in microwave techniques. The circular guide has the advantage of greater power-handling capacity and lower attenuation for a given cutoff wavelength, but it is also larger and heavier. Also, the polarization of the transmitted wave can be altered by minor irregularities of the wall surface of the circular guide, whereas the rectangular cross section definitely fixes the polarization.

The wave of lowest frequency or the dominant mode in the circular waveguide is the TE_{11} mode. The subscripts that describe the modes in the circular waveguide are different than for the rectangular waveguide. For the circular waveguide, the first subscript, "m," indicates the number of full-wave variations of the radial component of the electric field around the circumference of the guide. The second subscript, "n," indicates the number of half-wave variations across a diameter. Also, the second subscript indicates the number of diameters that can be drawn perpendicular to all electric field lines, and in the case of TE_{01} waves, it indicates the half-wave variations of the electric field across a radius of the guide. Illustrations of various circular waveguide modes are shown in Fig. 3–36.

Before investigating some applications of waveguides in microwaves, a review of some of the important points covered thus far, along with a comparison of waveguides with coaxial or two-conductor lines, is in order.

1. The characteristic impedance and phase velocity are essentially independent of frequency in the TEM mode of propagation. Lossless transmission lines carrying TEM waves are nondispersive.

2. Propagation occurs in a waveguide only above a critical frequency called the cutoff frequency. Thus, a waveguide carrying a TE or TM wave may be considered to be a high-pass filter as far as the particular mode is concerned. At frequencies far above the cutoff point, the propagation characteristics approach those of free space; i.e., the waveguide wavelength approaches the free, unbounded wavelength in the medium contained inside the waveguide.

3. In the waveguide, the phase and group velocities and the wave impedance of TE and TM waves are functions of frequency. Therefore, the waveguide is a dispersive medium in that it tends to dissipate waves of different frequencies that may be applied simultaneously at a common point. This dispersion phenomenon

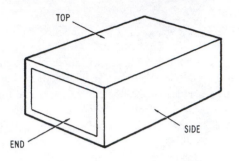

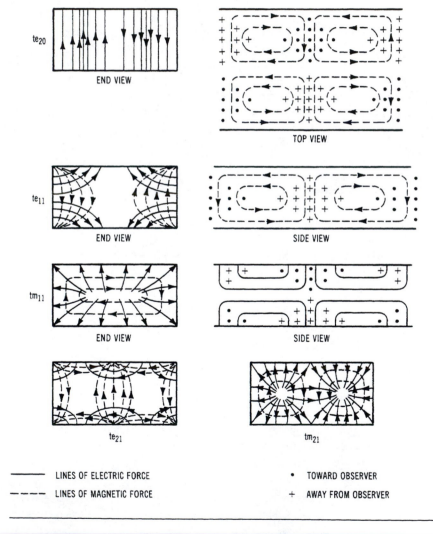

Figure 3–35. Field configurations of higher order modes in the rectangular waveguide.

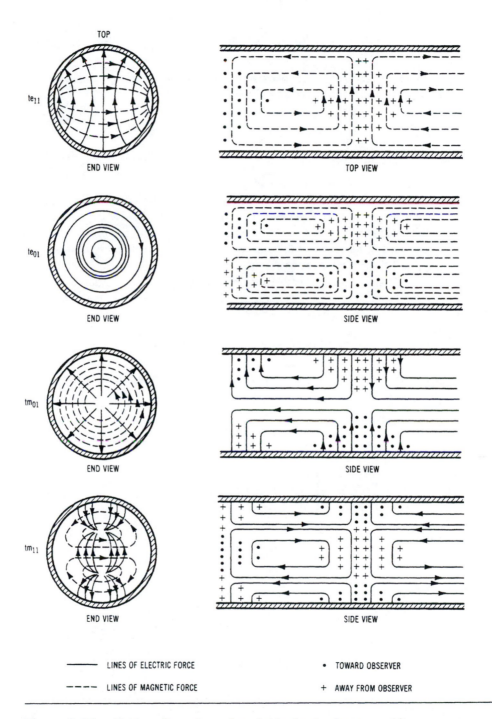

Figure 3–36.   Field configurations of modes in the circular waveguide.

causes distortion in waves that have modulation components extending over a wide frequency range.

4.   TEM waves cannot exist in closed waveguides without inner conductors, but TE and TM waves may exist as higher order modes on two-conductor or coaxial lines if the operating frequency is sufficiently high.

3.5.2   Junctions and Tees

Our first application will be with junctions and tees. These elements play an important part in waveguide circuitry and techniques. The two most common forms of junction are the E-plane and the H-plane tees. The former is also called a *series tee*, because the axis of the side arm is parallel to the E-field of the main transmission line; the latter is referred to as a *shunt tee*, because the axis of the side arm is parallel to the planes of the magnetic field of the main transmission line.

A combination of both tees forms a hybrid waveguide junction called a *magic tee*, shown in Fig. 3–37. The term is sometimes reserved for the hybrid junction in which matching structures have been introduced to improve the match of the E and H arms to the junction.

The characteristics of the hybrid in which the E and H arms are symmetrically placed are such that energy applied to either arm divides equally between arms 1 and 2, and no energy emerges from the opposite arm. If the fields entering arms 1 and 2 are equal in

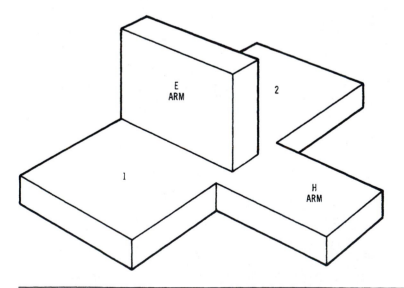

Figure 3–37.   The magic tee.

amplitude and of the same polarity, the net field in the E arm is zero, and the total energy emerges from the H arm.

At the junction, each arm is terminated effectively by the two other arms of equal impedance. Therefore, discontinuities are inevitable unless special precautions, such as compensated junctions, are observed. One manner in which the magic tee can be matched is by the use of tuning rods, whereby one rod is placed normal to the E field in the series arm, and the other is placed normal to the E field in the shunt arm. The precise location and configuration of the various matching structures are usually determined empirically. The magic tee also can be used as a phase shifter when the series and shunt arms are terminated with adjustable short circuits. Furthermore, the device has been used with microwave receivers where crystal detectors are placed in arms 1 and 2. The signal frequency enters the H arm, while the local oscillator signal is fed into the E arm. The tee provides isolation between sources of the two signals mixed in the crystals. The magic tee also is employed in microwave phase-measuring systems and various microwave bridge circuits.

3.5.3 Attenuators

A very useful waveguide component is the waveguide *attenuator*. It consists of an attenuator plate supported within the guide either by metal rods or by moving the plate in and out of the guide at a fixed location. In the latter method, a resistance card is inserted through a longitudinal slot in the top (center) of the waveguide, composing what is known as a *flap attenuator*. To reduce leakage, absorbent material is placed next to the slot.

Cross sections of the two types of waveguide attenuators are shown in Fig. 3–38. A drive mechanism is attached to the struts to move the vane in and out of the electric field to obtain the variable attenuation characteristic. Also, a dial is attached to provide accurate calibration of the attenuator. Waveguide attenuators can be matched by various means; the resistive transformer method, shown in the diagram, is only one such method. A match can also be obtained simply by tapering the vane at each end—a one-half wavelength taper provides a good match. In addition to the frequency sensitivity of the variable glass-vane attenuator, considerable phase shift is encountered since the moving vane is a dielectric material. (A much improved version of a waveguide attenuator is the *rotary vane attenuator* by virtue of its precision and direct-reading feature. It has a calibrated range of 50 dB, accurate within 2% of the decibel reading at any frequency in the waveguide band. The phase-shift variation of the attenuator is less than one degree between 0- and 50-dB variations.)

3.5.4 Couplers

A useful form of hybrid known as a *directional coupler* is constructed by placing an auxiliary section of uniform waveguide along the narrow or wide dimension of the main guide with appropriately located apertures connecting the two.

The *Bethe hole coupler*, shown in Fig. 3–39A, consists of a single-coupling aperture in the wide dimension of the guide. The electric field that leaks into the auxiliary guide is actually a combination of two fields: the fringing electric field through the hole and the field

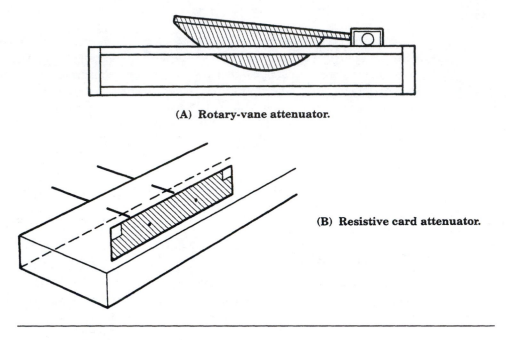

(A) Rotary-vane attenuator.

(B) Resistive card attenuator.

Figure 3–38.   Waveguide attenuators.

across the hole caused by the flow of charge through it. This charge flow is caused by the transverse component of the magnetic field of the wave. The directions of the component electric fields in the auxiliary guide are such that the fields tend to cancel in the forward direction from the aperture and reinforce in the reverse. The magnetic coupling is a function of the angle between the guide axis, whereas the electric coupling is essentially independent of the angle. Therefore, the maximum directional property can be obtained by an optimum setting of the angle between the auxiliary guide and the main guide, as shown in Fig. 3–39A. The coupling and directivity are sensitive to changes in frequency.

The *two-hole directional coupler* is illustrated in the cross-section diagram in Fig. 3–39B. The two auxiliary holes are placed one-quarter wavelength apart. The signal that travels back to the first hole from the second hole is 180° out of phase. The two signals tend to cancel in the reverse direction; in the forward direction, they are in phase and reinforce each other. The coupling and directional properties are impaired by off-frequency operation when the distance between the holes is no longer one-quarter wavelength.

The coupling factor is defined as the ratio, expressed in decibels, of the power entering the main line input to the power output of the auxiliary guide:

$$C = 10 \log \frac{P_i}{P_o} \qquad (3.22)$$

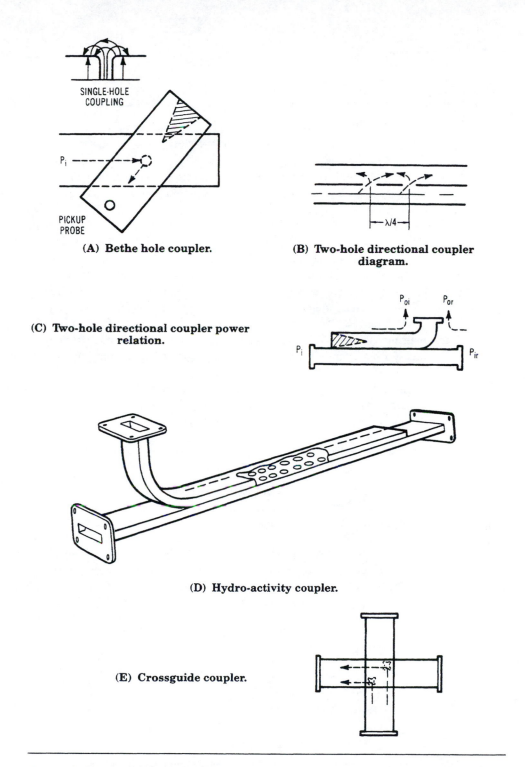

(A) Bethe hole coupler.

(B) Two-hole directional coupler diagram.

(C) Two-hole directional coupler power relation.

(D) Hydro-activity coupler.

(E) Crossguide coupler.

Figure 3–39. Waveguide couplers.

From the definition and illustration of Fig. 3–39C, the directivity is given by

$$D = 10 \log \frac{P_{oi}}{P_{or}} \qquad (3.23)$$

Directivity is defined as the ratio, in dB, of the powers out of the auxiliary guide when a given amount of power is alternately applied in the forward and reverse directions to the main guide.

The leakage signal due to the intrinsic directivity of the coupling holes, the reflected signals from the internal termination, and any discontinuities at the output flange of the coupler determine the total directivity signal.

Multihole couplers operate on the same basic principle as does the two-hole coupler. The coupling array providing high directivity is illustrated in Fig. 3–39D. The coupling versus frequency variations are ± 0.5 dB over the entire waveguide frequency range. Coupling is obtained through a series of graduated holes accurately machined along the broad or narrow face of the waveguide. In couplers designed for high-power operation, the holes usually are drilled in the narrow dimension of the waveguide.

In applications where a multihole coupler is not required, the less expensive and more compact crossguide coupler, shown in Fig. 3–39E, is used.

3.6   Chapter Summary

In the beginning of this chapter we stated that a microwave transmission line was much more than something to attach two points together. Throughout the chapter we have covered the full gamut of these microwave transmission lines: general theory, with inductance, capacitance, conductance, and resistance; reflective, VSWR, and characteristic impedance; terminations on microwave transmission lines; and coaxial cable structures, both flexible and semirigid. The topics of stripline, microstrip, and waveguide (both rectangular and circular) were covered to illustrate how they apply to microwave circuits.

These types of transmission lines verify our definition of transmission lines as devices that transfer energy from one point to another with a minimum amount of loss.

3.7   References

1.   Brown, R. G., Sharpe, R. A., and Hughes, W. *Lines, Waves, and Antennas*. New York: Ronald Press, 1961.

2.   Howe, H., Jr. Stripline Circuit Design. Dedham, MA: Artech House, 1974.

3.   Lance, A. L. *Introduction to Microwave Theory and Measurements*. New York: McGraw-Hill, 1964.

4.   Laverghetta, T. S. Microwave Measurements and Techniques. Dedham, MA: Artech House, 1976.

5. Olyphant, M., Jr., and Ball, J. H. *Dielectric and Packaging Factor in Stripline Performance*. Cu-Tip #2, 3M Company, St. Paul, MN.

6. Presser, A. "Properties of Microstrip Line." *Microwaves*, March, 1968.

7. Reintjes, F. J., and Coate, G. T. *Principles of Radar*. New York: McGraw-Hill, 1952.

8. Richardson, J. K. "Graphical Design of Strip-line Directional Couplers." *Microwaves*, October, 1967.

Questions

3.1 General Theory

1. Define *characteristic impedance*.
2. How was Figure 3–2 arrived at?
3. How was the standard 50Ω characteristic impedance determined?
4. What is VSWR?
5. Is a high VSWR or low VSWR desirable for low loss circuits? Why?
6. What is *return loss*?
7. Which circuit has the better VSWR, one with a return loss of 2 dB or one with a return loss of 26 dB? Why?
8. Describe the parameters of a matched condition on a transmission line.
9. Which circuit has the larger VSWR, one with $Z_L = 5Z_0$ or $Z_L = Z_0/5$? Why?
10. Why is the VSWR infinite on a transmission line terminated with a reactive load?
11. Compare the parameters for transmission lines terminated with a short and an open circuit.

3.2 Coaxial Cables

12. Define *coaxial cable*.
13. What factors determine the characteristic impedance of a coaxial cable?
14. How does the attenuation of a coaxial cable vary with frequency?
15. Why is the dissipation factor (tan d) used to determine attenuation in a coaxial cable?
16. Which material used for dielectrics slows the microwave energy the most?
17. What factors determine the capacitance/ft. in a coaxial transmission line?
18. What happens when the cut-off frequency is exceeded in a coaxial cable?
19. Describe pulse overshoot. Why is this undesirable?
20. Name two types of cable noise.
21. What is a *flexible* cable?
22. What is a *semirigid* cable?
23. Which cable, flexible or semirigid, has the lower losses at microwave frequencies?

3.3   Stripline

24. Define *stripline*.
25. What is the ground plane spacing in stripline?
26. Define t/b, s/b, and w/b ratios for stripline.
27. Name two types of stripline couplers and sketch their layouts.
28. Define *odd* and *even* mode impedances in stripline.

3.4   Microstrip

29. Why is effective dielectric constant used for microstrip?
30. What is the ground plane spacing for microstrip?
31. Define *filling factor*.

3.5   Waveguide

32. Describe the "a" dimension in waveguide.
33. Define TE Mode.
34. Define TM Mode.
35. What is the *dominant mode*?
36. In the expression TE_{mn}, what are m and n?
37. In a magic tee, describe what happens when energy is put into the E arm.
38. In a magic tee, describe what happens when equal energy is placed in ports 1 and 2.
39. Describe the difference between a rotary-vane attenuator and a resistive card attenuator.
40. Name three types of waveguide couplers.

Problems

3.1   General Theory

1. A transmission line has a VSWR of 3.2:1. What is the reflection coefficient and return loss?
2. The reflection coefficient on a transmission line is 0.415. What is the VSWR?
3. A return loss of -8.2 dB is read. Our VSWR cannot exceed 2.5:1. Is this return loss condition acceptable?
4. The characteristic impedance of a transmission line is 45Ω. The line is terminated with a 78Ω impedance. What are the VSWR, reflection coefficient, and return loss for this line?
5. The load impedance on a 75Ω transmission line goes from 225Ω to 25Ω. What is the range of VSWR for this variation in Z_L?
6. A 60Ω transmission line can tolerate no more than a 2.65:1 VSWR on the line. What is the maximum value of load impedance that can be used?

3.2   Coaxial Cables

7.   A coaxial cable has been found with no markings on it. If it is measured and found that D = 0.160″, d = 0.080″, and the dielectric is Teflon®, what is its impedance?

8.   A piece of RG-8 coaxial cable is to be used for a 2.5 GHz application. The cable is to be 50 feet long, and the attenuation cannot be more than 8 dB. Can RG-8 cable be used for this application? Explain.

9.   Find the velocity of propagation when materials with relative dielectric constants of 2.95, 3.2, and 6.0 are used.

10.   For RG-9 and RG-174, find the capacitance and inductance per foot. (Use a Teflon® dielectric.)

11.   A piece of RG-196 is intended to be used up to 15 GHz. Will this be beyond the cable's cut-off frequency, or can it be used?

12.   What is the 90% Higher Mode Cut-Off Frequency for the cable in problem 11?

3.3   Stripline

13.   What is the width of stripline on a piece of microwave material 0.030″ thick with 2-oz. copper if the impedance is to be 65Ω? (ϵ_r = 2.95)

14.   We need a stripline side-coupled coupler with 12 dB of coupling using material with ϵ_r = 2.5. What are the values of "s" and "w"? (Material is 0.20″ thick.)

15.   We need a stripline broadside coupler with 7 dB of coupling using material with a dielectric constant of 2.5 (0.030″ thick). What are the values of "s" and "w"?

3.4   Microstrip

16.   We have a microstrip circuit operating at 3.5 GHz on a material with ϵ_r = 2.6. The material is 0.030″ thick, and the trace to be etched is for a 45Ω line. What is the length of a quarter-wave line in inches?

17.   A 3-section microstrip power divider is to be designed with impedances of 70Ω, 85Ω, and 93Ω, respectively. If we operate at 4 GHz using material with ϵ_r = 2.6 and 0.060″ thick, how long will the quarter-wavelengths of each section be in centimeters?

3.5   Waveguide

18.   What is the cut-off wavelength for waveguide in the dominant mode where the dimensions are 1.00″ × 0.500″?

4 Microwave Components

Objective

To present the components used in microwave circuits and systems and explain the operation of each so that the reader may use this information to create specific circuits. The components to be covered will be couplers, hybrids, detectors, mixers, filters, circulators and isolators, attenuators, and chip components. Each will be explained and placed in representative applications.

Key Terms

Amplitude Balance
Attenuator
Bandpass Filter
Chip Capacitor
Chip Inductor
Chip Resistor
Circulator
Continuously Variable Attenuator
Conversion Loss
Cut-off Frequency
Detector

Directional Coupler
Directivity
Highpass Filter
Insertion Loss
Isolation
Isolator
Lowpass Filter
Mixer
Noise Figure
Phase Balance
Quadrature Hybrid

Rejection   Step Attenuator
Ripple   TSS
Sensitivity

4.1   Introduction

Thus far we have covered the materials that microwave components are made of and the transmission lines that connect them together. Now, let us cover the components themselves.

What do we mean by *microwave components?* They are the individual pieces, like parts of a puzzle, that make up a complete system. Each has its own function and its own specific task to perform, making the overall system operate the way it was designed. Just as in the case of a puzzle where each piece must fit together in the proper place to produce a final picture, so also the microwave component must be put in the proper place for the microwave system to produce its final "picture."

We will now cover eight types of microwave components. For each we will describe the component, provide a sample data sheet, define the terms on that sheet, and list applications for that component. Those types we will cover are the following:

- Directional couplers
- Quadrature hybrids
- Detectors
- Mixers
- Filters
- Circulators and isolators
- Attenuators
- Chip capacitors, resistors, and inductors

4.2   Directional Couplers

The directional coupler is a very appropriate component with which to begin our discussion. It is the basis from which some components evolve (quadrature hybrids) and is an integral part of others (mixers and attenuators). The directional coupler was mentioned earlier in Chapter 3 when we discussed stripline. Remember that we talked about side-coupled and broadside-coupled couplers when describing stripline applications. Fig. 4–1 shows both types of couplers as presented in Chapter 3 along with three-dimensional drawings to show the orientation of the lines. Notice that the broadside configuration has one line over the other, and the side-coupled device has these lines parallel. Between each line and surrounding each line is a dielectric material. The one point that should be evident about both of these couplers is that the lines never touch. One of the unique properties of directional couplers is that they provide action between two circuits without having any physical continuity between them.

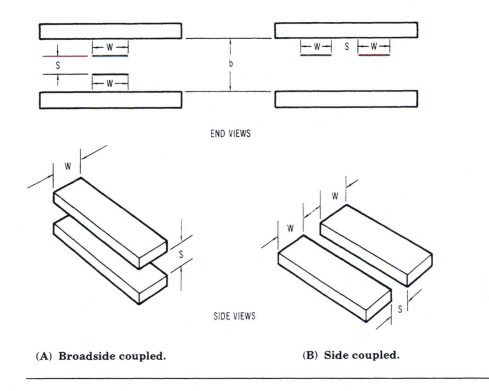

END VIEWS

SIDE VIEWS

(A) Broadside coupled.   **(B) Side coupled.**

Figure 4–1.   Directional couplers in stripline.

One exception to this is the branch-line coupler. This coupler has complete dc continuity but is a special case.

To help understand the directional coupler and why it can exhibit such properties as coupling and directivity, the basic coupler and the terms that go with it should be defined.

To begin our discussion, let us consider a coupler to contain two separate "circuits." First is the main-line circuit, and second is a coupled circuit. These circuits are shown in Fig. 4–2. The input is considered to be at port 1. The main-line circuit (ports 1 and 3) has the lowest loss, except for the quadrature hybrid coupler to be covered in the next section. In the quadrature hybrid coupler both circuits have equal loss. In all other couplers, however, the main line has the lowest loss. This loss can range anywhere from 0.1 dB up to 1.0 dB, depending on the coupling value of the device, which is called an *insertion loss*.

With microwave energy applied to the main-line circuit, a certain amount of energy will be impressed across the gap, S, into the coupled circuit. The amount of energy that crosses over between the two circuits depends on the spacing, S, between the two. The closer the circuits are together, the more energy is coupled between the lines (or circuits). Conversely, the farther they are apart, the less energy is coupled across. Also, the less energy coupled across to the coupled circuit, the less the effect on the

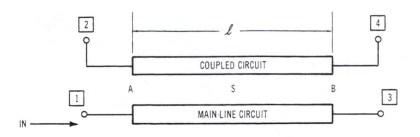

Figure 4–2.   Directional coupler circuit.

main line; that is, high values of coupling (20, 30, 40 dB, for example) have very low insertion losses. Low values (6 dB, for example) have higher main-line losses, so the effect on the main line is more than with a high-value coupler. The more energy taken in the coupler circuit, the less is available in the main line and, thus, a higher loss.

With the circuits shown in Fig. 4–2 energy could be coupled between the lines from either end. Since the coupling section is not really defined at this point, energy could probably be put in port 1 and result in a reasonable power reading at both ports 2 and 4 because, at this point, we have created two coupled lines, not a directional coupler. The term that makes all the difference in the world is the word *directional*. The parameter that brings about the property of being directional is the length of the coupling area, *l*. This length is set to be one quarter-wavelength at the frequency of operation or center frequency of the band of operation. Remember that in Section 3.1 on transmission lines, every quarter-wavelength a line showed opposite characteristics. If we had a low impedance on one level of a line, one quarter-wavelength away we would find a very high impedance. Conversely, a high resistance at one end results in a low resistance one quarter-wavelength away. If we apply these statements to the directional coupler, we can reason out how it becomes directional.

At the coupling end of the device (point A in Fig. 4–2), the energy is coupled very readily (the amount being dependent on the spacing of the lines). Thus, there is a low impedance to the microwave energy at this point. One quarter-wavelength away (point B), there will now be a high impedance to the energy in conformance with the theory previously discussed. This high impedance causes the majority of the input energy at port 1 to go straight through to port 3 with only an insertion loss and also to be coupled to port 2 where the low impedance is present. Very little energy will get to port 4. Thus, we have a directional transmission since the device will couple energy only in the forward port (port 2) and not the reverse port (port 4) when the microwave energy of the proper frequency is applied to port 1.

The same relationship will hold true if we make the input at port 3. The energy would readily be coupled to port 4 with very little present at port 2. The insertion loss of the device would be basically the same as before since it is now from port 3 to port 1 instead of from port 1 to port 3.

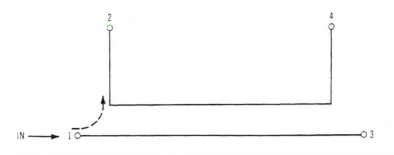

Figure 4–3. Directional coupler.

To summarize this discussion, let us refer to Fig. 4–3. The following designation can be put on the ports:

Port 1   Input
Port 2   Coupled port
Port 3   Straight through port (or insertion loss)
Port 4   Isolated port (isolated from the input)

One term in these designations has not appeared before in our discussions. That term is *isolated port*. We have previously said that this port has received very little microwave energy, but we have not referred to the port as being isolated. It is, in fact, truly an isolated port if the coupler is designed and operated properly. The actual energy at this port is a measure of how far the coupler deviates from an ideal design. The energy is, in actuality, a measure of the leakage present at the isolated port.

With the basic idea of a directional coupler in hand, we can now answer the question, What exactly is a directional coupler? You probably will agree that we can define this component as follows: A directional coupler is a component that allows two microwave circuits to be combined into one integrated system in one direction with the two completely isolated from each other in the opposite direction.

4.2.1 Data Sheets

To put all of the information covered up to this point into a practical perspective, we will now present a typical directional coupler data sheet and define the terms. Some of the terms have been touched on previously and will be covered again in a general way. Others have not been covered and will receive extensive definitions.

Model number	1000–20
Frequency	2 to 4 GHz
Coupling	20 dB
Coupling deviation	±1.0 dB
Insertion loss	0.25 dB (max)

Directivity	20 dB (max)
Input VSWR	1.2:1 (max)
Power	
Input	50 W
Reflected	10 W
Connector	SMA (female)
Size	3.0 in (76.2 mm) $\times$ 0.75 in (19.1 mm)
	$\times$ 0.375 in (9.5 mm)

Let us now take each of these terms and provide a definition for it.

Frequency   *Frequency* is the range of frequencies over which all of the specifications are valid. If the application is within this range, the coupling, loss, and all other parameters will be as indicated on the remainder of the data sheet. If the application is not within this frequency range, look for another coupler. This one will not work.

Coupling   *Coupling* is the amount, expressed in dB, that the coupled port is down from the input. Typical values are 6, 10, 20, and 30 dB. Special designs of virtually any value of coupling are possible by adjusting the spacing to the appropriate value.

Coupling Deviation   Since no device is going to be perfectly flat across even a narrow band of frequencies, there must be some allowances for any variations that occur. In our example we have allowed the coupling value to vary ± 1.0 dB. That is, the coupling figure in the data sheet can vary from 19 to 21 dB and still meet our specifications.

Insertion Loss   *Insertion loss* is the loss through the main line of the coupler. As we have previously mentioned, the loss depends on the value of coupling for the device. For high values of coupling (10, 20, 30 dB), the loss is small because very little energy is coupled from the main line. For low values (6 dB), the coupled port has much more effect on the main line and the loss is higher. This number also includes any copper losses through the device.

Directivity   *Directivity* is very important when performing reflection measurements. It is a measure of how well the coupler can isolate two signals, one forward and one reverse. In referring to Fig. 4–4 with the power in at port 1, we measure 20 dB down at port 2 and 50 dB down at port 4. The directivity would then be 30 dB, or the difference between the undesired coupling (ports 1 to 4) and the desired coupling (ports 1 to 2).

Input VSWR   *Input VSWR* is a measure of the input impedance compared to a common characteristic impedance (usually 50 ohms) of a system. This measurement tells how well the coupler will perform when connected to other components. A high VSWR will cause system problems when the coupler is inserted. The performance received will not be what was expected. A low VSWR is the desired parameter.

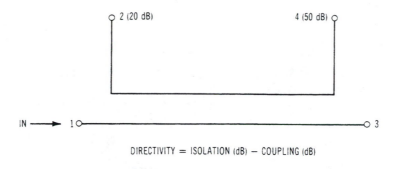

DIRECTIVITY = ISOLATION (dB) − COUPLING (dB)

Figure 4–4.   Directivity example.

Power   Two figures are shown on the data sheet—one for input (or forward) and one for reflected—because our example involves a 3-port coupler. So far, we have only shown 4-port couplers in our discussion. The 3-port coupler is a 4 port with the fourth port internally terminated. Fig. 4–5 shows a 3-port coupler, and from this figure it should be obvious why one power rating is higher than the other. With power applied to port 1, the normal coupling occurs, and the 50-watt rating is due to the construction and physical spacing of the coupler. The only consideration is the power rating of the component connected to port 2. In the reverse direction (power in at port 3), however, there is a terminating resistor to consider. These resistors are integrated into the overall coupler package and do not have much area to dissipate power. The reverse power rating on a 3-port coupler is always lower than that of the forward power.

Connectors   The term *connectors* is used here to emphasize the importance of having the proper connectors on the coupler. If the system is composed of SMA connectors, be sure the coupler has SMA connectors on it. Any other type of connector will require the use of adapters. Every time an adapter is used, there is an additional loss in the system and

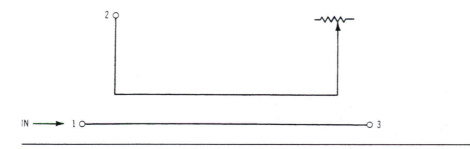

Figure 4–5.   Three-port coupler.

a VSWR setup at the junction of the adapter and the component. So don't add trouble to the system by introducing adapters to it. *Choose the proper connectors.*

Size   *Size* means just what it says—the physical size of the coupler. Most manufacturers give a variety of dimensions for couplers including mounting hole locations and coupler dimensions. These dimensions are usually given in both inches and millimeters.

4.2.2   Dual-Directional Couplers

Before covering applications of directional couplers, one special case should be discussed—the dual-directional coupler. This component is just what the name implies, two directional couplers. Fig. 4–6 shows the coupler, and you can see how the couplers are integrated back to back in a single package. This arrangement results in variations of the normal directional-coupler specifications. A comparison of a standard directional coupler and dual-directional coupler is shown here:

Specification	Directional Coupler	Dual-Directional Coupler
Coupling	20 dB	20 dB
Insertion loss	0.25 dB	0.75 dB
Directivity	20 dB	30 dB
VSWR	1.2:1	1.2:1
Power		
Input	50 W	50 W
Reflected	10 W	50 W

You can see that three areas have changed—insertion loss, directivity, and reflected power. The insertion loss has increased simply because a longer transmission line is on the main line for the energy to travel through. The directivity has improved because we are now taking advantage of the directivity (and isolation) of two couplers instead of only one. And, finally, the power rating in the reflected case is increased because we now are back to the basic 4-port coupler with no internal termination in the coupled path in a reverse direction as with the 3-port coupler. The terminations are both at the isolated ports.

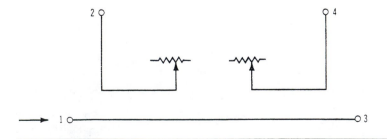

Figure 4–6.   Dual-directional coupler.

4.2.3   Coupler Applications

With the directional couplers and their associated terms explained, we will now present some applications of these components. Four applications are shown in Fig. 4–7. The first is a monitoring application (Fig. 4–7A). This application is used when a specific power level or frequency at a point in a system is needed. The directional coupler is ideal for this application because it can be inserted between the two circuits, as shown, and will not affect the operation of either. The insertion loss of the coupler will be virtually invisible when considering the overall system operation. Care must be taken to be sure that the coupling is of the proper level either to give the proper power meter reading or to trigger the frequency counter used.

A second application shown in Fig. 4–7B will be seen many times in the microwave testing chapter of this text (Chapter 10). The coupler is used in an automatic leveling circuit (ALC) when broad frequency ranges are swept and a leveled input to the circuit is needed. The coupler "samples" the microwave energy coming from the sweep generator. This energy is detected, and the dc voltage resulting is used to control internal circuitry in the sweeper to either increase or decrease the output level, as needed. Once again, a low insertion loss and good isolation make the coupler an ideal component for use in the application.

The third application (Fig. 4–7C) involves the dual-directional coupler. In this case the first meter (P1) reads the forward, or generator, power, and the second meter (P2) reads the reflected power from the load. By comparing the two power levels (P1/P2), a reading of the reflections from the load and, thus, the magnitude of the impedance of the load can be determined. This application takes advantage of the excellent directivity of the dual-directional coupler. This directivity is needed for the reflection application because the two signals must be kept separated to obtain an accurate reading.

The final application listed is one of power measurement (Fig. 4–7D). The coupler also could be used, once again, to measure output frequency. The coupler in this application is used as an attenuator to reduce the power level so that it is at a safe level to indicate the proper level on the power meter and yet not destroy the power sensor used with the power meter. The level to a counter must be reduced so that the front end of the frequency counter is not destroyed. The one consideration is that the terminating load resistor (R) in this application must have adequate power handling capability to survive the direct mainline power coming from the generator.

These are but four applications of directional couplers. As we progress through this chapter and the following chapters, many more applications will be observed and explained. Fig. 4–8 shows a directional coupler that is currently available.

4.3   Quadrature Hybrids

We referred to quadrature hybrids back in Chapter 2 when we were discussing microwave materials. At that time we covered thin Teflon® fiberglass laminates and their uses for microwave circuits. You will recall that the spacing between lines was so small that the normal side-coupled construction could not be used. Many facilities cannot hold good tolerances on

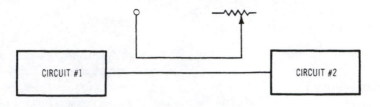

(A) Monitor.

(B) Leveling.

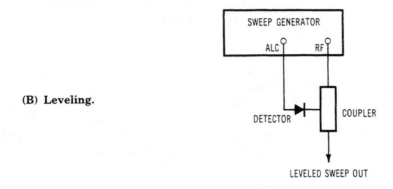

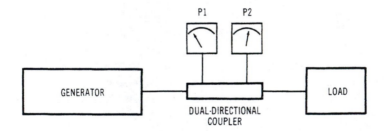

(C) Reflection measurement.

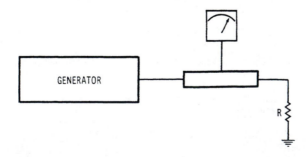

(D) Power measurement.

Figure 4–7. Directional coupler applications.

Figure 4–8.   Directional Coupler with a Frequency Range of 7–12.4 GHz *(Courtesy of Loral Narda).*

a 0.005-inch spacing for such a coupler. The broadside-coupled technique was thus the choice. Material of 0.005-inch thickness was used with the lines etched on both sides, and this etched material has been a very successful means of building this type of coupler.

The implications so far have been that the quadrature hybrid coupler is a special type of directional coupler, and the implications are absolutely right. Remember that the directional couplers we covered in the previous section had one port with a low loss (the main line) and the other port with a high loss (the coupled line). This loss was true for any coupler from 6 dB on up. The phasing of these two outputs was not considered because there was no constant relationship between them. The quadrature hybrid, however, is another matter. To see how different the hybrid is, consider the following definition: "A quadrature hybrid is a directional coupler whose two outputs are equal in amplitude and separated from one another by a constant 90° phase."

To illustrate these conditions, refer to Fig. 4–9. Notice that with an input at port 1, there is an equal split at ports 2 and 3 (−3 dB attenuation). There is also a 90° difference in phase between ports 2 and 3. This difference is where the term *quadrature* comes from. Anytime signals, objects, or graphical points are separated by 90° (or at right angles to one

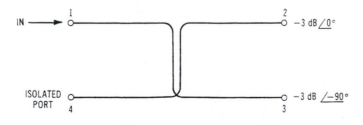

Figure 4–9.   Quadrature hybrid.

another), they are said to be *in quadrature*. So we now realize where one term comes from for the name of a component.

The other term that needs to be understood is *hybrid*. This term usually implies a combination or cross between two types of material. Plants, for example, are crossbred to produce hybrids, or some early radios were hybrids because they contained both transistors and tubes. What we are speaking about in the quadrature hybrid is a combination, or a junction. The hybrid, or hybrid junction, is a four-terminal network with two conjugate pairs of terminals. By a conjugate pair, we mean a pair of terminals oriented in such a way that an input applied to one of the branches will have no effect on the other. That is, two branches (or terminals) form a conjugate pair if they are isolated from one another. Ports 1 and 4 are a conjugate pair in Fig. 4–9 as are ports 2 and 3.

4.3.1   Data Sheets

When a device has two conjugate pairs (hybrid junction), power flowing into any terminal of the junction (port 1, for example) does not appear at its conjugate (port 4), but divides equally between the other two terminals (ports 2 and 3). Thus, the definition of quadrature hybrids can now be understood as presented earlier. Fig. 4–10 shows pictures of actual quadrature hybrids. To further understand the quadrature hybrid, consider the data sheet here.

Model number	HY-1
Frequency	2 to 4 GHz
Isolation	20 dB (min)
VSWR	1.2:1 (max)
Insertion loss	0.25 dB
Amplitude balance	±0.5 dB
Phase balance	±1.5°
Connectors	SMA

The data sheet terms will now be defined to help you understand what the sheet is saying and how to adjust the component to your applications.

Frequency   *Frequency* is the band of frequencies over which all of the remaining specifications apply. Below and above this range the specifications are not valid.

Isolation   *Isolation* tells you how good a conjugate pair is formed by the terminals. The isolation between terminals is very important to consider if the hybrid is going to be used to drive two active devices, such as transistors or diodes (covered under applications). If the isolation is not high enough, an imbalance in one device will affect both of them and may eventually cause one, or both, to be destroyed.

VSWR   The *VSWR* is a measure of how well the input of the coupler is matched to the system characteristic impedance (Z_0 is 50 ohms, normally). A low VSWR means a good

Figure 4–10   Quadrature hybrid's *(Courtesy of Loral Narda and Anaren Microwave).*

match; conversely, a high VSWR is a bad match. A 1.2:1 VSWR can be considered to be a good match for a quadrature hybrid.

Insertion Loss   Fig. 4–11 shows the *insertion loss* of a quadrature hybrid. It is the difference between the average output level and the ideal 3-dB mark. The internal losses that make up the insertion loss are copper losses, the input and output VSWRs, and that loss due to an imperfect coupler dumping some energy into the isolated port.

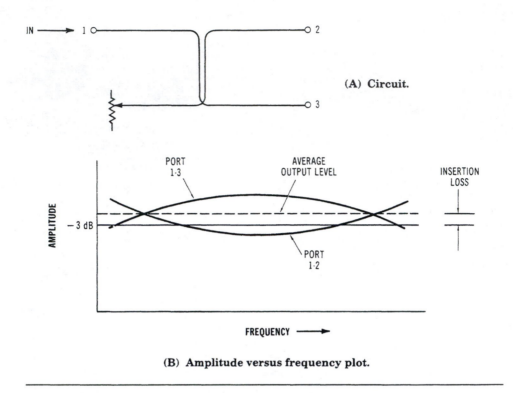

(A) Circuit.

(B) Amplitude versus frequency plot.

Figure 4–11. Insertion loss of a quadrature hybrid.

Amplitude Balance   This figure is a comparison of the power levels of the two output ports of the hybrid. Ideally, there should be no difference in amplitude between the two outputs, but in the real world there is some variation. This figure should be as low as possible so that each circuit driven by the hybrid will receive the same amount of microwave energy.

Phase Balance   The quadrature hybrid is designed to have a 90° phase separation between the output port, as we have previously stated. Because of variations within the device, however, this phase will vary with frequency. The phase balance figure shown on the data sheet is a measure of how well the phase difference between the two output ports tracks over the frequency band of operation.

Connectors   This term tells what connectors are on the device. If the connectors of the quadrature hybrid do not match those on the rest of the system, adapters will have to be used. Adapters result in increased losses, and each adapter has a VSWR that will cause problems within the overall system.

4.3.2 Applications

With the data sheet explained, we will now discuss applications of quadrature hybrids. Fig. 4–12 shows four common applications. The first is as a mixer. The properties of the hybrid junction make it ideally suited for this application. The conjugate ports are RF input, LO (local oscillator) input, and the two diode ports. For a mixer to perform properly, the RF and LO ports must be isolated and the diodes must receive the same level of energy. From our previous discussions, you can see how the quadrature hybrid provides all these requirements with ease.

Fig. 4–12B is similar to the mixer application, but there is only one input. The RF input is applied to the hybrid, and both diodes receive an equal amount of the RF input. The video output is then taken off the anode sides of the diodes.

Figs. 4–12C and D show applications that take advantage of the operation of quadrature hybrids when put back to back. If two hybrids are connected together and one port is chosen as an input, the diagonal port will have the same level at the output, with a small insertion loss. If we put two amplifiers in between the hybrids, as in Fig. 4–12C, we will combine the parameters (gain) of the amplifiers, and our amplifier will be additive and isolated. The same effect can be realized by placing *pin diodes* (to be discussed in Chapter 6) between two hybrids. In one case the diodes will be conducting, and the input signal will be at output number 1. In the opposite case the diodes will be turned off, and all the energy will be reflected to output number 2. Anytime an arrangement such as in Figure 4.12 (C&D) is used, it is termed a "constant impedance" device. This is because any disruption between the two quadrature hybrids (open or shorting of a transistor or diode) will not be seen at the input or output of the overall device because of the isolation properties of the hybrid, as previously explained. Thus, the input and output never see a change in impedance or match because no additional reflections are present. It displays only a *constant impedance*.

There are many different ways of using a quadrature hybrid coupler in microwaves. Any place that needs an isolated port and even amplitudes with a quadrature phase (90°) is where the quadrature hybrid finds a home.

4.4   Detectors

With detectors, the terms reliability and simplicity have a close relationship. The fewer parts involved and the simpler the design, the less that can go wrong—thus, a more reliable system. The *microwave detector* is one of the microwave components based on this idea of simplicity. Fig. 4–13 shows a block diagram of a detector. As can be seen from the figure, the detector takes an rf signal, rectifies it with a diode, filters out the high-frequency ripple, and sends out a dc signal proportional to the rf input level. This component is very useful when a microwave signal needs to be displayed on a conventional oscilloscope. Applications of the microwave detector will be covered later in this section.

Let us now examine the detector block diagram in more detail. The first block is an input matching circuit. Its purpose is to match the diode impedance to the system characteristic impedance ($Z_0 = 50$ ohms). Fig. 4–14 shows a plot of rf impedance of a Schottk diode used

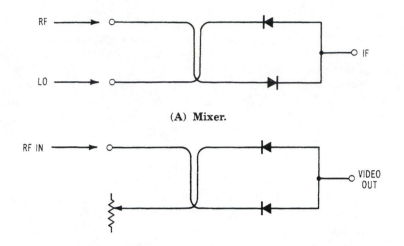

(A) Mixer.

(B) Matched detectors.

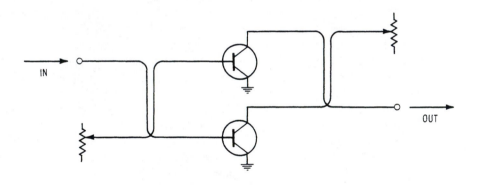

(C) Combining amplifiers.

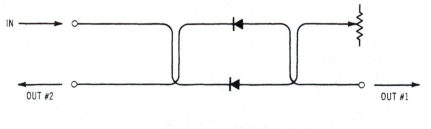

(D) Spst pin switch.

Figure 4–12. Quadrature hybrid applications.

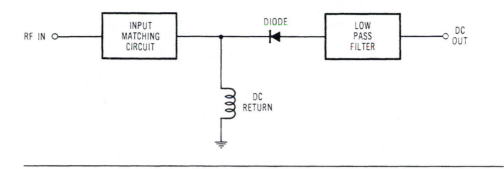

Figure 4–13. Microwave detector.

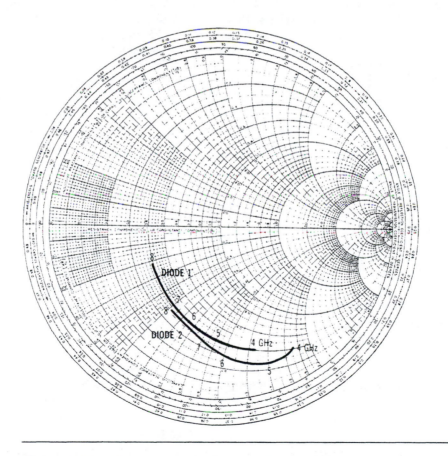

Figure 4–14. Detector diode impedance plots.

for microwave detectors on a Smith chart. The *Smith chart* indicates the impedance of a device. When *normalized* (the center of the chart is 1.0) as shown in Fig. 4–14, any imaginable impedance may be determined because this chart may be multiplied by the particular impedance. When we make the chart a 50-ohm chart, for example, all the numbers within the chart are multiplied by 50, and the impedance may be read directly. The plot in Fig. 4–14 goes from 4 to 8 GHz and shows two different diodes. If we are using a 50-ohm system for diode number 1, the impedance goes from $20 - j62.5$ ($0.4 - j1.25$) ohms at 4 GHz to $20.5 - j15.5$ ($0.41 - j0.31$) ohms at 8 GHz. For diode number 2 the impedance goes from $20 - j87.5$ ($0.4 - j1.75$) ohms at 4 GHz to $17.5 - j25$ ($0.35 - j0.5$) ohms at 8 GHz. These impedances would need to be matched to 50 ohms to ensure a maximum transfer of power from the rf source to the diode.

The diode is, of course, the heart of the detector. As we have mentioned, the Schottky diode is used usually for such an application because of its excellent rf characteristics, including high sensitivity. The Schottky diode will be covered in detail in Chapter 6.

The dc return noted in Fig. 4–13 serves a dual function. It provides a ground for the diode current to return to and also is used as an rf choke so that none of the rf energy going to the diode is shunted to ground. Instead, the diode receives all the power coming through the matching circuit.

The low-pass filter at the detector output is designed to remove the high-frequency ripple produced as a result of the rectification process within the detector. This filter could be a simple shunt capacitance or a series of capacitance and inductance, depending on the particular application of the detector.

4.4.1   Data Sheets

To aid in understanding the microwave detector, let us present a typical data sheet and explain the terms it contains.

Characteristics	*Specifications*
Model number	DET-1
Frequency	1 to 18 GHz
Frequency response	±1.25 dB
Sensitivity	1500 mV/mW
Tangential sensitivity (TSS)	−52 dBm (min)
Impedance	50 ohms
Maximum input	200 mW
Polarity	Negative (see options)
VSWR	1.5:1
Connectors	Input—SMA(M)
	Output—BNC(F)

Options: 1A: Positive Output, 50 ohms
2A: Negative Output, 75 ohms
2B: Negative Output, 75 ohms

Frequency   *Frequency* refers to the frequency range over which the specifications listed in the data sheet are valid.

Frequency Response   *Frequency response* is used to indicate the rf performance of the detector. It is sometimes given in dB/octave or in one figure for overall performance. Basically, it is a measure of the variation in sensitivity of the detector, measured in decibels.

Sensitivity   To understand what is meant by sensitivity, refer to Fig. 4–15. This curve is not the curve for our sample data sheet but a typical curve of a commercially available detector. The two factors affecting sensitivity are the input power and the output voltage. We, therefore, define sensitivity as how much voltage is produced for a given input power level. Thus, the parameters millivolt/milliwatt come about. One additional factor must be considered when speaking of sensitivity. That factor is the load resistance, R_L. You can see in

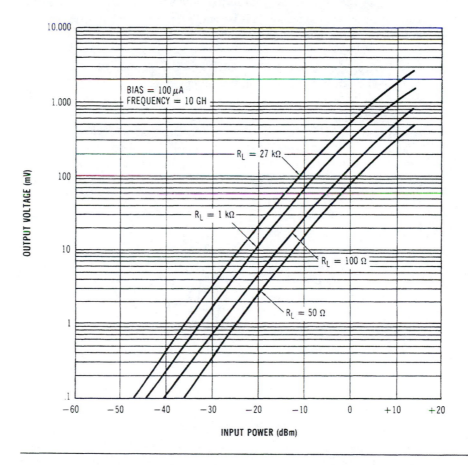

Figure 4–15.   Detector sensitivity curve.

Fig. 4–15 that four curves for sensitivity are generated with four values of R_L. Notice how output voltage will vary if, for example, we have a circuit that produces a -10 dBm input level and if RL is 100 ohms; we would have 30 millivolts at the output. If the resistance is increased tenfold to 1 kilohm, we would have 80 millivolts. Thus, the value of load resistance is of prime importance when specifying the sensitivity (or mV/mW) of a detector.

Tangential Signal Sensitivity (TSS)   TSS is a direct measure of the signal-to-noise voltage. The measurement is carried out with a pulse signal. The level of this pulse signal is adjusted so that the highest noise peaks observed on a scope with no signal are the same level as the lowest noise peaks when a signal is present. This pulse signal corresponds to a signal-to-noise ratio of approximately 2.5, which becomes 8 dB when it is plugged into the noise figure formula: $NF_{dB} = 20 \log 10$ (noise ratio). The log is multiplied by 20 since we are discussing voltage and not power. The presentation at TSS is shown in Fig. 4–16. To summarize TSS we can say that it is the rf input signal level that produces an 8-dB video output-to-noise voltage ratio. It is basically the lowest input level where a usable output voltage can be expected.

Impedance   *Impedance* tells what the rf input impedance of the detector is. In our case it is 50 ohms, which does not mean that all detectors are 50 ohms. The detector to be used should be checked to ensure it is the proper input impedance for your system.

Maximum Input Power   This term is the maximum level of power that may be applied to the rf input of the detector without damaging the diode. You can see from our original block diagram of the detector that there is a diode in series with the input. Anytime there is a semiconductor oriented this way, care must be taken with regard to input power level.

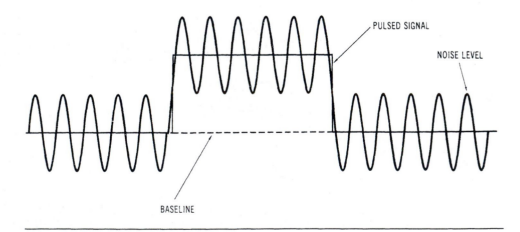

Figure 4–16. TSS display.

Polarity   This is the polarity of the output signal, either positive or negative. The options listed on the bottom of the data sheet allow for both polarities with different input impedances. This term is important because it will have to be connected to some external circuit, and that circuit will probably require a specific polarity input for its proper operation.

VSWR   This term tells how well the input matching circuit is working. It tells how much of the input power is getting to the diode and how much is being reflected back to the input because of mismatches. The lower this value is, the more efficiently the input matching circuit is functioning.

Connectors   This term is added in to once again emphasize the importance of the proper connectors. To pay for a detector that has a low input VSWR and then have to put a couple of adapters on the unit to use it defeats the whole purpose of obtaining a good match because all adapters have a VSWR and an insertion loss. Make an effort to buy detectors with the proper connectors for your system.

4.4.2   Applications

With the microwave detector described and its parameters defined, we will now investigate a few applications. Fig. 4–17 shows an actual detector that could be used in these applications.

Fig. 4–18 shows three applications where the microwave detector finds wide usage. Fig. 4–18A is the same application we cited for directional couplers—a circuit monitor. The microwave signal coming down the line is sampled by the coupler, and the detector provides a dc voltage proportional to the power on the line. This dc voltage is used to register on a monitoring device (meter, light, etc.) to give an indication of what is happening on the main line.

Figure 4–17.   Small Detector
(Courtesy of Loral Narda).

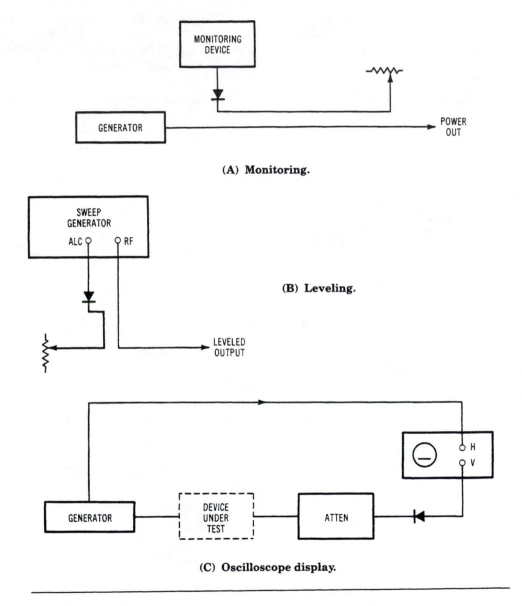

(A) Monitoring.

(B) Leveling.

(C) Oscilloscope display.

Figure 4–18. Detector applications.

Fig. 4–18B also was presented in our section on directional couplers. The application is referred to as leveling. Once again, a coupler and detector work together to sample the main line rf; the detector converts the rf to dc to activate the leveling circuits. This dc voltage causes either the rf power out of the source to be attenuated or the attenuation to be removed to raise the power level. The result of this instantaneous pattern of action and reaction of coupler and detector is that the rf power remains level at the system output. This ALC (automatic leveling circuit) was referred to previously and is used in many test setups.

The application shown in Fig. 4.18C is using the detector to display microwave energy on an oscilloscope. Most conventional oscilloscopes do not come anywhere near having the capability to directly display a microwave signal, although the response is getting higher for modern-day scopes. Also, storage scopes are available that will display the lower end of the microwave spectrum. Network analyzers can be used but they generally are not standard equipment in every laboratory. So Fig. 4.18C is an inexpensive and easy answer. A simple combination of an attenuator and a detector does the job very nicely. The attenuator is of great importance in the setup, since you should keep below a maximum power level. The attenuator ensures that you do, and it also keeps you in the linear range of the sensitivity curve of the detector (Fig. 4–15), giving much better results.

It should be evident from our discussion that this highly reliable and simple microwave component has some critical jobs to perform within the microwave spectrum. It performs these jobs very well and does so time and time again because of its basic idea of "the simpler the better."

4.5   Mixers

Every component presented so far has required a single microwave signal for its operation. The microwave mixer, however, requires two signals to perform its required function. These two signals are termed RF (radio frequency) and LO (local oscillator). Fig. 4–19 is a schematic representation of a microwave mixer showing both the necessary inputs and the resulting output, the IF (intermediate frequency).

Fig. 4–20A is a basic block diagram showing the three fundamental parts of a mixer—the input coupling network, the diode circuitry, and the output filtering network. Fig. 4–20B shows a mixer that is commercially available. The input coupling network is the circuitry that allows the RF and LO signals to be combined properly before being applied to the diode portion. This network could be a directional coupler, a transformer, or a quadrature hybrid; its prime feature should be that, under operating conditions, it combines the two signals while providing isolation between them. The diode circuitry consists of the diode(s), any matching circuitry used, and a dc return for the diodes; this area is where the actual mixing process occurs. When the RF and LO signals are applied to the diode circuitry, many frequency combinations of the two are generated. The signals are at different levels, but the ones that are most prominent are each of the originals and their sum and difference. In other words, if the RF signal is called f_1 and the LO signal called f_2, the output would have f_1, f_2, $f_1 + f_2$, $f_1 - f_2$, and every combination of fundamentals and harmonics ($2f_1$, $2f_2$, $2(f_1 + f_2)$, etc.).

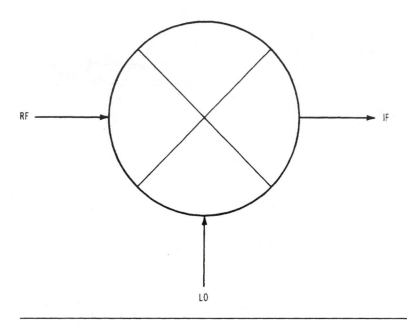

RF ──────────►

IF

LO

Figure 4–19. Mixer schematic.

These combinations of frequencies are the reason why the final block is needed in a mixer—the output filtering network. Usually, the upper or lower sideband (that is, $f_1 + f_2$ or $f_1 - f_2$) is the desired output; everything else must be eliminated to produce the desired and necessary clean signal. The filtering usually is accomplished by a low-pass network that might be as simple as a shunt capacitor or as complex as a large number of low-pass filter poles; it could be a series of trap circuits that would eliminate only specific unwanted signals. For certain applications, the filtering could be performed by a bandpass or high-pass filter. The type of filtering and device used depends, of course, upon the individual situation.

The RF signal to the mixer is the primary signal; that is, it is the one that initiates the operations. This signal can be compared to the input of an amplifier, a coupler, or a detector. It is usually at a considerably lower level (-20 dBm, for example) than the LO signal, and the RF signal should be as clean a signal as possible (low spurious and harmonic content).

The LO signal provides the RF bias for the mixer and turns the diodes on inside the mixer. It is at a much higher level than the RF input and can range from 0 dBm up to $+23$ dBm. There is usually an optimum level of power for the LO that produces a minimum conversion loss. (Conversion loss is the loss between the RF input and the IF output. It gets its name from being a loss generated as a signal is converted from one frequency to another.) This level should not be too high or too low. If it is too low, the diodes will not be biased correctly, and the conversion loss will be high. If the level is too high, the diodes will be overdriven, and the conversion

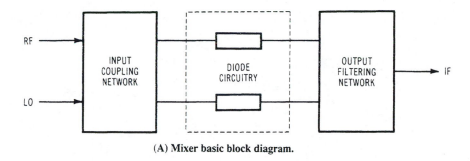

(A) Mixer basic block diagram.

(B) Image Rejection Mixer

Figure 4–20.   Microwave mixer diagram and sample mixer *(Courtesy of Anaren Microwave).*

loss will once again be too high. Also, an excessive LO level can damage the diodes.

The IF signal is the desired output signal. It is usually $f_1 - f_2$ but could also be $f_1 + f_2$. The level of this signal will depend upon the characteristics of the mixer itself.

4.5.1   Single-Diode Mixers

The simplest form of mixer is the *single-diode mixer*. Fig. 4–21 is a diagram of the single-diode mixer. It fulfills the requirements of being a mixer by containing the

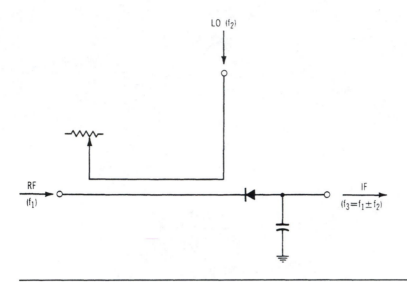

Figure 4–21. Single-diode mixer.

three main sections referred to earlier. It has an input coupling network (a directional coupler), a diode circuit (the single diode), and an output filter network (the shunt capacitor that acts as a simple low-pass filter). The mixer obtains RF to LO isolation because of the characteristics of the directional coupler. The RF is applied to the main line of the coupler and the LO to the isolated port. In this way they can be combined at the diode but are isolated from one another. This is true because the coupler is a reciprocal device.

The advantage of such a circuit can be seen in its simplicity. It is simple to design and construct since only a coupler, one diode, and one capacitor are needed. Because of this simplicity, it stands to reason that it is also a very low cost circuit.

As simple and low cost as the single-diode mixer is, the disadvantages quickly cancel any advantages we have listed. The component is a narrow-band device since only one diode is used. This use of only one diode also means that the noise terms within the mixer are not cancelled. The overall noise figure of the mixer, thus, is high. The conversion loss of the mixer also is high because of inefficiencies within the coupling structure. Actually, the directional coupler shown in Fig. 4–21 is the optimum coupling structure for this type of mixer and still yields higher than normal conversion loss. Other methods that may be used are capacitive or inductive probes on a transmission line or even a simple "T" connection. All these, however, result in worse performances than the directional coupler.

The single-diode mixer does not find many applications in the microwave field anymore. The limitations previously listed have diminished its usage greatly.

4.5.2   Balanced Mixers

Since the single-diode mixer does not do the job, there is a "better mousetrap" to overcome the disadvantages of the single-diode mixer and provide the necessary performance, the *balanced mixer*.

Fig. 4–22 is a comparison of the single-diode mixer and the balanced mixer. The improvements are readily seen. The input to the balanced mixer is a quadrature hybrid instead of a directional coupler. Remember that the quadrature hybrid had the property of having conjugate pairs; that is, when power flows into one terminal of a hybrid junction, the power is isolated from the conjugate terminal, and the power divides equally between the other two terminals. Because of this hybrid junction, if we apply the RF signal to its terminal, it will be isolated from the LO port and divided equally for application to the circuits of the diodes. The same properties are true considering the LO port input. This improvement contributes greatly to the *balanced* properties of the mixer.

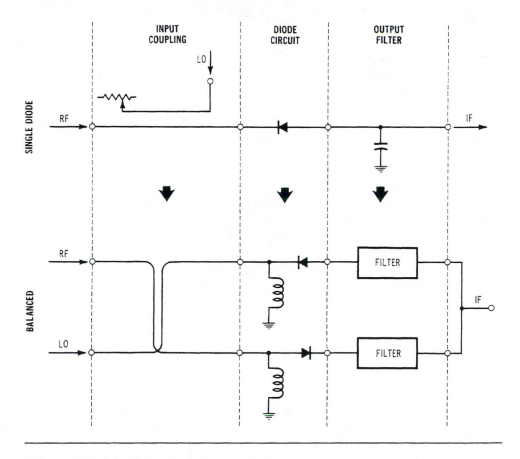

Figure 4–22.   Single-diode and balanced mixer.

The second area that is different is the diode circuits. In the single-diode mixer there is, of course, only one diode. In the balanced mixer there are two diode circuits. We refer to this area as diode circuits instead of simply diodes because there are times when more than one diode is used in each arm. These may be a matched pair or even four diodes per circuit. These diode circuits, called *quads* when four diodes are used, provide matched operations that reduce conversion loss and also the noise figure of the mixer.

The third section is the output filtering. In the single-diode mixer a shunt capacitor works quite adequately. In the balanced case there is a low-pass filter for each leg of the device. Many times there are a number of filter sections for each leg (two sections, for example), and one common section is added when the two legs are tied together at the IF output.

There are many advantages to using a balanced mixer for microwave applications. They have very good noise characteristics, low conversion loss, and good LO-RF isolation, and they can be used for wideband applications because of the quadrature hybrid broad-band (usually octave and greater) characteristics.

The only real disadvantage of balanced mixers (if you can call it that) is that the circuit is more complex to fabricate. The diode circuits and input networks are much more sophisticated than the single-diode mixer, and more time must be taken to build such a device. Their performance is so superior, however, that this disadvantage is far outweighed. The balanced mixer is used in many microwave applications. It is by far the number-one choice for microwave systems in use today. It is considered to be a workhorse component for the microwave industry, and it is surely a fine workhorse.

4.5.3 Image-Rejection Mixers

There are some times when the balanced mixer will not quite fill the bill for a particular application. You may need a specific IF signal and need it to be much cleaner (spurious and harmonic free) than the balanced configuration can give. What is needed at this point is an *image-rejection mixer*. This mixer performs the task its name implies; it rejects the unwanted image (sideband) coming out of a mixer. If we have an RF signal (f_1) and an LO signal (f_2) applied to a mixer, we will have a sum ($f_1 + f_2$) and a difference ($f_1 - f_2$). The value of these will depend on the frequency of the LO—that is, whether it is above or below the incoming RF. If we consider the most common case, where the LO is below the RF in frequency, we will want a signal that is $f_1 - f_2$. There will, however, be the ever-present image (or upper sideband) at the output also. Filtering will reduce this signal but will not reduce it far enough for many applications, such as for use in a spectrum analyzer where distinct signals must be measured. To be sure the unwanted image is rejected, the conventional balanced mixer must be abandoned, and an image-rejection mixer must be used.

Fig. 4–23 shows a block diagram of an image-rejection mixer. It consists of an input quadrature hybrid coupler for the RF signal and an in-phase power divider for the LO signal. This combination sets up the proper phasing for each of the balanced mixers used in this circuit.

Up to this point we have simply combined two balanced mixers and fed them with the proper phase to produce two IF outputs—IF 1 and IF 2. These two IF outputs are

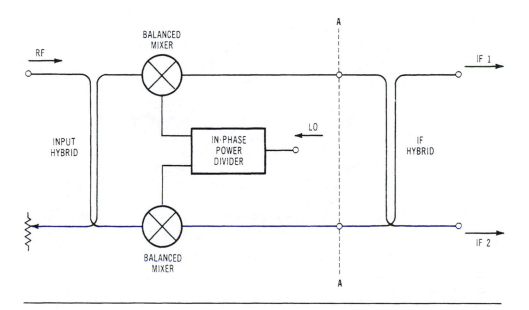

Figure 4–23.   Image-rejection mixer.

all of the circuitry to the left of the line AA. What really makes the image rejection mixer tick is what is to the right of the line AA—the IF quadrature hybrid labeled IF HYBRID. It is designed to operate at the IF frequency, whether it be 30 MHz, 400 MHz, or 2 GHz. One output (IF 1) produces the upper-image frequency; the other (IF 2) produces the lower-image frequency. The quadrature characteristic (90° phase relationship) results in only one of these frequencies being present at each port when the other port is terminated. It is very important that the image port not wanted be terminated.

The rejection of the unwanted image depends on a balanced condition. Balanced condition means both phase and amplitude coming from the two balanced mixers and within the hybrid itself are balanced. Remember that we mentioned both phase and amplitude balance when we discussed quadrature hybrids and how important it is to have good balance. Here is one of the applications that bears out that statement. Fig. 4–24 shows two graphs indicating how the rejection of the image is decreased (made worse) by increasing imbalances in both amplitude and phase. It is easy to see that it does not take much imbalance in amplitude to decrease the rejection of the image. The phase imbalance is not quite as critical but still must be held as small as possible to obtain reasonable rejection. These graphs come from the following expressions:

Amplitude

$$\text{Rejection (dB)} = 20 \log \left(\frac{V_1 + V_2}{V_1 - V_2} \right) \tag{4.1}$$

where, V_1 is the amplitude of signal 1 to the IF hybrid,

V_2 is the amplitude of signal 2 to the IF hybrid,

θ is the phase error between 1 and 2 (assume $\theta = 0$).

Phase—if $V_1 = V_2$,

$$\text{Rejection (dB)} = 20 \log \left(\frac{\sqrt{1 + \cos \theta}}{\sqrt{1 - \cos \theta}} \right) \tag{4.2}$$

The main advantage of the image-rejection mixer is that it does away with the unwanted signals. Doing away with these signals is of great value when single-frequency receivers are used or in a spectrum analyzer front-end application. These applications demand a clean IF signal for system operation.

The only disadvantage would be its complexity to design and fabricate. If the band of operation is not too wide, however, this would not be too large a problem.

4.5.4 Data Sheets

To help understand the microwave mixer, we will present a typical data sheet and explain the terms. Since the balanced mixer is the most widely used type, we will present a sheet characterizing such a component.

Model number	M102
Frequency (RF & LO)	0.5 to 18 GHz
Frequency (IF)	2 to 5000 MHz
Conversion loss	9.5 dB (max)
Isolation (RF/LO)	20 dB (min)
Noise figure	10 dB (max)
VSWR: LO	2.5:1
RF	2.0:1
LO power	+13 dBm
Connectors	SMA

The following terms are the primary parameters that characterize a mixer and its operation: conversion loss, isolation, noise figure, and LO power. Because they are the primary parameters, we will provide definitions for only these terms. The other terms have all been defined many times through the text.

Conversion Loss   *Conversion loss* , as previously mentioned, is the level of the IF output as compared to the level of the RF input. Usually expressed in decibels, it can be considered as a measure of the efficiency of the mixer; that is, it indicates how efficiently the two input frequencies are combined to form the IF output.

Isolation   It may be any one of three combinations of measurements referred to by this term: RF/LO, RF/IF, or LO/IF. In each case, *isolation* is a measure of the undesired frequency at an auxiliary port. An example would be as follows: the RF/LO

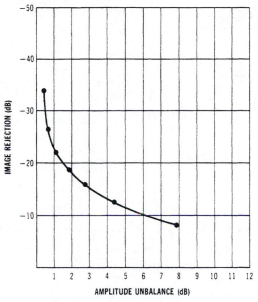

(A) Image rejection versus amplitude unbalance.

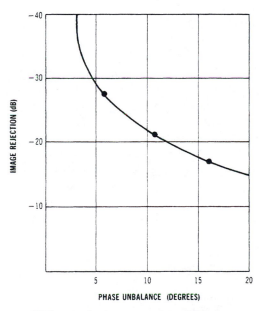

(B) Image rejection versus phase difference.

Figure 4–24. Amplitude and phase balance for image-rejection mixers.

isolation figure, in decibels, is the amount of RF power measured at the LO port compared to that at the RF port. (The other two isolation figures are found the same way.) The RF/LO isolation is a measure of the operation of the input coupling network (a quadrature hybrid, for example). Any isolation between the inputs and the IF output is a measure of diode efficiency. This isolation is controlled primarily by matching all LO drive levels.

Noise Figure   *Noise figure* and conversion loss are two of the most important and closely related mixer parameters. The primary requirement for a mixer is to provide the maximum IF output power with the minimum RF input power while generating the least amount of noise. Every network, whether a mixer or other component, has a certain amount of internal noise associated with its operation. A comparison of this level of noise to that of the desired signal in the network is called the *signal-to-noise ratio*. There are *input* and *output* signal-to-noise ratios for each network under consideration. The noise figure measures the reduction of a signal-to-noise ratio by the network under consideration (for example, the balanced mixer). It is a term expressed in decibels that compares the input and output signal-to-noise ratios. Expressed mathematically, it is

$$NF = 10 \log \frac{S_i/N_i}{S_o/N_o} \tag{4.3}$$

where, NF is the noise figure in decibels,

S_i/N_i is the input signal-to-noise ratio,

S_o/N_o is the output signal-to-noise ratio.

At times, noise figure may be referred to as a "figure of merit" of a device. In a broad sense, such evaluation is appropriate. The noise figure of a device is an indication of how its internal noise is handled in the process of trying to produce as nearly a noise-free output as possible. Effective execution of such a task could be described as a measure of the merit of a device.

LO Power   *LO power* is the amount, usually expressed in decibels in reference to a milliwatt (dBm), necessary to provide proper operation of the mixer. This power level is used to bias the diodes in a nonlinear region so that mixing action occurs. Power levels either above or below this level will result in less than predicted results for the mixer. Too low a level will not bias the diodes properly. Too high a level, aside from improperly biasing the diodes, could cause damage.

4.5.5   Applications

The mixer finds many applications in today's microwave systems. The first application is in a receiver front end; Fig. 4–25 shows a typical instance of how the mixer is incorporated. It is used to convert the high-frequency input signal to a lower frequency to obtain the transmitted information; it is much easier to operate on this lower frequency than on

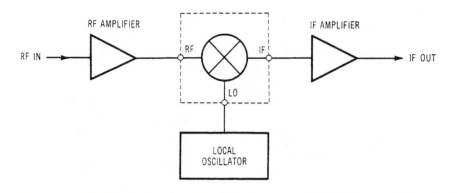

Figure 4–25. Mixer application.

the input signal as it appears at the antenna. The same concept is used in microwave network analyzers. The input gigahertz signals are converted down to a set frequency in the kilohertz range for easy operation and greater accuracy. In applications of this type, it is very important to have low-noise mixers with a high degree of isolation and minimum conversion loss. Other utilizations are as an up-converter in a transmitter (to increase the frequency for transmission), as a balanced modulator (which incorporates a 180° hybrid and reversed diodes in a balanced mixer), as a single-sideband modulator, and in many instances where signal processing must be accomplished. One application in testing is to use a mixer to down-convert a signal for noise-figure measurements. Some noise-figure meters have set frequencies as inputs; therefore, an IF signal sometimes needs to be converted to a lower frequency.

To summarize this component, we could say that the key to understanding mixers in use in today's microwave systems is having a knowledge of the balanced mixer. From this basic circuit you can build more complex mixing schemes, and you are equipped to use the mixer as a valuable tool in needed test setups.

4.6   Filters

Filters are used whenever there is something to be passed and something to get rid of. An oil filter in a car is supposed to pass the oil and trap any dirt or other particles that may damage the engine, a filter in a furnace is supposed to pass the warm air and trap dust and dirt in the air, and a filter on a faucet is supposed to let the water through and trap any particles in the water line. The list could go on and on, but the purpose still remains the same. A filter is designed to remove something that is unwanted and pass on what is wanted.

A microwave filter serves the same purpose. It is designed to pass a certain band of frequencies and reject the rest. The low frequencies may be rejected with a *high-pass filter*, the high frequencies may be rejected with a *low-pass filter*, or all frequencies except a specific band may be rejected with a *bandpass filter*. Whatever the need, there is a filter to take care of it. Let us now investigate the three types of filters just described.

4.6.1   Bandpass Filters

The first filter we will look at is the *bandpass filter*. We cover this one first because it is the component that has the largest number of applications. This filter, of course, does what the name says—it passes a band of frequencies. It passes these frequencies and rejects every-thing below and above this band. Fig. 4–26 shows a bandpass filter response. This re-sponse shows an area termed *passband* that is the area of minimum loss in the filter. On ei-ther side of the passband the response falls off sharply. These dropoffs are termed *skirts* of the filter, and their slope depends on the amount of attenuation present as a function of fre-quency. That is, if we have a frequency only a few megahertz away from the passband and we have 50 to 60 dB of attenuation, the skirts are very sharp. If the attenuation is only 10 or 20 dB, the skirts have a gentle slope. These attenuation (or rejection) points are usually at the 60-dB points of the response, and a band of frequencies is given for this attenuation.

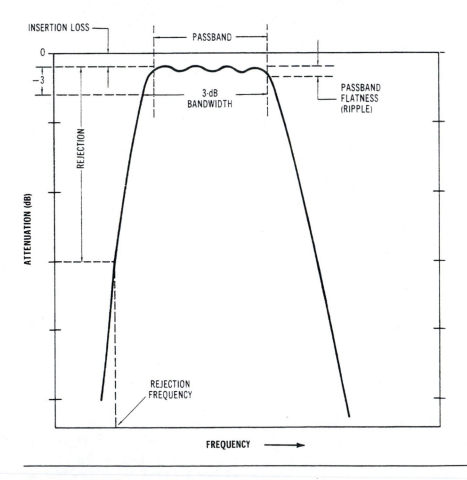

Figure 4–26.   Bandpass filter response.

A 3-dB bandwidth is also given to further characterize the filter. The ratio of the 60-dB width to the 3-dB bandwidth is called the *shape factor* of the filter. If, for example, the 60-dB bandwidth were 1.2 GHz and the 3-dB bandwidth were 0.48 GHz (480 MHz), the shape factor of this particular filter would be 2.5. Sometimes the shape factor is referenced to a different bandwidth other than 60 dB. There may be reference made to 30 dB and 3 dB. Be sure to know what reference a manufacturer is using when looking at a filter for your applications.

In the beginning of this chapter we said that we were going to present each of the components and show a typical data sheet with the terms explained. We are going to deviate from that statement for the filters only because filters are very specialized components and a catalog that gives a data sheet for a specific filter cannot be found. Instead, a range of values is presented for all the filters a manufacturer can build. For this reason we will present specific terms that apply to each filter and explain them.

For the bandpass filter these terms will be *insertion loss, ripple, passband,* and *rejection.*

Insertion Loss   This loss is the internal loss through the filter in the band to be passed. No device will have zero loss through it simply because there is energy passing through a transmission line. There is, therefore, a small loss, even in the band to be passed. This insertion loss increases with an increase in bandwidth of the filter and also as the number of poles used to construct the filter increases. This increase stands to reason since a wider bandwidth is obtained by adding poles. The number of poles interact to form the filter response and increase the loss because there are more elements (transmission lines) for the energy to pass through. Fig. 4–27 is an exploded view of the entire response shown in Fig. 4–26 and shows the insertion loss.

Ripple   We mentioned above how the individual poles of a filter interact with one another to form the overall filter response. If everything interacted perfectly, we would get a perfectly flat response for every filter from the low-frequency end of the band to the high end. Unfortunately, the poles do not interact perfectly, and the response develops variations in amplitude across the band that correspond to the poles within the filter. These variations are termed the passband *ripple* of the filter and must be kept to a minimum. Strive to keep this figure in the ±0.1-dB range for good filter operation. Fig. 4–27 shows the filter ripple and how it relates to insertion loss.

Passband   The *passband* is the band of frequencies that are affected only by the insertion loss of the filter. These frequencies are the ones the filter is designed to pass virtually unaffected. Passband is different from a bandwidth figure in that the passband involves only the insertion loss. A bandwidth figure would have a decibel number attached to it. In Fig. 4–26, for example, there is a 3-dB bandwidth indicated. This band is obtained by moving 3 dB down from the insertion loss and reading the frequencies. This type of bandwidth could be a 1-dB bandwidth, 2-dB bandwidth, or even a ripple bandwidth. It must be remembered, however, that it is a *bandwidth;* the numbers associated with the insertion loss are for a *passband.*

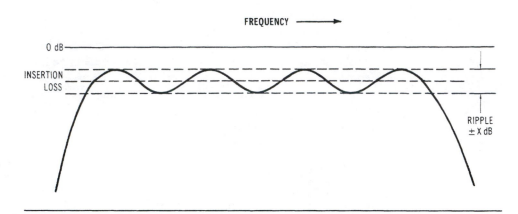

Figure 4–27.   Insertion loss and ripple for bandpass filter.

Rejection   This term tells how much the undesired frequencies are attenuated. The number that comes up most frequently is 60 dB because, as previously mentioned, the 60-dB bandwidth is commonly used to specify the shape factor of the filter and thus needs to be characterized. If a certain frequency is unwanted in a system, simply go to the filter response, find that frequency, and find the amount the signal is attenuated (or rejected).

The bandpass filter may be fabricated in many ways. Three methods are shown in Fig. 4–28. The first (Fig. 4–28A) is a commonly used method where space is not the most critical item. This configuration, however, may be folded to save a certain amount of space. The second filter (Fig. 4–28B) is very widely used in *microwave integrated circuit (MIC)* work. This filter is easily fabricated and can be used in either microstrip or stripline. The third type (Figure 4.28C) used to be one that had limited usage. This was because good grounding techniques for the resonators were very difficult to achieve using Teflon®-based circuit boards. Now, with excellent plated-through-hole technology, this filter is finding more and more applications.

The bandpass filter finds many applications in the field of microwaves. It can be used to reduce noise and harmonic content of a system by limiting the bandwidth of the signals seen by that system. They are many times placed at the input of a receiver to improve its selectivity.

The bandpass filter is used many times following a frequency multiplier. The multiplier will produce a variety of signals called a "comb." Then, one signal must be selected for your system by the bandpass filter. A similar application is at the output of the generator of a swept test setup. Unwanted signals can cause a transistor amplifier, for example, to encounter problems—especially when lower frequencies could cause excessive gain, oscillations, and destruction of the device.

We have said at the beginning of this section that we would not present a typical data sheet for filters because each one was a special item, so special specification sheets to completely define the filter are needed. Appendix G has such a sheet for the bandpass, low-pass, and high-pass filters. Study these sheets and see how you might fill in each of the

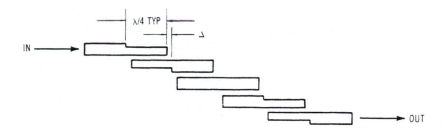

(A) **Side-coupled half-wave resonator bandpass filter.**

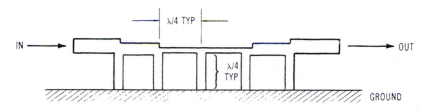

(B) **Short-circuited quarter-wave stub bandpass filter.**

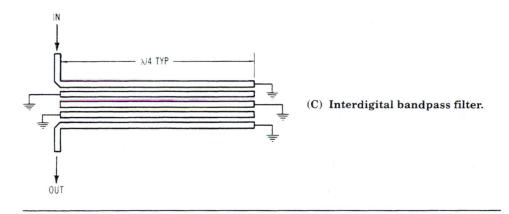

(C) **Interdigital bandpass filter.**

Figure 4–28. Bandpass filters.

sections. Be very cautious, however, because it is not an easy task to specify a component that will interface exactly with a system. Consider both the job of the filter in the overall system and the parameters of the system itself.

4.6.2 Low-Pass Filters

The second filter to be covered is the *low-pass filter*. This filter passes all frequencies below a certain frequency and rejects those above. Fig. 4–29 shows a low-pass filter

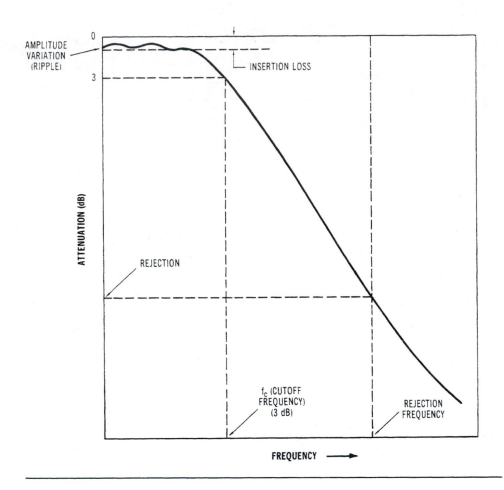

Figure 4–29. Low-pass filter response.

response. The *passband* of this filter can be considered from zero to the portion termed *cut-off frequency* (f_c). Generally, we think of the zero frequency as dc. The filter will, in fact, pass dc since it consists of series inductances and shunt capacitances. This equivalent circuit is shown in Fig. 4–30. You can see that with such a configuration dc could be passed easily through this filter.

The rejection point shown in Fig. 4–29 is usually on the order of 60 dB. This figure, of course, will vary with the specific application. Anywhere near 60 dB of rejection may not be needed, but the cut-off frequency may have to be very precise, and 30-dB rejection is needed soon after the cut-off frequency. There could be any number of combinations. Remember to take each application as an individual case to be treated as a unique situation, and you should not have any problems specifying low-pass filters.

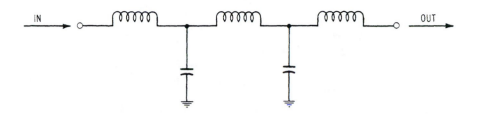

Figure 4–30.   Low-pass filter equivalent circuit.

One term that has been used a couple of times and that has not as yet been defined is *cut-off frequency* (f_c). This frequency is when the response falls 3 dB (half power below the insertion loss of the filter), to where the filter begins to take on the characteristics it is designed to have. Below this frequency the filter has low loss and is considered to be in its passband. Above this frequency there is ever-increasing loss as we approach the rejection frequency and an ultimate (or maximum) attenuation. Remember that if you want to pass all frequencies below 1.0 GHz, for example, do not specify a lowpass filter as having $f_c =$ 1.0 GHz because at 1.0 GHz the response will be down 3 dB. You should specify the filter to have a cut-off frequency *above* 1.0 GHz so that the required frequencies will be in the low-loss passband.

The other terms associated with low-pass filters are the same as for a bandpass filter. Looking at the response, you can see it is basically a bandpass filter with no frequencies below the passband. Such terms as insertion loss, ripple, rejection, and VSWR are the same as in the case of the bandpass filter. As in the case of all components, areas such as power handling capability and the proper connectors must be considered always. When working in a low-power system (receiver application, for example), the power handling capability is not that critical. However, for high-power transmitter usage, it is a number-one consideration.

The proper connectors are always a necessity. Anytime the need for adapters can be eliminated, potential problems are being lessened. Adapters always have a certain amount of loss and their own VSWR that contributes to the loss and the VSWRs in the system.

Two methods of fabricating low-pass filters are shown in Fig. 4–31. The first (Fig. 4–31A) is the conventional type of low pass we presented in Fig. 4–30, using series inductance and shunt capacitance. The narrow lines are the inductor, and the large areas are shunt capacitors. This filter may be constructed in either stripline or microstrip.

The second type is called an *elliptic-function low-pass filter*. You can see that it is more complex in that there is a combination of both inductance and capacitance in the shunt legs. This combination can be seen as a narrow line connecting to a wide section for each of the poles in the physical circuit layout. This filter performs basically the same as the previous low-pass filter discussed earlier when we considered the passband. The rejection curve produces a series of peaks and valleys once the rejection frequency is passed. This series of peaks and valleys is different from the previously discussed filter, which had a

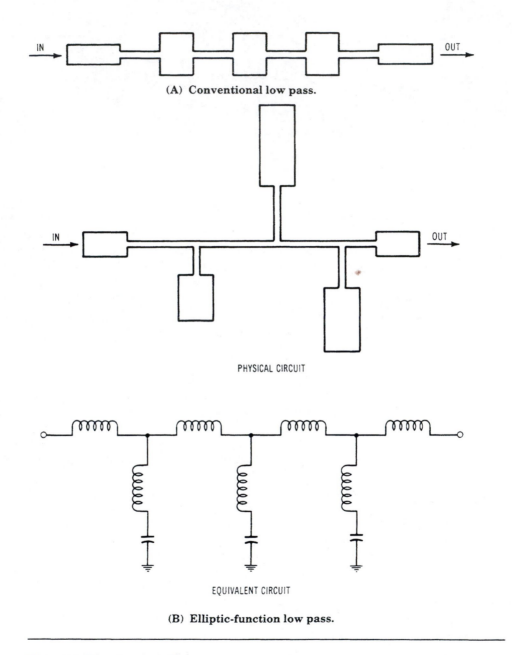

(A) Conventional low pass.

PHYSICAL CIRCUIT

EQUIVALENT CIRCUIT

(B) Elliptic-function low pass.

Figure 4–31. Low-pass filter.

continuous curve for attenuation until the ultimate attenuation point was reached. There are times when a conventional low pass has to follow an elliptic-function low pass simply to clear up the peak responses that are characteristic of this device. This type of filter has been presented as a comparison to show some low-pass possibilities but not to say that the elliptic-function filter does not have any applications. On the contrary, there are times when this type of filter will work very well.

One application of the low-pass filter has been mentioned previously in Section 4.5, when we covered mixers. This application is to filter out all of the mixer products generated within the device and allow only the IF ($f_1 - f_2$) to appear at the output. This filter could be a bandpass filter also, but there is less complexity involved when using a low pass. This filter also provides a good match between the mixer circuitry and the outside world at the IF frequency. This interconnection with other components in a system is much more compatible than are other filters.

A second application for a low-pass filter is for removal of those pesky signals called spurious responses and harmonics. There are times when a circuit, for one reason or another, produces a variety of extra signals. These extra signals are very prevalent when high-power active circuits are being designed. The low-pass filter eliminates the higher frequency signals that could cause problems further down the line in a system.

These are but two of many applications for low-pass filters. Anytime there are unwanted signals present that are higher in frequency than the desired output, a low-pass filter is the component to do the job.

4.6.3   High-Pass Filters

The third filter to be covered is the high pass. This filter could be considered the mirror image of the low-pass filter just discussed. It is designed to attenuate the frequencies up to a cut-off frequency and then pass the frequencies *above* this point. This design does not mean, of course, that the filter has 0.2-dB loss from just beyond the cut-off frequency to blue light. The attenuation will increase at higher and higher frequencies beyond a reasonable point. Up to this point, however, the filter will show good performance.

The response of a high-pass filter is shown in Fig. 4–32. You can readily see the rejection point with the rejection frequency (f_R), the cut-off point with the cut-off frequency (f_c), the insertion loss, and ripple of the filter.

A typical high-pass filter is shown in Fig. 4–33. The actual construction is shown in 4–33A, with the equivalent circuit in 4–33B. The series capacitors may be put in the filter in one of two ways. First, they may be chip capacitors. These are ceramic dielectric capacitors designed for microwave use. They will be covered in detail later in this chapter. The second method is the one shown in Fig. 4–33A. This method is an overlap type of construction in which a trace is printed on both sides of a thin dielectric board. You may recall from Chapter 2 how we were speaking of 0.003-inch and 0.005-inch material for making couplers. This same type of material can be used for the high-pass filter. This type of construction is designed for stripline applications. The overlap construction could not be achieved using a microstrip medium. With the capacitors printed on the board (both sides) and the inductors grounded, a high-pass filter is available for use.

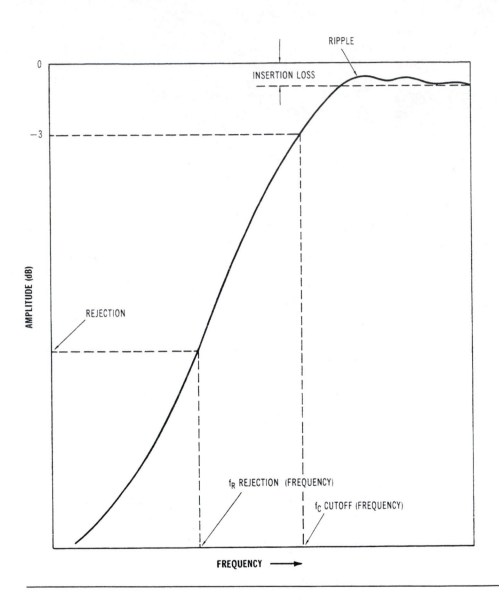

Figure 4–32.   High-pass filter response.

The high-pass filter does not find as many applications as do the more popular bandpass and low-pass filters. One particular instance does stand out, however; the high-pass filter can be very useful when low-frequency signals may be disrupting the operation of the system. The high-pass filter removes any low-frequency signals present at the input of a device or section, thereby preventing spurious responses or a microwave transistor breaking into low-frequency oscillations. This model also holds true for test setups; the filter can be

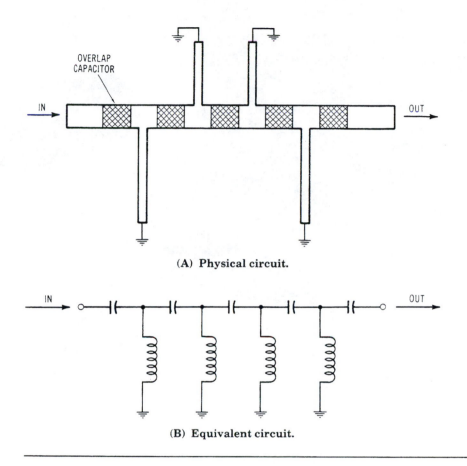

(A) Physical circuit.

(B) Equivalent circuit.

Figure 4–33.   High-pass filters.

put at the output of a generator to ensure a clean output signal that is free of low-frequency problems. To eliminate the low-frequency portion of a signal, high-pass filters are the appropriate devices.

We have covered three basic and very useful types of filters. They have specific applications as well as those that can be used for the same, or similar, applications. Regardless of their designated applications, they all have these properties in common: They will pass the desired frequencies and reject those that are not desired. Fig. 4–34 shows a series of filters available in a variety of sizes and shapes.

4.7   Circulators and Isolators

In Section 4.2 we discussed a device that passed microwave energy in a forward direction but provided isolation in the reverse direction. This device was termed a *directional*

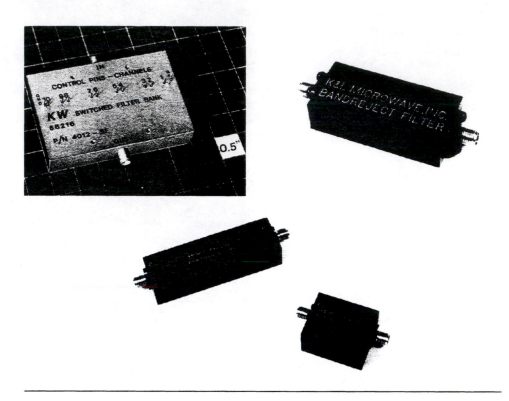

Figure 4–34. Microwave filters *(Courtesy of KW Microwave Corp.).*

device—more specifically, a directional coupler. We will now cover a device with the same type of forward and reverse characteristics that is called a *nonreciprocal* device. This device is the circulator (or isolator). You might ask why two devices that accomplish the same purpose are called two different names. To answer this question we can say that these devices may have their forward and reverse characteristics in common, but the similarity ends there because of their construction and internal characteristics. We will expand on these differences as we get further into the topic of circulators and isolators.

Probably the best way to describe a nonreciprocal device is to say that it provides a one-way street for microwave energy. To illustrate this statement, refer to Fig. 4–35. This figure shows a common piece of equipment used at sporting events, concerts, etc. It is a turnstile. These devices will turn only one way. When going in, the "in" direction is the only way to proceed. This device is nonreciprocal in that it allows movement in one direction and not in reverse. The only way to get out through this particular turnstile would be to jump over it.

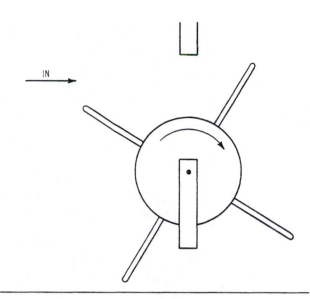

Figure 4–35. Circulator
illustration.

Now compare this type of operation to a microwave circulator. Fig. 4–36 shows such a circulator. With an input at port 1, most of the energy will appear at port 2. There will be a small insertion loss on the order of 0.3 to 0.5 dB, but basically all the energy at the input will be at port 2. If the load at port 2 is matched to the system characteristic impedance, there will be no reflections at that point. If reflections are present, they will be sent from port 2 to port 3, and they will not be felt back at port 1. In this way we have a nonreciprocal device when considering ports 1 and 2 (or if we look at ports 2 and 3 or ports 3 and 1).

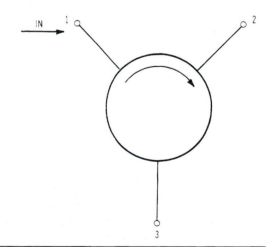

Figure 4–36. Microwave
circulator.

In each case, there is forward power but nothing in reverse. The microwave energy circulates around the device to the proper point. Thus, the name *circulator* can very readily be seen to apply to this device. Remember that when we say nothing is transmitted in the reverse direction, we are speaking from a theoretical point of view. In actuality, there is some small amount of leakage in the reverse direction since it is impossible to build a circulator or any other component that is perfect. Therefore, when we speak of the reverse direction we are referring to some finite value of isolation expressed in decibels.

The nonreciprocal action of a circulator is not caused by mechanical gearing, as in the turnstile. Rather, it is brought about by a gyromagnetic action. This action is set up by an interaction of the fields of the microwave energy and the internal magnetic field of the circulator. The structure that causes this interaction is shown in Fig. 4–37. The circuit itself is of stripline construction, with the printed circuit board in the center and ground plates on both sides. In between the circuit and grounds are ferrite "pucks." The term *puck* is used since it very closely resembles the shape of a hockey puck. This small unit is what determines the characteristics of the circulator.

On either side of the stripline package containing the circuit, ferrite pucks, and ground plates are pole pieces and magnets. As we mentioned, this magnetic field interacts with the electric fields of the microwave energy to produce the gyromagnetic action and, thus, the circulation of energy.

The magnetic field can be either a clockwise or counterclockwise direction, depending upon the orientation of the magnets. The direction of rotation of the microwave energy results in the terminology *right-hand circular* (for clockwise) or *left-hand circular* (for counterclockwise). Different configurations may be made using either a right-hand circular, a left-hand circular, or a combination of both. The type used depends on the particular application. The most widely used type is the right-hand circular.

The shield shown in Fig. 4–37 is placed around the entire device to both protect it from outside effects as well as keeping any external circuitry from being affected by the internal magnetic field of the circulator. The circuit itself is three printed lines joined in the center to form a star arrangement. Using a meter to check between ports shows that they are all shorted together. Remember that in the directional coupler there was no dc continuity. In the circulator there is complete dc continuity. The nonreciprocal action takes place because of the rf and not the dc properties of the device. As a direct comparison, we can say that the directional coupler has both rf and dc isolation but the circulator has rf isolation but not dc.

Before getting into a data sheet for circulators it would be appropriate to distinguish between a *circulator* and an *isolator*. The circulator, as we have mentioned, is a three-port device that allows energy to be passed in a forward direction and provides isolation in the reverse direction. Energy can be transferred between any two adjacent ports. An isolator is a two-port device consisting of a circulator with its third port internally terminated. This unit is used exclusively for providing isolation between circuits and improving the match between them. Fig. 4–38 shows schematic representations of a circulator and an isolator.

Operation of the isolator (and also the circulator) is very highly dependent on the termination within the device (or on the third port of a circulator). The maximum

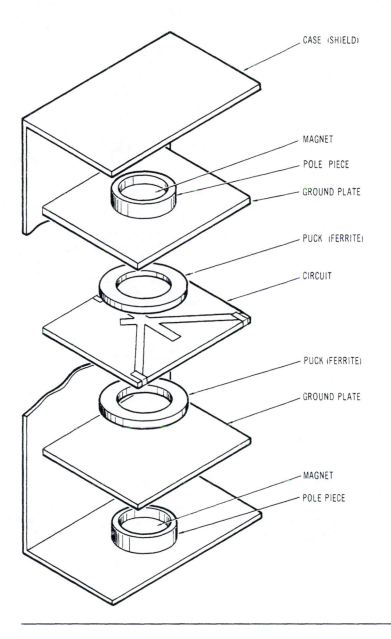

CASE (SHIELD)

MAGNET

POLE PIECE

GROUND PLATE

PUCK (FERRITE)

CIRCUIT

PUCK (FERRITE)

GROUND PLATE

MAGNET

POLE PIECE

Figure 4–37. Circulator construction.

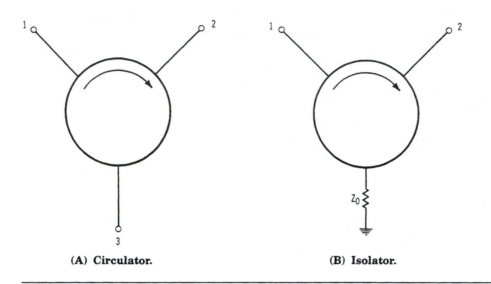

(A) Circulator.   (B) Isolator.

Figure 4–38.   Circulator and isolator schematics.

isolation of the device depends on how good a termination there is with respect to the characteristic impedance (Z_0). To see just how important the termination is, refer to Fig. 4–39. As an example, if the termination on an isolator or the load on the third port of a circulator produced 1.5:1 VSWR, the maximum isolation possible would be on the order of 14 dB. Similarly, a VSWR of 1.1:1 would increase the isolation to in excess of 25 dB. So you can begin to appreciate how important that third port really is. The advantages of circulators could be completely neutralized if just any old load were slapped on the third port or if proper concern for the termination of an isolator were not shown.

One other consequence of an improper match is the input power rating. A large mismatch would cause multiple reflections, and thus less input power could be handled. Fig. 4–40 shows a curve of input power rating as a function of load VSWR. This rating is a multiplier of the actual termination rating. With a one-half watt termination and a 2:1 mismatch, for example, approximately 4 watts of input power maximum could be applied. Other values can be read from the curve to see how fast the rating decreases on VSWR increases.

4.7.1   Data Sheets

Typical data sheets for a circulator and isolator follow. Most of the specifications are similar. There are some, however, that are different simply because one device is a circulator with three ports and one is an isolator with an internal termination.

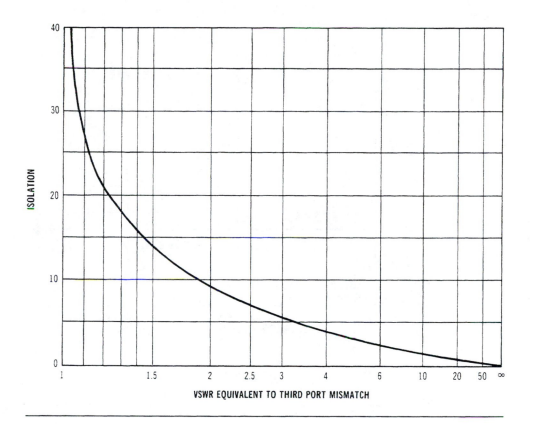

Figure 4–39.   Circulator and isolator isolation curves.

	Circulator	*Isolator*
Frequency	2 to 4 GHz	2 to 4 GHz
Insertion Loss	0.5 dB (max)	0.5 dB (max)
Isolation	18 dB (min)	18 dB (min)
VSWR	1.30:1 (max)	1.3:1 (max)
Power	125 W (cw)	1.0 W (cw)
	600 W (peak)	600 W (peak)
Connectors	SMA	SMA

The major difference between the circular and isolator shown in the two data sheets is in the power handling capability. The circulator has a 125 W cw rating, and the isolator is rated only at 1.0 W. This power handling difference is because of the internal resistor that is being used to terminate the isolator. Its power rating may only be one-half or one-quarter watt. Therefore, as we have previously seen from Fig. 4–39, with a good VSWR on the terminated port, the power handling is in the order of 1.0 W. The circulator, on the other

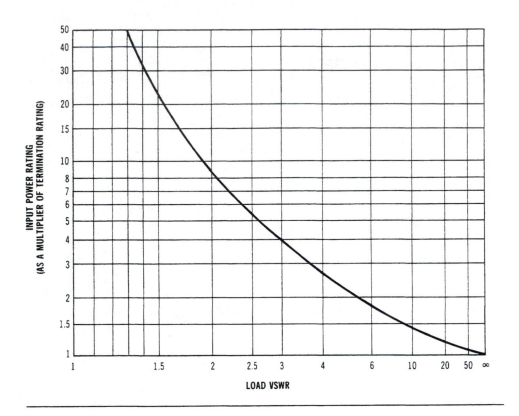

Figure 4–40.   Power rating versus VSWR.

hand, does not have the internal resistor. The component itself, therefore, can handle considerably more power than the isolator. It should be pointed out, however, that to feed 100 W of power into a circulator something should be on the third port that will handle a considerable amount of power, or any reflections from the output port will make that third port component look and smell like the french fries were in the grease too long. So be careful when reading a data sheet. For this case the power rating is for the device *only*. Your application may be of the same type. Double-check all power ratings. This checking will save time, money, and tempers later on.

We have previously said that the prime purpose of an isolator is to provide isolation between circuits and improve the match between them. For a graphic example of this type of application, refer to Fig. 4–41. Fig. 4–41 is a chart that shows the reduction of VSWR by use of a circulator or isolator. To see how this chart, and thus an isolator, works to match two circuits, let us consider an example. Suppose we have a load with an 8:1 VSWR. This VSWR is pretty high and intolerable in most cases. If we use our isolator from out of the data sheet example and show 18 dB of isolation, you can see that the circuit driving the

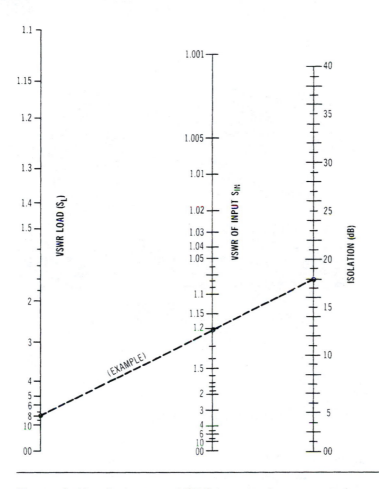

Figure 4–41.   Reduction in VSWR by a circulator or an isolator.

isolator, and eventually the 8:1 VSWR circuit, sees a 1.2:1 VSWR instead of the gigantic 8:1 actually at the load. This difference in VSWR makes a tremendous difference in the operation of the circuits involved.

One type of isolator should be mentioned to round out our discussion, a *drop-in* isolator. There seems to be a difference of opinion as to just what constitutes the label of drop-in for an isolator (or circulator). Some people classify them as coaxial devices with the connectors removed and tabs soldered to them for connections. Others believe that the only true drop-in is one with no case but simply the inside circuitry with tab connections. In a sense both definitions are correct. There is more to it, however, than simply removing connectors and connecting tabs. There is a matter of impedance matching when going from a connector to a tab output. Internal tuning must be done to

Figure 4–42.   Coaxial isolators and circulators *(Courtesy of TRAK Microwave).*

accomplish impedance matching. The most accurate definition of a drop-in would thus be an isolator or circulator with no coaxial connectors at any port. Tabs for stripline or microstrip usage are used for input and output connections. For a comparison of coaxial and drop-in devices, refer to Fig. 4–42 for coaxial circulators and isolators and Fig. 4–43 for drop-ins.

4.7.2 Applications

Some of the applications of circulators and isolators have been covered throughout the text. Probably the most common use for the three-port isolator is the diplexer shown in Fig. 4–44. The transmitter is connected to port 1, the antenna to port 2, and the receiver to port 3. This arrangement allows simultaneous transmission and reception since there is isolation between the transmitter and the receiver. The signal from the transmitter can go only to the antenna, and the signal received at the antenna can go only to the receiver. Notice that both the transmitter and the receiver should be well matched to the circulator to ensure maximum isolation between the two systems.

As previously mentioned, isolators find their widest applications as devices used to match two circuits. They do, however, have other uses. They can be used to reduce local

Figure 4–43. Drop-in isolators and circulators *(Courtesy of TRAK Microwave).*

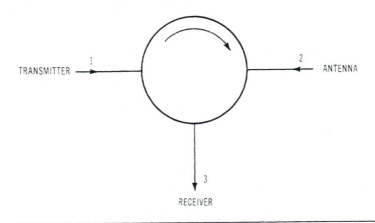

TRANSMITTER → 1

2 ← ANTENNA

3
RECEIVER

Figure 4–44. Circulator application diplexer.

oscillator radiation back into a receiver antenna (Fig. 4–45A), they isolate the output of an oscillator to reduce frequency pulling (Fig. 4–45B), and they are used commonly at the output of a generator in a test circuit to ensure that the generator power and frequency remain stable throughout the testing (Fig. 4–45C).

In conclusion, we can say that the circulator (or isolator) is a unique component. We began this section by relating it to a directional coupler, but you can now see that there are distinct differences.

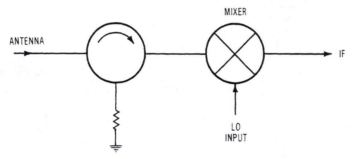

(A) Local oscillator radiation reduction.

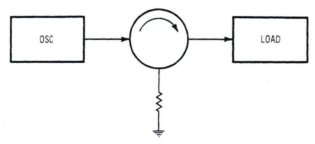

(B) Oscillator pulling reduction.

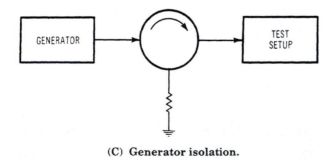

(C) Generator isolation.

Figure 4–45. Isolator applications.

Directional Coupler	Circulator (Isolator)
Directional device	Nonreciprocal device
No dc continuity	Dc continuity
Two separate circuits	One single circuit
(main line and coupled)	
Operates by coupling action	Operates by gyromagnetic action

4.8   Attenuators

When thinking of attenuating something (like a microwave signal), we should think of lessening the value of that particular item. There are many times when we have to lessen the value, or level, of a microwave signal to allow the rest of the system to work properly. For use in microwaves there are two types of attenuators: *fixed* attenuators, commonly called pads; and *variables,* both continuously variable and step attenuators. Both of these types will be discussed in detail.

4.8.1   Fixed Attenuator

The *fixed attenuator* is probably the most recognizable type of attenuator. Those are the ones in breadboard test setups or in a small box on somebody's bench just waiting to be used. Fig. 4–46 shows some typical fixed attenuators. You have undoubtedly seen these many times, but have you ever wondered what was inside of them to give the precise amount of attenuation they say they will? The configuration used is the same type that would be used to attenuate a low-frequency signal. They can either be a "T" or "π" configuration, as shown in Fig. 4–47. The configuration is the same as for low-frequency applications, but the components are very different. The resistors used are not carbon resistors with colored bands on them but film resistors that are designed to operate at microwave frequencies. These resistors employ a nichrome alloy film on a high-quality ceramic substrate to ensure a firmly bonded film with low-temperature coefficients. This type of construction makes the resistors extremely stable at high frequencies. The skin effect, discussed earlier, of the resistive conductor using this structure is excellent and, as such, is a big plus for microwave applications.

You can see from Fig. 4–47 how the resistors are placed in series on the center conductor or in shunt, contacting both the center and outer conductor. In this way the "T" configuration with one shunt flanked by two series resistors and the "π" configuration with one series flanked by two shunt resistors can be fabricated. It is also very easy to see why there is a limited power handling capability for the standard fixed attenuator. Whether a "T" or a "π" configuration is used, a series resistor is in each. This resistor has a limited power handling capability, usually less than 1 W unless an elaborate heat sinking is provided. This resistor is, thus, the limiting factor in the amount of power the component can handle. There are power attenuators available, but they are usually very large and require much heat sinking, as we have previously mentioned.

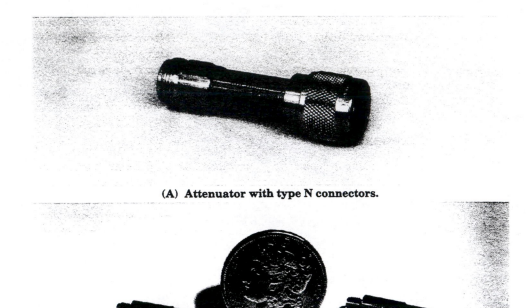

(A) Attenuator with type N connectors.

(B) Attenuator with type SMA connectors.

Figure 4–46. Fixed attenuators *(Courtesy of Loral Narda)*.

4.8.2 Fixed Attenuator Data Sheets

A typical data sheet for a standard attenuator and a power attenuator follow.

	Standard	*Power*
Frequency	dc to 12.4 GHz	dc to 8 GHz
Attenuation	20 dB	20 dB
Accuracy	± 0.5 dB	± 0.75 dB
VSWR	1.20:1 (dc to 8 GHz)	1.2:1 (dc to 4 GHz)
	1.25:1 (8 to 12.4 GHz)	1.3:1 (4 to 8 GHz)
Input power	2 W (cw)	50 W (cw)
Connectors	SMA	Type N
Size	0.95 inch long, 0.30-inch dia.	4.75 inches long, 1.7-inch dia.

Relevant terms will be explained with distinctions and differences between the standard and power attenuators following the explanations.

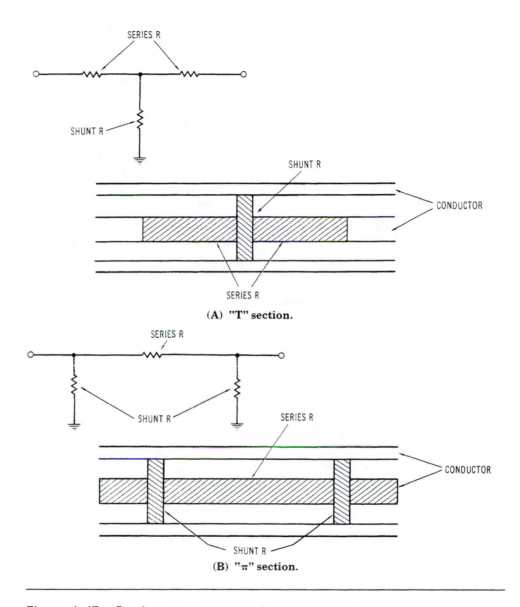

(A) "T" section.

(B) "π" section.

Figure 4–47. Fixed-attenuator construction.

Attenuation   *Attenuation* is the figure, in decibels, that tells how much power is lost between the input and the output of the attenuator. That is, it is how much the input power is lessened by the time it gets to the output. If we express it mathematically, it would be attenuation (dB) = 10 log (P_{out}/P_{in}).

Accuracy   *Accuracy* is a measure of how close the attenuator is to the value of attenuation (20 dB in this case) specified in the data sheet. This measurement is usually a plus or minus ($\pm$) value that gives a range over which the attenuation of the attenuator can vary.

VSWR   This term is mentioned only to clarify when the VSWR holds true. It is the voltage standing-wave ratio (or input match) of the attenuator when the output port is terminated in the characteristic impedance, Z_0 (usually 50 ohms).

Input Power   This term is probably the most important parameter to consider when working with any power from ½ watt and up. Check this figure and leave a good safety margin to keep from suddenly detecting a strange odor in the lab, followed by a sinking feeling in your stomach caused by finding that your $100 pad has just been destroyed and that there are no more in the lab.

There are some very obvious differences in the previous data sheet when comparing the standard attenuator with the power attenuator. First, the accuracy is not as good (±0.75 compared to ±0.5 dB). Also, the VSWR is higher for the power version than for the standard device as well as a frequency range of dc to 8 GHz for the power unit as opposed to dc to 12.4 GHz for the standard device. When needed, the standard attenuator is available from dc to 18 GHz. The next two areas of difference between the devices are very obvious when you look at the data sheets. They are the input power figure (2 W versus 50 W) and the size. These two, of course, are directly related since the larger resistance area can handle more power as well as incorporate the best sinking capabilities.

4.8.3   Fixed Attenuator Applications

Applications of the fixed attenuator are numerous throughout the microwave industry. Basically, all the applications can be broken into two distinct categories: reduction in signal level and impedance matching of a source and load. The categories can be broken down as follows:

Reduction of signal levels:

1.   Operation of a detector in its square-law range for most efficient operations.
2.   Testing of devices in their small signal range—getting the input level down low enough for proper operation.
3.   Reduction of a high-power signal to a level compatible with sensitive power measuring equipment—power sensor and thermistor mounts.

Impedance matching of sources and load:

1. Reduction of signal variations as a function of frequency. The variations we are speaking of in this case are caused by a high VSWR—the attenuator provides a reduction in these variations and a better match.

2. Reduction in frequency "pulling" (that is, changing the source frequency by changing the load) of solid-state sources by high-reflection loads—this application is similar to the circulator application in the preceding section.

4.8.4   Variable Attenuators

There are times when you may not know what value of attenuator is needed or you may need more than one value of attenuation. For cases like this, the variable attenuator is used. We will cover two types of variable attenuators: *step* and *continuously variable*.

Step Attenuators   The step attenuator consists of a series of fixed attenuators mechanically arranged to provide stepping in discrete increments. This arrangement is sometimes called a *turret*. These attenuators are mounted on the rotating element with stationary coaxial contacts for the input and output of the device. Attenuators of this type have many advantages:

1. Wide frequency range—dc to above 18 GHz.
2. Repeatability—within 0.05 dB is common; within 0.03 dB is possible.
3. Wide attenuation range—0 to 100 dB.
4. Long life—with the simple construction technique used, the attenuators will not be stressed and will last.
5. Manual or programmed operation—this quality has become very important in a step attenuator in this age of automatic test setups.

Step attenuators can come in 0.1-dB, 1.0-dB, and 10-dB steps. By combining all three of these steps, 0 to 100 dB in 0.1-dB steps can be obtained. A typical data sheet for a step attenuator is shown here:

Frequency	dc to 12.4 GHz
Attenuation range	0 to 60 dB (10-dB steps)
Accuracy	±1.5 dB
Insertion loss	1.0 dB
VSWR	1.5:1
Power	2 W (cw)
	100 W (peak)
Connectors	Type SMA

In the step attenuator data sheet, the attenuation range tells the maximum and minimum attenuations obtainable with the attenuator (60 and 0 dB, respectively) and the steps that can be taken with this component (10-dB steps). The accuracy tells how close each step actually is to each value. For example, 10-dB steps in our case could be anywhere from 8.5 to 11.5 dB. When making precise measurements, this type of information is vital. The insertion loss of the attenuator is how much loss is read from the input to the output when the attenuator is set at 0 dB.

The VSWR of the attenuator is a measure of the match of the attenuator when it is stepped through its attenuation range. Remember, just as with the fixed attenuators, it is measured with its output terminated in its characteristic impedance (Z_0).

The power figures listed were for cw and peak pulse. These figures guarantee the long life previously mentioned. If the power figures are observed, the attenuator will last. If they are not observed, you will end up with a high-priced paperweight with a strange, charred odor.

One feature not discussed as yet is repeatability. This feature allows coming back to an attenuation value and being assured that it is at the same value as originally set. This value stands to reason since the fixed attenuator will remain the same no matter how long the step attenuator is switched to another value. When it is switched back to an original value, the attenuation should be highly repeatable. The only variation should be that of the input or output coaxial contacts.

Continuously Variable Attenuators   There are some times when neither a step attenuator nor a fixed attenuator will do an adequate job in your setup or system. At this point a continuously variable attenuator should be used. Many different types of continuously variable attenuators are available for use. Some of them are as follows:

- Piston
- Variable card
- "T"
- Variable coupler
- Low loss cutoff
- Lossy wall

Each of these will be briefly explained to give an idea of the different ways that attenuation may be varied.

The *piston attenuator* is shown in Fig. 4–48. This type of attenuator is also referred to as a waveguide below cutoff attenuator because the cylindrical waveguide used is operating at a frequency lower than it was designed for (waveguide *below* cutoff). The attenuator operates in the same manner as any piston except it is hollow, whereas most pistons (as in an automobile) are solid. The use of a cylindrical waveguide has advantages and disadvantages. A much higher power handling capability is obtainable over a coaxial configuration. This higher power handling capability is definitely an advantage when working in high-power applications. The minimum insertion loss, however, is much higher for this type of attenuator since it is operating outside its intended range of operation. Insertion loss can run as high as 15 to 20 dB, but once this loss is overcome, the piston attenuator is one of the most accurate around. Values of 0.001

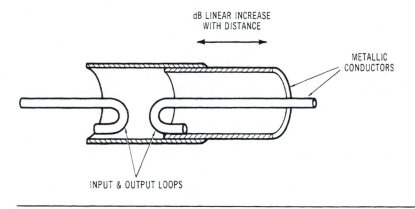

Figure 4–48.   Piston attenuator.

dB/10 dB of attenuation over a 60-db range are not uncommon. A good input/output match is obtained by inductive loops within the respective waveguides. This excellent match is attained over the entire range of attenuation because of inductive loop coupling.

A second type of continuously variable attenuator, shown in Fig. 4–49, is the *variable-card attenuator*. It consists of a resistance card attached to a substrate. A wiper arm varies

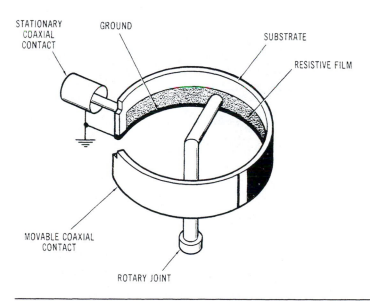

Figure 4–49.   Variable-card attenuator.

the resistance between the input and the output and thus changes the amount of attenuation. This action is very similar to a potentiometer used in a dc circuit. This type of sliding contact attenuator can have variations in resistance, so the VSWR is not always the best. Most of the time the mismatch is not serious enough to cause any problems with any attached circuitry. The insertion loss of the variable-card attenuator is on the order of 4 dB, which is a very reasonable figure.

The "T" attenuator is shown in Fig. 4–50. This attenuator is the same configuration used when we discussed fixed attenuation. It contains two series resistors and one shunt resistor. In the case of the variable attenuator, however, all three resistors are varied simultaneously to result in the appropriate attenuation and still maintain a good input/ output VSWR.

The *variable-coupler attenuator* shown in Fig. 4–51 is one that operates differently from those previously discussed. The previous attenuators all dissipated power within the units to provide attenuation. The variable coupler operates by varying the amount of coupling between two lines. The insertion loss of this type of attenuator is about 5 dB because of the mechanical restrictions concerning how close the two lines can be. With an input at port 1, there is a certain amount of energy coupled to port 2. This amount depends on the

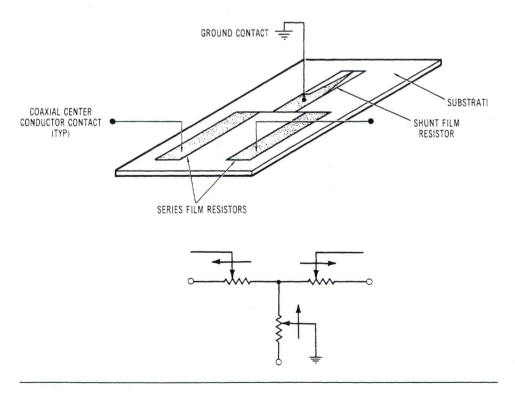

Figure 4–50.   "T" attenuator.

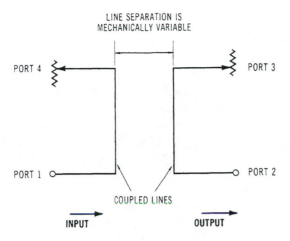

Figure 4–51.   Variable-coupler attenuator.

mechanical spacing between the lines. The amount of power this type of attenuator can handle depends on the external load at port 4. The input/output VSWRs are excellent with this attenuator because it exhibits the qualities of a directional coupler while being an efficient attenuator.

Fig. 4–52 shows an attenuator that makes use of another component we have discussed earlier—the quadrature hybrid. This figure shows the *low-loss cutoff attenuator*. The attenuator uses two hybrids (to divide and re-sum energy) with mechanically coupled variable-cutoff waveguide sections between them. By mechanically varying the cutoff point of the waveguide, the attenuation at the desired frequencies can be varied. The

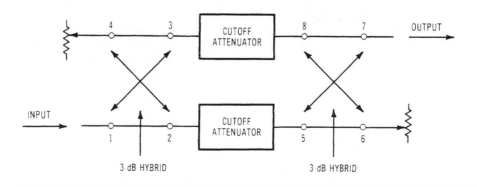

Figure 4–52.   Low-loss cutoff attenuator.

insertion loss of such a device is very low. With the attenuator set at zero, the insertion loss will be on the order of 1.0 dB. The input/output VSWRs on this type of attenuator are excellent. There are times when this type of attenuator is termed a *constant impedance* device, as previously explained.

Fig. 4–53 shows the only attenuator that makes sense, and its operation may be viewed easily. This attenuator is the *lossy-wall attenuator*. It is made of stripline construction with a section of its outer conductor replaced by a lossy type of laminate material. The position of this lossy material is varied to vary the attenuation. With zero attenuation (insertion loss generally below 1.0 dB), there is simply a stripline conductor with low-loss dielectrics above and below it. As the lossy material is moved over the transmission line, the attenuation increases until the stripline conductor is completely covered with lossy material. At this point the attenuator exhibits its maximum attenuation.

These attenuators are generally used above 2 GHz because of the availability of the appropriate lossy materials. Power handling capability of this type of attenuator is limited to approximately 10 W because it operates on a dissipation principle.

A typical data sheet for a general, continuously variable attenuator is shown here.

Frequency	2 to 4 GHz
Attenuation range	6 to 120 dB
Accuracy	±1.5 dB
VSWR	2.0:1 (6 to 15 dB), 1.5:1 (15 to 30 dB), 1.2:1 (30 to 120 dB)

The terms used on this data sheet are identical to those presented for the step attenuator. The only variation that should be noted is that of VSWR. For low values of attenuation the VSWR is high. As the attenuation increases, the VSWR decreases because as the attenuation increases, there is more power absorbed, so loss is reflected. Therefore, there will be a lower VSWR.

4.9   Chip Components

In Chapter 1 we discussed microwave operations and how everyday distributed components could not be used because of the skin effect. We referred to chip components at that time and again in Chapter 2 when the benefits of ceramic material were discussed. This section will investigate further the components used every day in microwave: chip capacitors, chip resistors, and chip inductors.

4.9.1   Chip Capacitors

Probably the most recognizable chip components in microwave circuits are the *chip capacitors*. Whenever a microstrip amplifier is used, there are chips used as coupling and bypass capacitors. They also are used as capacitive elements in high-pass and bandpass filters. What makes these components so remarkable is their size. There are capacitors available that are as small as 0.010 inch square. The most common

SIDE VIEW

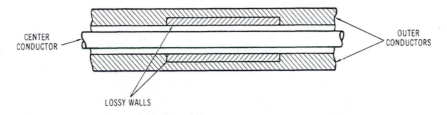

CENTER CONDUCTOR

OUTER CONDUCTORS

LOSSY WALLS

END VIEWS

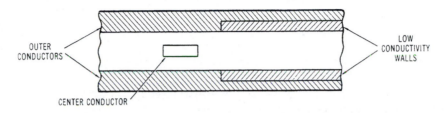

OUTER CONDUCTORS

LOW CONDUCTIVITY WALLS

CENTER CONDUCTOR

(A) Minimum attenuator.

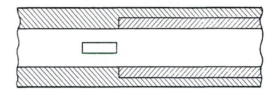

(B) Small attenuator.

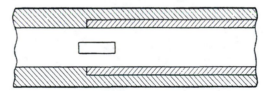

(C) High attenuator.

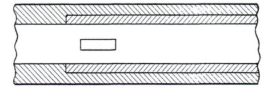

(D) Maximum attenuator.

Figure 4–53. Lossy-wall attenuator.

181

types, however, are 0.020 inch square, 0.030 inch square, and 0.050 inch square. These are cube-shaped capacitors called *pellets*. This type of capacitor is shown in Fig. 4–54 along with the same type of capacitor with leads. The figure contains the pellet (or chip), microstrip, radial wire, radial ribbon, narrow axial ribbon, axial ribbon, and axial wire. The device used for your application will depend on what type of circuitry you are using. The capacitors you see most often are those shown in Fig. 4–55. These are the pellet and microstrip types of devices. (The only difference between a pellet and a chip is that the chip has palladium silver terminations so that the length of the chip is increased 0.003 inch to 0.005 inch to allow for the solder.)

The capacitors we have talked about so far will work up into the 10- to 15-GHz range very satisfactorily. Be sure to check the data sheet for any series or parallel resonances that may be present to see whether they will interfere with your system. Capacitors are also available to operate up to 50 GHz. These are shown in Fig. 4–56 and are called millimeter-wavelength capacitors. They range in size from as small as 0.018 inch square to larger for higher capacitors (up to 1800 pF). Surface mount capacitors are also available that range in size from 0.063″ × 0.031″ to 0.360″ × 0.400″. Thicknesses are from 0.035″ to 0.071″, with values of capacitance from 1.0 pf

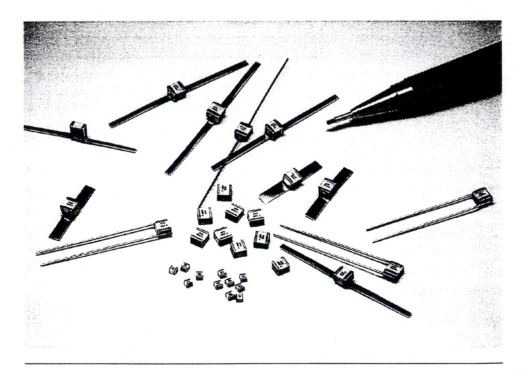

Figure 4–54.   Pellet and microstrip capacitors *(Courtesy of American Technical Ceramics).*

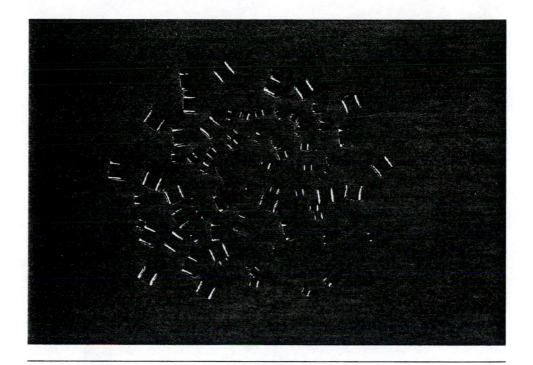

Figure 4–55.   Chip capacitors *(Courtesy of American Technical Ceramics).*

to 18 µf. Specific manufacturers should be consulted for a particular surface mount application.

The components we have been talking about so far are very small compared to the standard everyday capacitor used for filtering of dc lines and coupling of low-frequency circuits. What is it that makes these components operate so well at microwave frequencies yet be so small? You will recall from basic ac theory that with a single-plate capacitor (two electrodes with a dielectric between them), the capacitance would be found as follows:

$$C = \frac{A\epsilon}{t} \tag{4.4}$$

where, C is the capacitance in farads,
   A is the area of the plates,
   ϵ is the dielectric constant of the dielectric material,
   t is the thickness of the dielectric.

The single-plate capacitor is shown in Fig. 4–57A. To increase the capacitance, it is necessary to increase the dielectric constant (ϵ), increase the area of the electrodes (A), and reduce the dielectric thickness (t). All of these are done in the structure shown in

Figure 4–56.   Millimeter-wavelength capacitors *(Courtesy of American Technical Ceramics).*

Fig. 4–57B. You cannot see it in the figure, but remember that we discussed earlier using ceramic for a dielectric. Using ceramic material will fulfill the first requirement of increasing the dielectric constant. The area of the electrodes is increased because there are so many more electrodes in this interdigital configuration, that is, a series of stacked electrodes that are alternately offset from one another. Finally, the effective thickness of the dielectric is reduced because the electrodes are much closer together. So you can see how the miniature ceramic microwave chip capacitors shown in Figs. 4–54, 4–55, and 4–56 are possible.

Now that we have the chip capacitor, we must have a means of attaching it to our circuit. This attachment involves more than slapping some solder on the component and circuit, then forgetting it. To understand completely how all of the attachments are made, you must first understand that there are other orientations of chip capacitors, as shown in Fig. 4–58, and that these chips could not possibly be attached in the same way as those previously presented. Therefore, a variety of methods are needed. These methods are as follows:

1. *Soldering.* Soldering may be dip soldering or paste and preform soldering.
   A. Dip Soldering. In dip soldering the chip terminations and substrate pads are pre-tinned with solder. The chip and the substrate are heated gradually so that

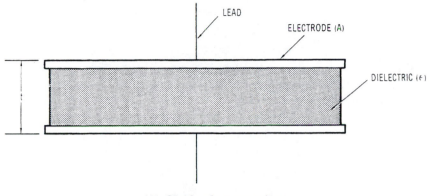

(A) Single-plate capacitor.

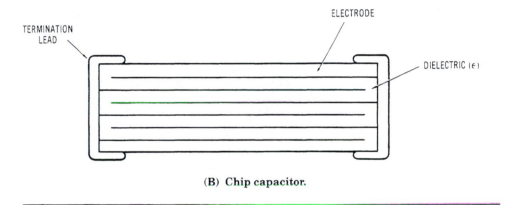

(B) Chip capacitor.

Figure 4–57.   Evolution of a chip capacitor.

the solder will flow freely and bond firmly. Temperatures used will depend on the type of solder but are generally between 400 and 500° F (204 to 260°C).

B.   Paste and Preforms. The solder paste is applied to the substrate by a screening or preform method. The chip and substrate are then heated to allow the paste to flow and bond. Temperatures are usually between 500 and 600°F (260 to 310°C).

2.   *Ultrasonic Bonding.* This bonding is a cold bonding (no temperature process is needed) method usually used to join dissimilar metals or metals of different thicknesses. The ultrasonic bonds are produced by applying pressure to the surfaces to be joined with a transducer tip vibrating at a rate of approximately 60 kHz. The force of the scrubbing action causes molecular mingling of the surfaces in contact, thereby forming a bond. A preform material is normally placed between the chip and the conductor to ensure the bond.

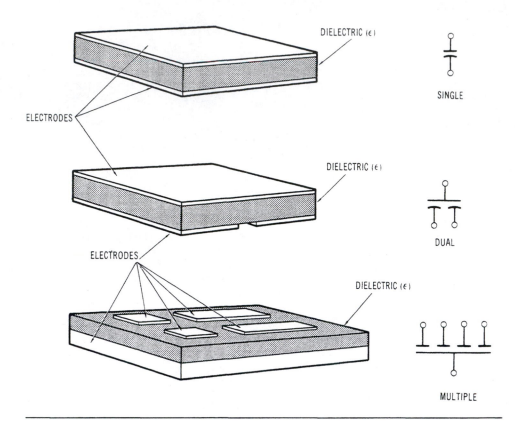

Figure 4–58.   Chip capacitors.

3. *Thermo-Compression Bonding.* This process is a heat-pressure bonding process. The chip is pressed onto a heated substrate (or circuit). A preform material, such as AuSn (gold-tin), is once again used at the interface of the chip and the substrate. The temperature range for this type of bonding is 700 to 750°F (375 to 400°C).

4. *Conductive Epoxy.* This bonding is a very popular method of connecting chip capacitors to substrate. The epoxy is a one- or two-part paste material that is applied in the area of 300°F (149°C). Temperature and curing times may vary with different types of epoxy.

5. *Nonconductive Epoxy.* The chip is bonded to the substrate with this epoxy, but electrical connection is completed using gold or aluminum bonding methods. Two of these methods are shown in Fig. 4–59. You can see that both bonding wires and ribbon leads are acceptable means of making connection between the chip and its associated substrate.

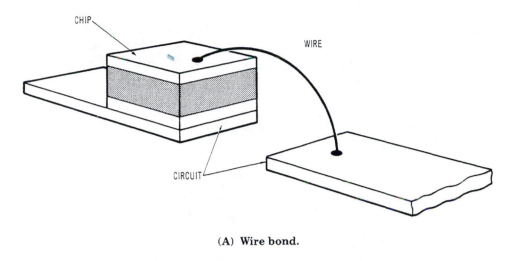

(A) Wire bond.

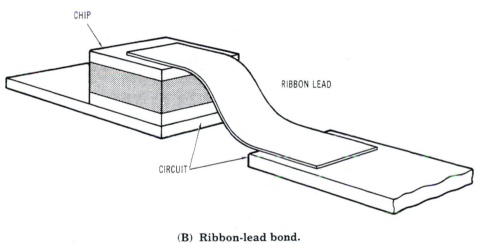

(B) Ribbon-lead bond.

Figure 4–59. Chip bonding methods.

The type of attachment used for capacitor applications will depend upon the composition the substrate conductor (gold, silver, palladium silver, etc.) and the material of the capacitor terminals (gold, silver, etc.). Check with the capacitor vendor before attaching the chip to ensure obtaining a strong, reliable, lasting bond of the chip capacitor to the substrate.

4.9.2   Chip Resistors

The *chip resistor* is a microwave component that finds many applications as a terminating resistor or attenuator. One reason why is because of the number of sizes of the resistor available. The sizes range from 0.040 × 0.020 inch up to 0.1 × 0.25 inch for low-power devices (½ W) and from 0.1 × 0.2 inch to 1.0 × 1.0 inch for high-power devices (1 kW for the 1-inch × 1-inch type). The size selected depends upon the power needed to handle and the real estate available to you.

The resistors mentioned come with a variety of lead configurations. Fig. 4–60 shows four common configurations. All but the last configuration show a side view with the ceramic base clear and the metallized contacts (terminations) darkened. The first configuration is probably the one that is the most common. It is the ceramic base with two metallized contacts, or terminations, on the top of the device. These two contacts are connected to a substrate by a variety of means that will be covered later in this section. Normally, you do not see the terminations on this device since it usually is turned upside down and attached to the substrate. All that is seen is a small, white ceramic rectangle sitting on the circuit.

Fig. 4–60B is similar to the first configuration except that one termination is wrapped completely around the resistor. This type of resistor is commonly used for terminations to ground. The contact on the top only is attached to the circuit, and the wraparound contact is connected to ground. A typical usage for this configuration is a 50-ohm termination of an isolated port of a coupler.

Fig. 4–60C has both contacts wrapped around the ceramic of the substrate, allowing contact to the circuit either way. Once the resistor is connected to the circuit, continuity readings and other measurements can be taken easily. The final configuration is the same

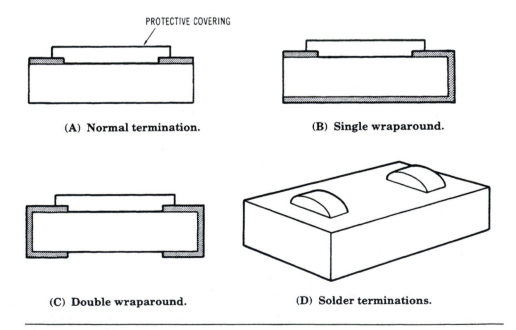

PROTECTIVE COVERING

(A) **Normal termination.**   (B) **Single wraparound.**

(C) **Double wraparound.**   (D) **Solder terminations.**

Figure 4–60.   Chip resistor lead configurations.

as the first one except solder "bumps" are used for the connection rather than metallized contacts.

Resistance values are obtained by either using a nichrome film or a thin film depositing technique. With these means a resistive layer with a specified ohms/square is deposited on a ceramic base. Metallized contacts of such composition as electroplated silver over nickel, platinum gold, or gold are attached to the resistant layer, and a protective coating is constructed over the resistive layer so that its resistive properties will be isolated from scratches, chips, and scrapes that would change the value. Normally, the ceramic base used is the same material we discussed in Chapter 2, alumina.

Just as in the case of the chip capacitors, the chip resistor must be mounted, or attached, to a substrate with a circuit on it. Two recommended methods are *epoxy* and *soldering*.

Epoxy   Two-part, room temperature setting compositions are recommended. A curing process the same as with the chip capacitor is accomplished once the epoxy has been applied to the resistor. Temperatures should not go above 150°C since chip damage may result. Any epoxy containing solvents should be avoided, and the epoxy used should be a thin line. Any excess should be removed before hardening and curing.

Solder   The "reflow technique," which requires pre-tinning, is the preferred method.

Pre-Tinning Chips   Two methods are available, dipping and soldering. The *dipping method* is preferred, and a $^{60}/_{40}$ or $^{63}/_{37}$ tin-lead solder should be used. Dip the chip into a solder pot held at approximately 220°C (426°F). Good wetting occurs in about 2 to 5 seconds, depending on the chip. Remove the excess flux with isopropyl alcohol or other suitable solvent. With the *soldering method* a 35-watt (maximum) soldering iron with a small chisel point is recommended. Apply solder to iron and transfer to the chip. Good wetting occurs in about 3 to 8 seconds, depending on the chip. The same solders and removal of flux as in the dipping process should be used. Due to contamination that may have been picked up in handling or storage, the solder may not cover the entire termination area. A light abrasion (like an ink eraser) prior to soldering will remove this contamination.

Attachment to Circuit   With *small chips*, apply the flux to the circuit. Reflow the solder on the chip and circuit by applying a 35-watt iron to the circuit terminals. Heat should not be applied for more than 15 seconds. Ideally, heat should be applied to all the solderable pads at the same time. When applying heat to all the pads is not possible, start with the pad that has the most solder. During reflow, the chip will settle down toward the circuit board due to the downward pressure. All the joints may need to be reflowed again to put the chip into its final position against the board. Apply the minimum amount of heat necessary for reflow. With *large chips* that have higher power, use the same procedure. A 50-watt iron applied for a longer duration may be necessary to attain adequate reflow. Units with conductive bases may be soldered to a mounting block or a heat sink in a similar manner. A pre-heat stage is recommended. Soldering heat should be applied to the mounting block with a controlled heat and time cycle to achieve a flow. Heat should be removed within 5 seconds of achieving a flow.

Once again, the type of attachment used for the chip resistor depends on your application and need.

4.9.3   Chip Inductors

Chip inductors are components that, up until a few years ago, were not thought of as chip components. When an inductance was needed for a microwave circuit, the designer would place a narrow transmission line in the circuit similar to those discussed in the previous section on filters. However, this has changed a great deal. Chip inductors now available give inductance from 1.6 to 550 nH in a size ranging from 0.047 inch square to 0.117 inch square and only a maximum of 0.025 inch thick. The basic package is shown in Fig. 4–61.

The entire inductor is vacuum deposited on the alumina chip (or substrate) in such a way as to form continuous turns, with the number depending on the required inductance. Fig. 4–62 shows how just such an inductor is constructed, using a six-step sequence of mask changes and alternating depositions of aluminum and ceramic thin films. The procedures followed were these:

1.   With mask 1 in place, aluminum was deposited for the first termination pad and the first half-turn of the winding.

2.   Mask 2 was used, and ceramic material was deposited to partially insulate the first half-turn.

3.   Mask 3 was called up, and aluminum was deposited to obtain the second half-turn.

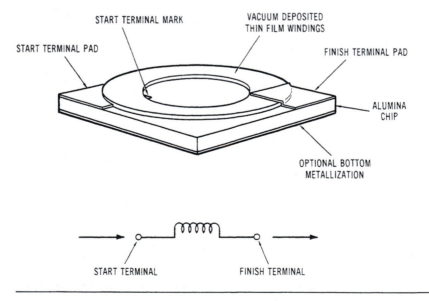

Figure 4–61.   Chip inductor.

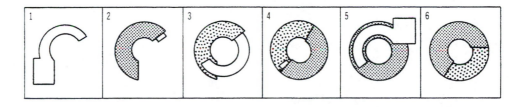

Figure 4–62.   Chip inductor construction.

4. Mask 4 was set in place, and ceramic material was deposited to partially insulate the second half-turn. For a specific inductance value, half-turns are added until the proper inductance value is reached.

5. Mask 6 was called up, and ceramic material was deposited to insulate the last half-turn, which will also hermetically seal the inductor.

This process makes it possible to deposit over 100 identical inductors on a single 2-inch × 2-inch alumina substrate.

Wire bonding is the most frequently used method for mounting the chip inductors. Aluminum or gold metallization is available on the terminal pads. These metallizations make ultrasonic or thermocompression bonding processes natural choices.

Solder may also be used if a 50% indium plus 50% lead alloy composition is ordered on the leads. Epoxy may also be used for both terminal connection (conductive epoxy) and for chip attachment to a substrate (nonconductive epoxy). The metallization on the bottom of the chip is optional. The chip itself may be soldered or ultrasonically bonded if metallization is used.

With processes and materials progressing, we may safely say that the chip inductor is a component that will improve in the future and find wide applications throughout the microwave industry.

4.10   Chapter Summary

We have covered many components in this chapter. The devices discussed can be considered to be the backbone of the microwave industry. The combinations of couplers, hybrids, mixers, and filters make up the complex systems now in use. Each component has unique properties that make it different from any other component, and each performs its assigned task to make larger assemblies operational. The chip components have made smaller and smaller assemblies a reality and will continue to do so.

If you become familiar with the operations of each of these components, you will have a firm grasp on the heart and soul of microwaves and will be able to handle virtually any problems that may arise in a microwave system.

4.11   References

1.   *Coaxial and Waveguide Components*. Hewlett-Packard Co., Palo Alto, CA, 1982.

2.   DePalma, H. A. "Manufacturing Process Produces Multilayer MICs." *Microwaves,* October, 1981.

3.   Howe, H., Jr. *Stripline Circuit Design*. Dedham, MA: Artech House, 1974.

4.   Laverghetta, T. S. *Microwave Measurements and Techniques*. Dedham, MA: Artech House, 1976.

5.   *Microwave Catalog*. Anaren Microwaves, Syracuse, NY, 1982.

6.   *Microwave Coaxial Connectors*. Omni Spectra Connectors, Waltham, MA, 1982.

7.   *RF and Microwave Components*. Anzac Division Adams Russell, Burlington, MA, 1982.

8.   *RF and Microwave Filters*. K & L Microwave, Salisbury, MD, 1982.

9.   *ThinCo Catalog*. ThinCo Corp., Hatboro, PA, 1980.

10.   *Understanding Chip Capacitors*. Johanson Dielectrics, Burbank, CA.

Questions

4.2   Directional Couplers

1.   Define *directional coupler*.
2.   Define *insertion loss*.
3.   What makes a directional coupler directional?
4.   Draw a directional coupler, label the ports, and identify the insertion loss port, the coupled port, and the isolated port.
5.   What is *directivity*?
6.   Why is a specification for coupling deviation needed?
7.   Describe a dual-directional coupler.
8.   Name and describe two applications of directional couplers.

4.3   Quadrature Hybrids

9.   What does *quadrature* mean?
10.   Explain a *hybrid junction*.
11.   What is the insertion loss in a quadrature hybrid?
12.   Why is amplitude and phase balance so important in a quadrature hybrid?
13.   Name and describe two applications of a quadrature hybrid.

4.4   Detectors

14.   What is the purpose of the input matching circuit in a detector?
15.   Why is a dc return needed in a detector?
16.   What is another function of the dc return?

17. Explain detector sensitivity.
18. Explain Tangential Signal Sensitivity (TSS) and how it is measured.
19. Name and explain two applications of a detector.
20. Why are the applications of detectors and directional couplers so similar?

4.5 Mixers

21. Name the three elements that are necessary in order to have a mixer.
22. Explain the operation of the three elements in question 21.
23. Why is a balanced mixer better to use than a single diode mixer?
24. What is the purpose of an image rejection mixer?
25. Define *conversion loss*.
26. Why are noise figure and conversion loss so closely related in a mixer?
27. Name and explain two applications of a mixer.

4.6 Filters

28. What is the function of a filter?
29. What is the insertion loss of a filter?
30. Define *passband*.
31. Distinguish between passband and bandwidth in a bandpass filter.
32. What is a *rejection frequency*?
33. What is the cutoff frequency in a lowpass filter?
34. What is the difference between a conventional lowpass filter and an elliptic-function lowpass filter?
35. Explain how a lowpass filter can pass DC.
36. Does a highpass filter have an upper frequency limit? Explain.

4.7 Circulators and Isolators

37. What is a *nonreciprocal device*?
38. Distinguish between a right-hand circular and a left-hand circular circulator.
39. Describe the internal construction of a circulator.
40. Draw a schematic of a circulator and an isolator and explain their operation.
41. Explain the importance of the third port match in an isolator.
42. What is a *drop-in* isolator?
43. List one application of a circulator and one of an isolator.

4.8 Attenuators

44. Define *attenuation* as it applies to microwaves.
45. How is VSWR measured on an attenuator?
46. Name the two categories that fixed attenuators fall into.
47. Name the two types of variable attenuators.
48. Name and describe two types of continuously variable attenuators.

4.9   Chip Components

49.   How are chip capacitors built so small and still have a wide range of capacitance values?

50.   Name two lead configurations for chip resistors.

Problems

4.2   Directional Couplers

1.   A 5 mw signal is applied to a 20 dB coupler. How many watts are present at the coupled port?

2.   The insertion loss in problem 1 is 0.4 dB. What is the power level at the output of the straight-through port?

3.   A directional coupler has coupling of 15 dB. An input signal of 2 mw is applied. Power read at the isolated port is 8.8 μW. What is the directivity of the coupler?

4.4   Detectors

4.   A microwave detector is needed that will supply 100 mw when a −12 dBm signal is applied. What should R_L be for this detector? (Use charts in the text.)

5.   A detector is needed for a particular application. R_L is 100Ω, and the output voltage can be no greater than 50 mV or less than 1 mV. What is the range of input power to accomplish this? (Use charts in the text.)

4.7   Circulators and Isolators

6.   An isolator has a termination with a VSWR of 1.75:1. What is the maximum isolation expected from this device? (Use charts in text.)

7.   An isolator is characterized as having an isolation figure of 25 dB. What is the VSWR of the internal termination? (Use charts in text.)

8.   An isolator has an internal termination rated at 0.75 watts. There is a 2.5:1 mismatch. What is the maximum input power that can be applied? (Use charts in text.)

9.   An isolator is used in a circuit that has a mismatch of 5:1. The isolator has 20 dB of isolation. What does the input circuit see instead of the 5:1 mismatch? (Use charts in text.)

5

Microwave Transistors

Objective

To provide an insight into the world of microwave transistors. Topics such as bipolar and gallium arsenide transistors, as well as High Electron Mobility Transistors (HEMTs), will be introduced, explained, and categorized as far as frequency of use and applications.

Key Terms

Alloy Transistor
Bipolar
Diffusion
Epitaxial
Field-Effect
GaAs FET
Gain Contours
HEMT
Insertion Power Gain
Interdigital
Junction Transistor

Linearity
Mesa
MESFET
Metallization
MOSFET
Noise Contours
Noise Figure
Oxidation
Planar
Unipolar

5.1   Introduction

The microwave transistor has probably accounted for some of the most significant advances in microwaves in recent years. The miniaturized and highly reliable systems of today have been made possible by high-technology wafer processing and bonding techniques. Areas such as low noise, linear power, and high power are now able to have their own special devices both in chip and packaged forms.

The transistors we will be covering will be bipolars, GaAs FETs, and the High Electron Mobility Transistors (HEMTs). The term *bipolar* is one that is probably familiar. It basically describes the structure of the device. That is, there are two charge carriers (electrons and holes) that account for transistor action. The *GaAs FET* may be somewhat unfamiliar. It stands for *gallium (Ga) arsenide (As) field-effect transistor*. The GaAs FET is a unipolar device in which only single-polarity carriers dominate and determine the transistor operation. The High Electron Mobility Transistor is a device which has been shown to outperform some low noise GaAs FETs in both noise performance and associated gain. Its construction is very similar to the GaAs FET but is not made up of only gallium and arsenic.

Before getting into specific details, examining a bit of transistor history might be interesting. This world of miniature marvels began back in 1948 when Drs. W. H. Brattain and J. Bardeen were studying the properties of germanium semiconductor rectifiers. During their studies they noticed that the conduction of a semiconductor diode could be controlled if they added a third electrode. This discovery resulted in the *point-contact transistor*. This early transistor is shown in Fig. 5–1. The collector and emitter wires are pressed into an n-type germanium material similar to the method used in early crystal sets using "cat's whiskers." The collector is often a phosphor bronze wire that has good conductivity. The p regions are introduced into the devices by a process called *forming*. This process

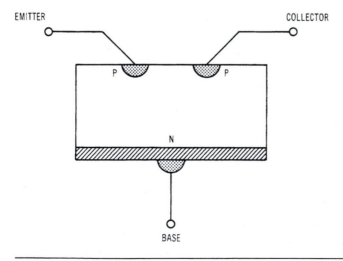

Figure 5–1.   Point-contact transistor.

involves the passage of a large current through the wires for a short period of time and into the germanium. The p-type germanium forms almost immediately at the contact point as a result of the diffusion of some acceptor impurities from the wire itself into the germanium. The resulting device in Fig. 5–1 is a pnp point-contact transistor.

The point-contact transistor was a major breakthrough at that time but had many drawbacks. It generated much more noise internally than the vacuum tubes it was designed to replace, it could not stand humidity and temperature since it was not hermetically sealed, and it was a rather fragile device to handle.

To counteract some of these drawbacks, W. Schockley of Bell Telephone Laboratories published a paper in 1949 examining the possibility of the now commonly known *junction transistor*. This device had much improved noise performance and was considerably more rugged than the point-contact devices.

The first junction transistors were prepared by a process called the *grown-junction technique*. This process resulted in devices that were mainly an npn structure. The *growing* of junctions is accomplished by starting with a small *seed* of material. The seed is touched to the surfaces of a bath or *melt* of molten semiconductor (germanium, silicon, etc.) and is then slowly pulled away from the metal. The size of the seed is increased (or grows) because the melt in the area of the seed sticks to the seed and freezes when it is withdrawn from the melt. The characteristics of the melt can be changed during the pulling operation by adding more or different impurities. The completed crystal is cut into transistors, and leads are welded to the proper areas. This device, as described, is also referred to as a *double-doped transistor*.

In 1951, a method was perfected that was termed the *alloy* or *fused-junction* method of transistor fabrication. The majority of this type of transistor are pnp devices. A wafer of base material (usually n-type) germanium is held in a jig between two pellets of impurity (indium, for example). The structure is heated until the impurity melts (155°C for indium), and the molten impurity penetrates the base material. During cooling, crystal regrowth results in the completed three-layer structure shown in Fig. 5–2. This device has a very high junction capacitance because of the fusing process. As a result, the alloy-junction transistor is not used at high frequencies.

In 1953, the shortcomings of the alloy-junction transistor, namely, the upper-frequency limit, were being worked on by a research group at Philco Corporation. As a result of this research the *surface-barrier transistor* was developed. This type of transistor is still widely used today, with many improvements, of course, on these early 1950s devices. Construction and operation were achieved by a manufacturing process that subjected a small piece of germanium to two jets of electrolyte, such as a solution of indium chloride, which strike the opposite faces of the germanium base slab. Etching of the semiconductor was achieved by passing a current through the electrolyte stream and the germanium. When the proper base thickness was achieved, current polarities were reversed, and the jets plated the metal emitter and collector contacts on either side of the base. Improvements in this procedure were made by the addition of impurity diffusion techniques. In this diffusion process, impurities of one conductivity type are made to diffuse at high temperature into the surface of a bar of the opposite type. The result is the formation of the familiar pn junction.

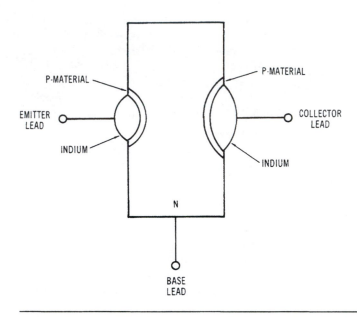

Figure 5–2.   Alloy-junction transistor.

In the *mesa* or *diffused-base transistor*, P_2O_5 (phosphorous pentoxide) may be diffused into an n-type silicon collector region to form the base. The emitter can be of aluminum alloyed into the base. Acid etching of appropriate areas results in a plateau or mesa. This type of device is shown in Fig. 5–3.

The *planar* and *epitaxial* types of construction are the ones that are used in microwave bipolar transistors. The *planar* construction uses diffusion and surface passivation that protect surfaces and junction edges from contamination. Leakage currents are low because SiO_2 (silicon dioxide), an insulator, is formed on the exposed surface. The *planar* structure is shown in Fig. 5–4.

The *epitaxial* transistor has a thin layer of low-conductivity material for its collector, with the remainder of the collector region of high conductivity. This structure is shown in Fig. 5–5. Both the *planar* and *epitaxial* devices are now used for low-noise and high-power application up into the 6- to 8-GHz range.

All of the devices referred to during our jaunt through transistor history have been bipolar devices. We should probably make our history lesson complete by mentioning another type of transistor that is finding a larger and larger place in the microwave fields. Although it was known in theory as far back as 1926 and thoroughly analyzed by Shockley in 1952, the *field-effect transistor* (FET) was not successfully fabricated for commercial use until 1961. It was not used for microwave applications until the mid 1970s, when gallium arsenide began to be the material that pushed the frequency limit well up into the upper microwave region. The *FET*, or *unipolar transistor*, exhibits certain characteristics superior to those of the junction device. These characteristics will be discussed in detail later in this

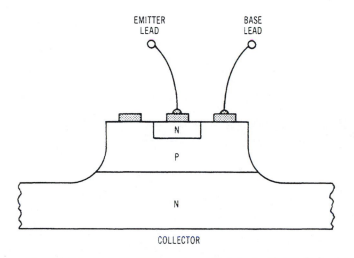

Figure 5–3.   Mesa transistor construction.

chapter. The FET was introduced at this point to emphasize its position in overall transistor history. The HEMT is a device which was introduced to the microwave industry in the late 1970s. It found limited applications at that time but has improved in the years that followed. The construction is similar to the FET in that it has a drain, gate, and source with a Schottky junction.

We have now gone through a brief summary of the transistor as it progressed from its early laboratory stages to the devices now available from a variety of manufacturers. It is now time to discuss in detail all of the devices mentioned.

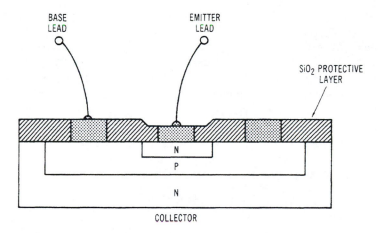

Figure 5–4.   Planar transistor construction.

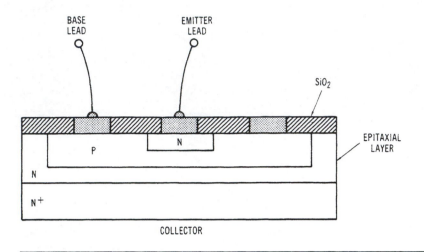

Figure 5–5.   Epitaxial transistor construction.

5.2   Bipolar Transistors

Most of us have seen an Olympic event called the *biathlon*. It involves a combination of cross-country skiing and shooting at targets along a prescribed route. Two skills are involved in this event. In 1976, the United States celebrated in great fashion the 200th birthday of the country, or its *bicentennial*. Finally, a person able to speak and/or write two languages is said to be *bilingual*. All these terms are recognizable and have undoubtedly been used more than once. Each of these terms has the prefix "bi" and is involved with the number two.

There is also a number two involved with the bipolar transistor. Most devices have only one means of energy movement. That means is the electron, or a majority carrier. In the bipolar transistor, however, two types of carriers cause the device to operate, majority and minority carriers (electrons and holes). You will recall from basic transistor courses how the electrons move and create vacancies in the atomic structure of the semiconductor material. This electron movement causes the effect of the vacancy, or hole, to move in the opposite direction. This combination of electrons and holes allows the bipolar transistor to operate.

Bipolar transistors find many applications in the microwave area. They are used for low-noise devices, for linear power amplification, and for power amplifiers (usually operating with common-base devices, Class C). Each of these bipolars has unique features that allow it to operate with low-noise characteristics, linear power, or high-power output. Let us look at each of these devices and see what these features are.

5.2.1   Low-Noise Transistors

The low-noise bipolar transistor finds its most prominent application as an amplifier in the front end of a microwave receiver. In this application a device that will amplify the low-level input signal and introduce a minimum amount of noise in the process is needed. To obtain this low-noise characteristic requires special construction techniques. As we describe this construction, two terms will be coming up, *planar* and *epitaxial*. These terms were mentioned previously in our introduction, but it would be beneficial for us to get a more detailed explanation of them at this point. You can understand how they would need explaining considering that the data sheet for a low-noise device would probably call it an *npn silicon planar epitaxial low-noise transistor.*

Remember from our previous discussions that *planar* construction uses diffusion and surface passivation that protects surfaces and junction edges from contamination. This construction makes the protected areas less prone to surface problems that sometimes occur when structures such as the *mesa* construction are used. The term *planar* is one that denotes that both emitter-base and base-collector junctions of the transistor intersect the device surface in a common plane. Perhaps, a better term for the structure is *co-planar*. The real significance of the *planar* structure, however, is that the fabrication technique of diffusing dopants (or impurities) through an oxide mask results in the junction being formed beneath a protective oxide layer. Thus, the previously mentioned isolation from surface problems. This protective oxide layer is emphasized in Fig. 5–6, where the unprotected *mesa* is compared to the protective *planar* construction.

The term *epitaxial* is actually a shortening of the term *epitaxial collector*. This term means that the collector region of the transistor is formed by the epitaxial technique rather than by the diffusion method, as mentioned earlier, for the base and emitter. The epitaxial layer is formed by condensing a single-crystal film of semiconductor material on a wafer of substance, usually the same material. Thus, an epitaxial device is one in which the collector region is formed on a low-resistivity silicon substrate. Epitaxial construction is shown in Fig. 5–7. Note that the epitaxial layer is not a different type of material. It is the same type material as the collector.

The base and emitter regions on the epitaxial device are diffused into the *epi*-layer, as it is termed. This diffusion can be noticed in Fig. 5–7 as well as the fact that the epitaxial structure can be thought of as the planar construction with an extra layer that separates the collector from the base and emitter structures. The epitaxial technique lends itself to precise tailoring of the collector region thickness and resistivity that improves device performance and uniformity.

The internal "geometry" of most low-noise transistors is an interdigital type of construction. This interdigital construction is illustrated in Fig. 5–8, with an actual device shown in Fig. 5–9. This geometry achieves the lowest base resistance without sacrificing gain. This low-base resistance high-gain combination results in a low noise figure. To understand why this comes about, consider what the noise figure of a transistor is.

The noise figure of a transistor is a measure of how much the signal-to-noise ratio (comparison of signal level to noise level) is degraded as the signal passes through the transistor. The degradation that does occur is due to the presence of three noise sources within the chip itself:

- Shot noise in the emitter-base circuit,
- Shot noise in the collector circuit,
- Thermal noise generated by the base resistance.

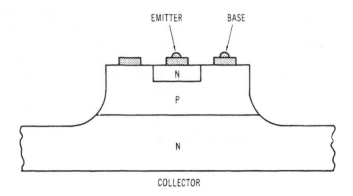

(A) Mesa construction.

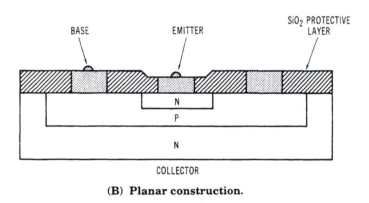

(B) Planar construction.

Figure 5–6. Mesa versus planar construction.

Shot noise is the noise generated at a semiconductor junction due to the current that is passing through it. It is directly proportional to the square root of the current applied to the junction. It is, therefore, logical to keep the current through the device as low as possible. Notice that when a low-noise transistor is presented later that the lower noise figures are achieved when the collector current (I_C) is low.

Thermal noise is the noise generated by the random thermal motion of charged particles (or electrons within the base region). Remember that we referred to the fact that the interdigital structure achieved the lowest base resistance without sacrificing gain. If we have a low resistance in the base region, we will have a small voltage dropped across that region and thus lower power dissipated. This lower power will mean very slight thermal activity in the base region and, therefore, low thermal noise.

With the information above, we can say that the best low-noise transistor should have a construction that offers an "efficient" junction structure and a low base resistance. By

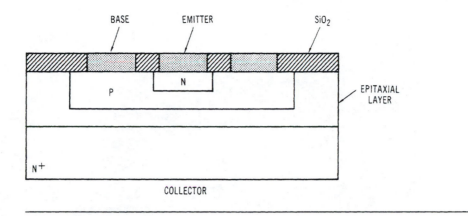

Figure 5–7. Epitaxial construction.

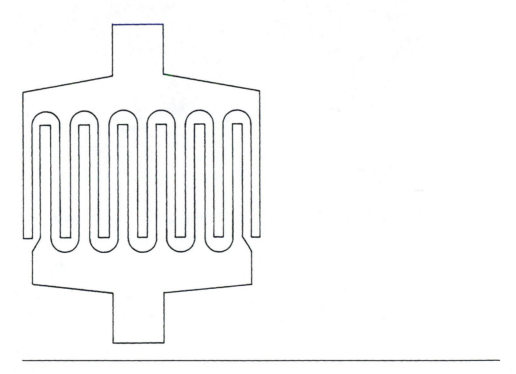

Figure 5–8. Interdigital construction.

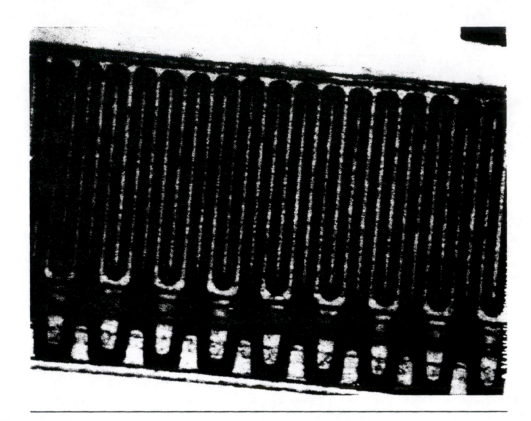

Figure 5–9.   Low-noise transistor *(Courtesy of M/A Com)*.

efficient junction, we mean one that will provide the pn junction action with a minimum amount of current or loss across that junction.

5.2.2   Low-Noise Transistor Construction

How would you make such a low-noise device for microwave operation? Let us play transistor designer for a while and construct a representative device. Fig. 5–10 shows the step-by-step process.

The process begins in Fig. 5–10A with an n-type epitaxially grown silicon layer. This layer ranges in thickness from 2 to 5 microns (0.00007874 inch to 0.00019685 inch). This silicon layer is the epitaxial layer shown previously in Fig. 5–7 and is supported by a substrate that is the collector.

A thermally grown oxide layer several thousand angstroms thick (1 angstrom equals 3.937×10^{-9} inches or 1×10^{-10} meters) is formed on the epi-layer, and the base contact cuts are defined and etched by photoresist techniques as shown in Fig. 5–10B. Through the open areas a heavily doped p-type (boron) diffusion is made as shown in Fig. 5–10C. Note

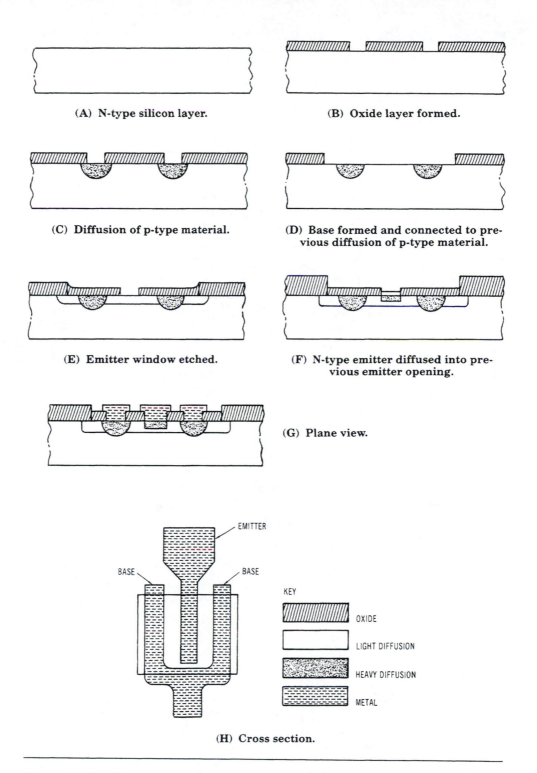

(A) N-type silicon layer.

(B) Oxide layer formed.

(C) Diffusion of p-type material.

(D) Base formed and connected to previous diffusion of p-type material.

(E) Emitter window etched.

(F) N-type emitter diffused into previous emitter opening.

(G) Plane view.

EMITTER

BASE

BASE

KEY

OXIDE

LIGHT DIFFUSION

HEAVY DIFFUSION

METAL

(H) Cross section.

Figure 5–10. Transistor construction.

that these areas, which are the base region, are of heavy diffusion so that a high concentration of p-type material will be available for each finger of the base construction. This high concentration lowers the metallization contact resistance and provides low-resistance contacts to the transistor base region. Remember that low base resistance is an important prerequisite for low-noise devices.

Fig. 5–10D shows the base area being cut into the previously deposited oxide layer. A precisely controlled amount of lightly doped p-type diffusions (boron once again) is inserted through the base area opening. This diffusion now forms the base and is automatically connected to the diffusion mode in Fig. 5–10C.

With the base of the transistor taken care of, the emitter is next to be produced. The emitter opening is defined and etched as shown in Fig. 5–10E. This opening must be located precisely between the two previous diffusions used for the base. The device is now completed by diffusing a shallow, heavily doped n-type emitter into the emitter opening etched previously, which can be seen in Fig. 5–10F.

To achieve good microwave performance in a device such as the one just described, the depth of the diffusions is kept very small. The total junction depth of the base of a 2-GHz transistor is on the order of only 0.3 micron (0.0000118 inch). The emitter will penetrate only about 0.2 micron (0.00000787 inch), and the base width, the difference between the emitter and base junction depths, will therefore be around 0.1 micron (0.000003937 inch).

Contact to the finished transistor is accomplished through metal fingers that align with the open p-base diffusion and the emitter fingers. These openings are made in two steps. First, the emitter oxide is "washed" away by a very short acid etch. Second, the p-base diffusion contact opening is defined and etched by photoresist techniques. The metallization is deposited in a film over the whole wafer area, and the fingers and bonding pads defined by chemical etching. A final cross section and plane view are shown in Figs. 5–10G and 5–10H. The pattern shown in Fig. 5–10H was also presented in Fig. 5–9, which showed the internal construction of an actual device.

Another method of transistor construction is *ion implantation*. This is a method of embedding the dopant into a wafer by accelerating the ions toward the wafer with a high electric field. The advantage of this process is that both the dopant dose and depth are controlled with much more precision. This precision results in both a much higher yield of devices and superior performance since the devices can be fabricated closer to the original design.

5.2.3   Low-Noise Transistor Data Sheet

With the low-noise transistor defined and constructed, it would probably be a good time to take a look at a typical data sheet. This data sheet is not from any specific transistor manufacturer but, rather, is a representative sample that takes the important parts of a variety of data sheets. When a low-noise bipolar transistor is needed, there are a few things to consider when making your choice.

Frequency   First, choose a device that is designed for your *frequency* of operation or over the band of frequencies needed. This selection process is sort of like picking out a new jacket. The first thing to do is find the rack that has your size. Then such things as style and color can be considered. It gets rather frustrating to find a jacket you really like and find out it does not come in your size. The same is true of transistor selection. One may be found that has a 1.5-dB noise figure but will only operate to 1 GHz. This limitation becomes a problem when your application is 2 to 3 GHz. So, the first thing to consider when selecting a transistor is its frequency range. Frequency range is usually one of the first items on the data sheet, and it will appear in an opening paragraph termed "Description" or "Description and Applications." It also may appear in a summary of specifications termed "Features." Wherever it appears, be sure to pick this up first. A paragraph entitled *description* would read something like this:

> The ABC-442 is a silicon bipolar transistor designed for use in low-noise, small-signal amplifiers up to 4 GHz. This device features excellent gain characteristics while maintaining its specified low noise figure over a broad range of frequencies. It is available in chip form as well as in our 70-mil and 100-mil packages.

Check out this paragraph on the transistor data sheet first. It will save time and aggravation later on.

Noise Figure   The next logical parameter to look at on the data sheet is, of course, the *noise figure*. This figure is sometimes called *spot noise figure* or *minimum noise figure*. When reading this number, be sure to read the conditions that go along with it. Simply saying that a transistor has a noise figure of 2.8 dB means little or nothing. If, however, the manufacturer says the device has a noise figure of 2.8 dB at 4 GHz with $V_{CE} = 10$ V and $I_C = 5$ mA, the device is characterized so that we can determine whether we can use it in a particular application. (V_{CE} is the collector-to-emitter voltage, and I_C is the collector current.) This type of detailed characterization is absolutely necessary since the design resulting from the use of a particular device depends on the S-parameter data that are taken at specific conditions ($V_{CE, IC}$). *S-parameters* are the parameters that define the transistor in reference to a 50-ohm system, and they are covered in Chapter 8. To have the proper matching circuits and ensure that the transistor will operate with the expected noise figure, these conditions must be duplicated. A typical curve of noise figure versus frequency for a microwave device is shown in Fig. 5–11.

Noise Contours   Some manufacturers use *noise contours* to show the noise figure of their devices. Noise contours are a graphical representation of the noise figure presented on a Smith chart, which is covered in Chapter 7. Fig. 5–12 shows just such a chart with various noise figures presented on it. Notice that there are different points where a specific noise figure can be achieved. The one parameter that may suffer, however, is the input VSWR if the input matching procedure is not treated carefully. The contours are derived by using the admittance parameter and optimum noise figures for the device and finding the circle center, the angle of the vector, and the radius of the circle. With this information, the admittance points can be picked from the chart for the noise figure desired, and the transistor can be *noise matched* for operation.

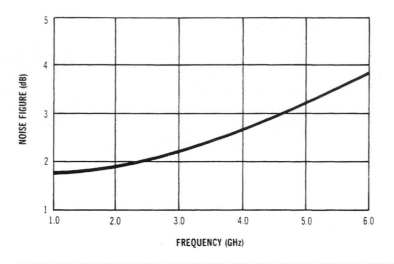

Figure 5–11.   Noise figure versus frequency.

Transistor Gain   The next parameter to be investigated is the *transistor gain*. With this parameter we are looking at two values: G_{max} and G_{NF}. G_{max} is the gain from *conjugately* matching the device for a specific gain. The noise figure obtained with this condition depends on the characteristics of the particular device used.

G_{NF} is the gain achieved when the transistor is matched specifically for the optimum noise figure at that particular frequency. This gain is significantly lower than the gain achieved at G_{max}. To illustrate, consider the following comparison of G_{max}, G_{NF}, and noise figure for three devices:

Device	Frequency	NF	G_{max}	G_{NF}
A	4 GHz	2.8 dB	12.0 dB	8.5 dB
B	4 GHz	2.7 dB	12.0 dB	8.0 dB
C	4 GHz	2.8 dB	12.2 dB	8.0 dB

This data is from three commercially available devices designed to provide basically the same performance. When designing for gain with these devices, considerably more gain is received than when designing for optimum noise figure. In most cases there is a difference of 4 dB in gain. This difference, of course, is a logical result since the design is around a conjugate match for power gain and around noise admittance parameters for a noise match. This type of matching does not guarantee a maximum transfer of power and thus will not guarantee a maximum amount of gain through the device.

Insertion Power Gain   One additional gain figure is sometimes given on data sheets— the *insertion power gain*, $|S_{21}|^2$. This gain is obtained by dropping the device into a

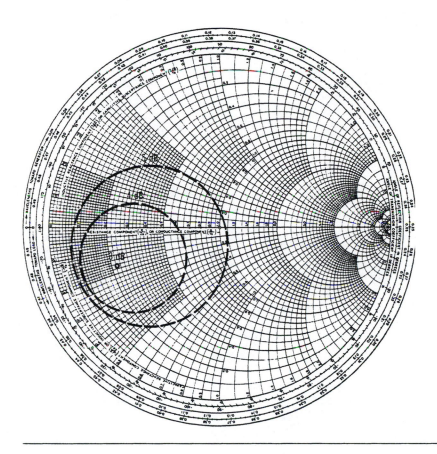

Figure 5–12.   Transistor-noise contours.

circuit with a 50-ohm line at the input and a 50-ohm line at the output. This gain is useful for wideband operations where a wide range of matching is not feasible using a conjugate match. The value of $|S_{21}|^2$ for the three previous devices is 7 dB, 8.2 dB, and 7 dB, respectively. About the same gain is received as with optimum noise figure with $|S_{21}|^2$. However, the good noise figure will not be received as when performing a specific noise match.

Gain Contours   *Gain contours* can be drawn just as we drew the noise contours. Fig. 5–13 shows these contours for one of the devices referred to previously. When we superimpose the gain and noise contours on one another, we have a picture of the capabilities of a device and can pick our operating point to comply with our specific requirements for noise and gain. The combination of noise and gain contours is shown in Fig. 5–14. One restriction should be made at this point concerning the gain contours. Note in Fig. 5–13 how

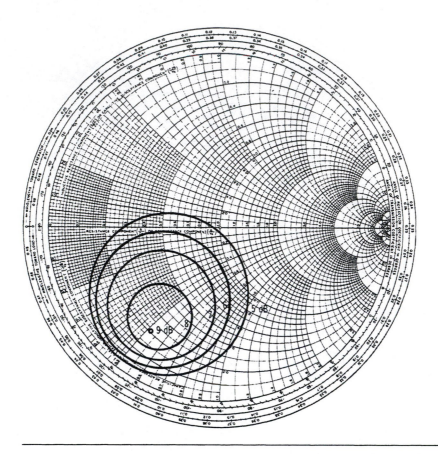

Figure 5–13.   Transistor-gain contours.

the circles are all inside the boundaries of the Smith chart. Within the boundaries means that the device will be unconditionally stable under these conditions. A gain contour that is not completely inside the chart tells that this condition has the possibility of becoming unstable and going into oscillation under certain conditions. So take care when designing to be sure to be in a stable area of the chart.

Current   One parameter we have mentioned previously as being important in low-noise operation is *current*. If we take the three transistors previously shown and list operating conditions, these conditions would be as follows:

Device	V_{CE}	I_C
A	10 V	5 mA
B	6 V	5 mA
C	10 V	4 mA

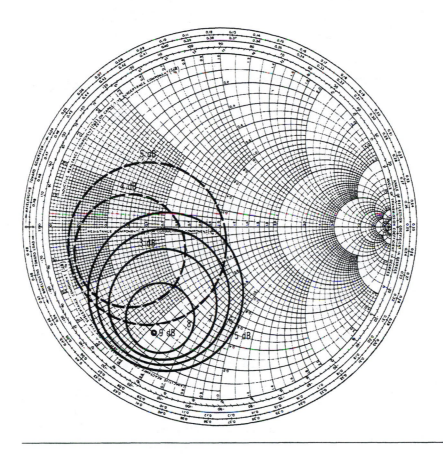

Figure 5–14.   Noise and gain contours.

To understand how important the operating current is to low-noise operation, consider Fig. 5–15. Notice that the noise figure of this device (which is device A in our example) is optimum (or lowest) at the 5-mA value. Above and below this point it increases in noise figure. This increase is why it is vitally important to note the operating conditions of a device and design with them in mind.

Package   One final area that should be looked at on a low-noise data sheet is the package used for the device. With the miniaturization of circuits and systems becoming more and more important, the size of the device used for these circuits and systems becomes important, too. Fig. 5–16 shows some of the packages that contain low-noise transistors. One item to watch in these packages is that some are in inches and some in millimeters. Fig. 5–16A is in both millimeters and inches, Fig. 5–16B is in inches only, and Fig. 5–16C is in millimeters only. Be sure to check the units. Another area to watch is the marking of base and collector leads. The drawings have a mitered lead on them to distinguish either the

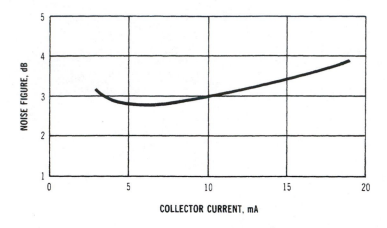

Figure 5–15.   Noise figure versus collector current.

base or the collector. This mitered lead is fine except when the leads get cut back to have the transistor put in the circuit. When this happens, the miter is lost and some other means of identifying leads must be relied on. There is usually a dot put on transistors at the collector to aid in this identification. Also, there is usually a name of the manufacturer printed on the top of the case. Note the orientation of this name *before* the leads are cut and go from there.

The low-noise transistor is one of the most unique devices. It provides a power gain, low noise figure, and does it all by using a minimum amount of current. You will have to search a long way before finding a device that will accomplish all of this in a package as small as what we have shown.

5.2.4 Linear Transistors

Remember from basic transistor courses that there are three classes of operations: Class A, B, and C. The class that is used depends on the biasing of the transistor. Class A is usually used for low-level stages where an increase in input results in a corresponding increase in output—up to a certain point. Class B would be used for a push-pull operation scheme where more power may be needed. Class C is almost exclusively used for high-power common-base operations where the increased efficiency is needed. These classes and usages are pretty much standard.

It is possible, however, to get a higher power output by biasing some transistors Class A. These are the microwave linear transistors. Devices are available that produce 7 W at 2 GHz or as much as 2 W at 3 GHz. This capability may be very impressive, or it may mean nothing without the understanding of what a linear power transistor does.

From basic transistor courses, remember that if a low-level transistor stage is built, the proper bias is applied, and an input signal is put into it, a certain output will be obtained,

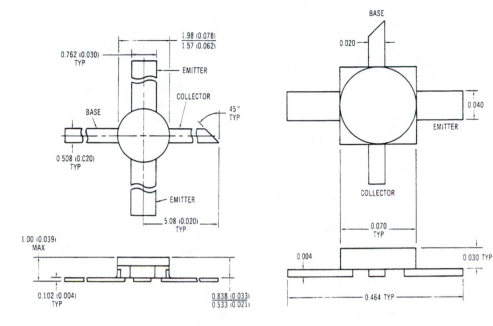

(A) Dimensions in both millimeters and inches.

(B) Dimensions in inches only.

(C) Dimensions in millimeters only.

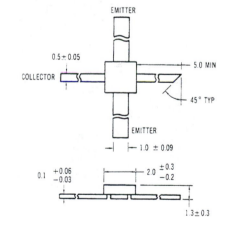

Figure 5–16. Low-noise transistor packages.

depending upon the characteristics of the transistor used. If the input is increased, the output will correspond. This will continue until a point termed the *compression point* is reached. At this point a 10-dB increase at the input will result in only a 9-dB increase at the output. This, appropriately enough, is actually called the *1-dB compression point*. When biased for Class-A operation, all circuits will exhibit this phenomenon.

Contrary to this Class-A operation is the operation of power stages. To obtain the efficiency needed, the devices are biased Class C. With Class-C operation, the input level can be run up with no significant output until the turn-on level is exceeded. The output level then jumps up to its necessary level and stays basically at that level if the input is further increased. With an input level that is required to vary over some power range, an output will be obtained only when that power is above the turn-on level of the power stage. Another method, therefore, must be available for taking care of such applications. This method, of course, is the linear power transistor.

The linear power transistor is constructed in such a manner that it will survive the higher collector currents necessary for power amplification in a Class-A mode. One point should be made clear, and that is the power levels for Class-A and Class-C operation. A single Class-C power stage will be able to produce in excess of 20 W at 2 GHz. The linear stage, on the other hand, will have a maximum of about 7 W at 2 GHz. This does not mean that power devices are not available that produce 7 W at Class C. They certainly are available and can be used for comparison of power and linear devices. The point is that the linear device is not a high-power device but one we could call an intermediate power device. We cannot make direct comparisons as far as maximum power levels are concerned.

5.2.5   Linear Transistor Data Sheets

To understand the linear transistor, let us look at a typical data sheet and explain some of the important parameters:

Frequency	1 to 2 GHz
Power output	1.6 W @ 2 GHz
VSWR mismatch	∞
Linearity	-0.2 dB, $+1.0$ dB

All these parameters have specific conditions associated with them. The conditions may be $V_{CE} = 20$ V, $I_E = 200$ mA, and frequency $= 2.0$ GHz. This indicates how the parameters were taken in the first place, and what conditions must be set up to duplicate the parameters.

Frequency   The first parameter, *frequency,* gives the range over which the transistor will operate. Frequency does not indicate, however, that the device will cover this entire band in one shot and have a response that will be flat within 0.5 dB. It says only that the frequency range given is the one the transistor was designed to handle. Fig. 5–17 is an impedance plot on a Smith chart for a typical linear device. (The Smith chart is covered in detail in Chapter 7.) The device shown in Fig. 5–17 has input and output impedances fairly close together between 1.0 and 2.0 GHz. These impedances mean that the device will

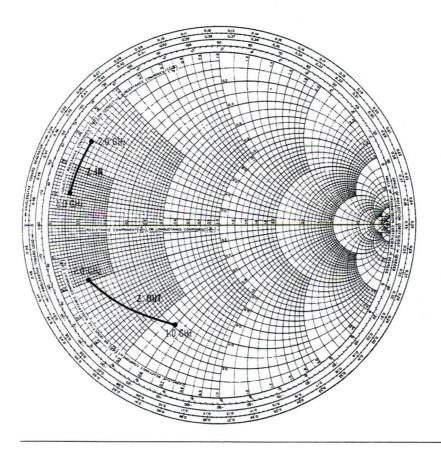

Figure 5–17.   Impedance data.

probably operate over a large portion of this band at one time. When the impedances vary greatly from one frequency to another, the band of operation decreases rapidly.

Power Output   *Power output* is the next important parameter to be checked when choosing a linear transistor. As we have said previously, certain conditions are spelled out when specifying parameters. A device cannot be said to deliver 1.7 W of power at 2 GHz without defining the conditions under which the 1.7 W was obtained. What is needed is such data as V_{CE}, I_E, and power input. To understand this need for such data further, consider the two curves in Fig. 5–18. The first curve (Fig. 5–18A) is power output versus frequency, and the power is relatively flat over the range of frequencies. Also, note that the test conditions are spelled out on the chart. The second curve (Fig. 5–18B) shows the importance of two of these conditions—V_{CE} and I_E. If we take the case of V_{CE} set at 15 V, we can get anywhere from 0.5 W to approximately 1.6 W by varying the emitter current (I_E) from 100 to 300 mA. If we increase V_{CE} to 20 V, the same range of current yields from 0.6

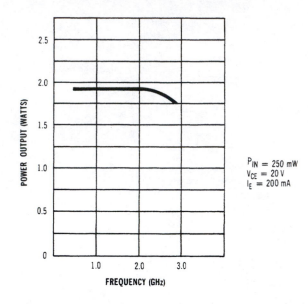

(A) Power output versus frequency.

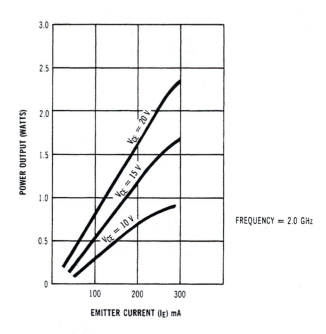

(B) Power output versus current.

Figure 5–18. Power output versus test conditions.

W to about 2.25 W. So you can see how important it is to specify the operating voltage and current to make the parameter repeatable. Therefore, the power output on the data sheet should always include test conditions, including frequency.

VSWR Mismatch   One of the most important properties of a device handling power is its ability to perform under mismatch conditions. If, for example, a device is producing 2 W of power to a load and a huge mismatch appears at that load, this 2 W (or a large majority of it) will be reflected back to the device. If the device is not equipped to handle this reflected power, it will be destroyed. (And you will have to explain to your boss why you have to spend some more money for a new transistor. This situation can be embarrassing, especially when you finally convinced your boss that this first device was the best thing since sliced bread.) So always check to see what sort of a mismatch tolerance the device exhibits. The example we have presented shows infinity as the specification for mismatch, meaning that a VSWR of infinity (or an open circuit) will not affect or damage the transistor.

Linearity   The final parameter on our typical data sheet is *linearity*. Linearity is a natural parameter to investigate since we are talking about linear devices. Linearity of a device is how close its output response tracks a straight line drawn between a low- and high-power end of its operating range. The data sheet device showed -0.2 to $+1.0$ dB of variation over this range. The conditions given for this parameter are usually frequency, power #1 (low end), and power #2 (high end). Typical numbers might be 2 GHz, $P_1 = 0.15$ W, $P_2 = 1.5$ W. If the response of P_{in} versus P_{out} is plotted and a straight line is drawn between P_1 and P_2, the variation from this line will be the linearity. The ideal figure, of course, would be ± 0 dB.

Case   One last area that should be considered when choosing a linear power transistor is the case, Fig. 5–19. Fig. 5–19A is a stripline package; Figs. 5–19B and C are similar to each other, with Fig. 5–19C using a stud for a heat sink to ensure proper operation of the transistor.

Notice the measurements of the devices in Fig. 5–19, and compare them to those for the low-noise transistors shown in Fig. 5–16. The linear devices are all considerably larger than the packages for low-noise devices because the linear devices must have larger areas to dissipate the heat that is generated within the higher power devices. Remember that we discussed currents in the area of 4 to 5 mA for low-noise devices, whereas the linear devices typically will draw in the area of 200 mA, which is 40 to 50 times more current than the low-noise transistor, and thus more power and heat are within the device, which must be removed. This additional power is why Fig. 5–19C has a stud attached to it, and Fig. 5–19A has the 0.250-inch × 0.800-inch plate on the bottom of it. The stud transfers heat from the chip to the baseplate that holds the components.

This microwave linear power transistor, then, is the device that produces an intermediate level of power amplification while operating in a linear, or Class-A, mode. Whenever a range of linear output power levels is needed, this transistor is the type to use.

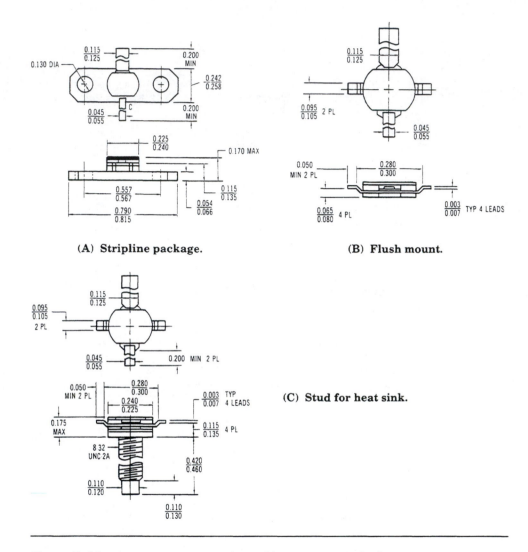

(A) **Stripline package.**   (B) **Flush mount.**

(C) **Stud for heat sink.**

Figure 5–19. Linear transistor packages (dimensions in inches).

5.2.6 Power Transistors

The *power transistor* is probably the most unassuming device you will ever see. These devices, which are usually about 0.300 inch by 0.250 inch with a flange placed under them, do not appear to be able to perform much of anything that is useful. But, as the old saying goes, "You can't judge a book by its cover." This marvelous device can produce as much as 20 W of power at 2 GHz and will do it continuously as long as the flange, mentioned previously, is firmly attached to an adequate heat sink. To further understand these dynamic little components, we will "build" one from the silicon wafer beginning to the protective cap epoxied on as the final step.

5.2.7   Power Transistor Construction

As mentioned, the power transistor begins with a silicon wafer. The silicon provides a medium for the transistor with certain properties that allow current to flow. These properties can be compared to a vacuum in a vacuum tube. For the tube the vacuum is the medium; for the transistor silicon is the medium. The wafers used are epitaxial. Remember from our discussions on low-noise devices that epitaxial is the "growing" of a substance on the same type of substance; that is, the growing of a layer of silicon on a silicon substrate. The silicon is grown in ingots 1 to 3 inches in diameter and from 6 inches to several feet in length. A single crystal "seed," ¼-inch square by several inches long, is used to start the growth. The seed is rotated and slowly withdrawn from a crucible of molten silicon that is doped to be n-type or p-type material by the addition of basic elements. Ingots of n-type doping use antimony or arsenic. The ingots are grown at temperatures approaching 1500°C. The dopants must all be distributed uniformly to prevent excess concentrations in one area and deficiencies in others. The ingots are sliced with a diamond saw into wafers to a thickness of around 0.015 inch. They are then mechanically and chemically polished to a mirror finish that reduces their thickness to around 0.012 inch.

The polished wafers are then placed in an epitaxial reactor at about 1200°C, and a doped film of silicon is grown on top of them by means of a chemical vapor deposition. The doped film may be as thin as 0.0001 inch (0.00025 cm) or as thick as 0.002 inch (0.005 cm). This film can be either n- or p-type, depending on the type of device being fabricated. The resulting material is a *silicon epitaxial wafer*, which is probably the most important section of the transistor since everything is constructed on top of this wafer. It is somewhat like the foundation for a building. The foundation must be right, or the entire building will crumble. Similarly, if the epitaxial wafer is not grown properly, the transistor will not operate properly. The wafer is now complete, and the next step in our building process is *oxidation* and *masking*.

Masking   The microwave power transistor requires a large number of narrow emitter images to obtain a high emitter perimeter to base-area ratio because a large current is drawn by power transistors, and there must be provisions within the device to handle this current. An absolute minimum practical width for the emitters is one micron (0.0001 cm, 0.000039 inch). The base, emitter, and their contact patterns are defined in an oxidated silicon wafer by a photoresist and etch process (print and etch). The one-micron geometries necessary for device operation must be maintained by this photoresist and etch process. The photoresists, which are polymers sensitive to exposure to ultraviolet light, are used to define the necessary pattern. These patterns may be either positive or negative.

When positive patterns are used, the exposed area is developed, and the resulting image is the same as that which was exposed. Thus, if a hole in the oxide to permit diffusion of the impurities is required, the mask would have clear areas with the pattern that is required. The remainder of the mask would be a dark field. This arrangement was once a very difficult item to align because the operator could not see through the dark field to align the mask. Now, however, there are see-through masks that are opaque to ultraviolet light, which makes the positive masks much more practical and useful.

Negative resists react, of course, in the opposite way of the positive mask. That is, the unexposed area is developed rather than the exposed area. Organic solvents are used to develop these resists and cause them to swell. When this swelling occurs, the openings shrink and all edge definition is lost. Various types of rinses can be used to attempt to reduce this swelling, but as the resist shrinks, its adhesion to the substrate is degraded, and some may even peel off. Therefore, it is desirable to use the positive resist for small geometries.

Thus, the positive resist has the following advantages for the small geometries used in power transistors.

- It does not cause swelling, so a much thicker resist can be used for better definition.

- The processing is easier since the positive resist does not have to be post baked at high temperatures to ensure adhesions.

Oxidation   The oxidation process is performed to protect the surface of the device. All diffusion takes place through "windows" cut in the oxide to allow the n-type dopants to diffuse into the silicon. As previously described, photosensitive material is applied to the top of the wafer. The photoresist is exposed when a light source is shown on the wafer, similar to the everyday process of developing a photograph from an instant camera. The mask (positive) shields the window from exposure. When developed, the unexposed window is removed since light is required to harden the photoresist.

Exposing the silicon for dopants to be diffused requires an etchant that attacks only the SiO_2 (silicon dioxide) and not the silicon or the hardened photoresist. That etchant is hydrofluoric acid (HF). Results of such an etchant and the oxidation process are shown in Fig. 5–20. This figure also shows the next phase of our power transistor construction—base and emitter diffusion.

Diffusion   Diffusion is officially defined as the movement of carriers from a region of high concentration to regions of low concentration. Diffusion can be likened to dropping a stone in the water and seeing the ripples move from the point of impact out until the water is smooth again. A high concentration (high amplitude) of water is in the center with a movement outward to the end of the large circle where a low concentration (low amplitude) is, which is the same general idea of diffusion in transistors. In the case of transistors where silicon and silicon dioxide (SiO_2) are concerned, doping impurities diffuse into the silicon at elevated temperatures to form the desired junctions. The same impurities penetrate the silicon dioxide much more slowly, so silicon dioxide on the surface of the silicon (shown in Fig. 5–20) acts as a mask to determine the areas into which diffusion occurs.

As seen in Fig. 5–20, base diffusion occurs before emitter diffusion. The base of an npn device is *boron,* and the emitter is *phosphorus*. Diffusion temperatures are high, as referred to earlier, and are on the order of 1000°C. This high temperature causes the dopant (either boron or phosphorus) to diffuse from areas of high concentration to those of low, which

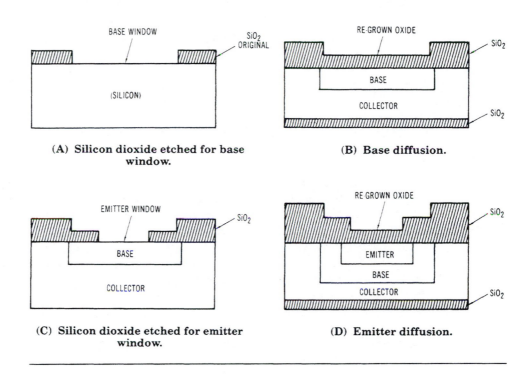

(A) Silicon dioxide etched for base window.

(B) Base diffusion.

(C) Silicon dioxide etched for emitter window.

(D) Emitter diffusion.

Figure 5–20.   Multiple oxidation.

satisfies the definition. Diffusion may be accelerated either by high temperatures or higher concentrations. As might be expected, the farther the dopant travels into the transistor, the lower the concentration. Whenever a high dopant concentration region exists (on the surface, for example), a low resistance exists, which results in a high concentration of carriers (electrons or holes).

The diffusion process, and how it occurs, determines the quality of the finished transistor. Once the base is diffused, the emitter/base junction must overtake the collector/base junction. That is, the base thickness must be controlled by controlling the separation of the emitter/base and collector/base junctions. This thickness will greatly affect the parameters of the transistor. By controlling the base thickness and the junction-area dimensions, the f_{max} (maximum frequency of oscillation) is improved greatly. Maximum frequency of oscillation is a measure of the ultimate frequency capability of the device, that is, how high in frequency the device will work. Obviously, when discussing microwave devices, this frequency will have to be as high as your application requires. In any case, it must be well within the microwave spectrum to be useful. The goal of this critical base region is to keep it thin. This thickness and the emitter depth can be controlled best when the doping densities are large, that is, the area of low resistivity where concentrations are high. Excellent operation occurs when the emitter depth is ½ to ¾ of that of the base.

Metallization   The final process before we put the device in a package and attach leads to it is that of *metallization*. There are two schools of thought on metallization. One says that aluminum is the best metal for microwave power transistors; the other says gold provides superior performances. There are advantages and disadvantages to both materials, and for that reason, both will be covered. The first to be covered will be aluminum. (This order does not mean that aluminum is any better or worse than gold. The only criteria used to make this decision was that "A" comes before "G" in the alphabet.)

The aluminum metallization technique is illustrated in Fig. 5–21. Note that the top portion of the figure is similar to that at the bottom of Fig. 5–20 after we finished our diffusion and deposition processes. Aluminum is evaporated on the wafer to something less than 2%. An electron-beam technique is used to achieve a uniform and pure aluminum layer. A photolithography process, described earlier, is used to define the final metal pattern needed. A patented etch process is used to etch away the unwanted aluminum. This technique is very powerful in that it allows extremely fine aluminum lines of less than 0.1 mil to be cut. In addition, the process is inherently very high yield and defect free. Its major advantages are finer line geometries and very high yield on large devices (permitting very high-power, high-frequency units).

Aluminum silicon metallization is another process employed in which there is an addition of 1 to 4% silicon to the aluminum. The advantages are the same as with the aluminum system, but there is the additional benefit of vastly improved yield and reliability on very high-frequency devices. Usually when a transistor is said to have aluminum metallization, it, in fact, has aluminum silicon metallization.

To those who prefer gold, the preceding discussion of improving aluminum would seem to be more "patching the wound" than "curing" the aluminum problem. Gold, it is said, offers distinct advantages over any aluminum system. It is less prone to electromigration (15 to 20 times better), it eliminates the corrosion problems encountered with aluminum, it can be etched easily (it does not undercut), and it virtually eliminates the microcracking associated with relatively brittle aluminum.

All of these advantages mentioned are applicable to pure gold—but pure gold cannot be deposited on bare silicon. Silicon and gold form a eutectic around 400°C, a temperature not compatible with transistor processing and assembly. Thus, a problem arises because the gold migrates into the silicon, and no clear boundary is formed between the two.

A high-temperature metal, therefore, must be used with gold to act as a barrier separating the gold from the silicon. This barrier metal should be a good barrier for gold (gold should not diffuse through it); it should make good, uniform electrical contact to the silicon; and it should adhere to the oxide of the device to allow for subsequent assembly.

In actual practice, no single element satisfies all of these conditions, so a multimetal system must be used. Some of the systems most commonly used as multimetal barriers follow.

Platinum-silicide/titanium/platinum/gold—Platinum silicide is used to provide good electrical contact, titanium is used for adherence to the oxide, platinum is used to provide a barrier to gold diffusion, and gold is used to carry high current densities. This system is primarily used in beam-lead applications. It is very complex and impractical for very small geometries because of the etching difficulties of both titanium and

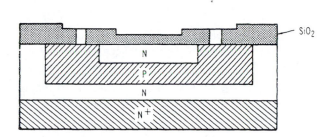

(A) Wafer with base and emitter contacts exposed for aluminum deposition.

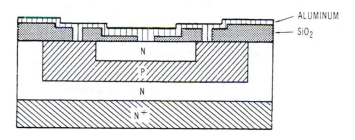

(B) Aluminum-covered wafer.

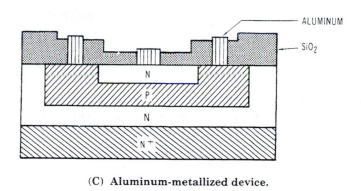

(C) Aluminum-metallized device.

Figure 5–21. Aluminum metallization technique.

platinum. This system, if it does work, is very reliable, if care is taken to passivate the titanium, which is worse than aluminum in terms of corrosion.

Platinum-silicide/molybdenum/gold—This metallization system has been used for several years in small-signal high-frequency transistors. The main problem in this system is the high electrical contact resistance. In high-power transistors, low input resistance is one of the essential features for achieving good performance. Platinum-silicide is used to minimize this problem. However, getting consistently low contact resistance still remains a problem to be solved.

Platinum-silicide/tungsten/gold—This system has been introduced for use with high-power, high-frequency transistors. Tungsten does not make good electrical contact to silicon; therefore, platinum-silicide has to be used to lower the contact resistance. Tungsten also does not adhere well to oxide. To overcome this problem, a very high sputtering energy, which tends to degrade the E-B junctions and, thereby, results in very poor yields, is required. To solve this problem, a fourth metal is required to "glue" the tungsten to the oxide. Tungsten, although better than titanium, is not an acceptable barrier to gold as accelerated tests at 600°C have proven. Moreover, tungsten is very hard to etch, as it undercuts the gold during etching, but it is impossible to quality inspect the seriousness of the damage. How far it has been undercut is clear only when the gold falls off the wafer. Obviously, this type of metallic barrier is not used very often.

Auromet®—This gold metallization system answers most of the problems encountered in gold metallization systems. It has excellent adhesion to oxide; it is a very good barrier to gold, as accelerated tests at 700°C for 24 hours have shown no significant interdiffusions; and it has excellent step coverage with no microcracking.

An additional composition of platinum-silicide/titanium/tungsten/gold has also been shown to be very successful. The platinum-silicide is used to provide a stable ohmic contact to silicon. The tungsten layer is also used as a barrier against gold diffusion into—and eutectic alloying with—silicon as well as providing adhesion of the gold to the silicon and silicon dioxide (SiO_2).

As we stated earlier, each of the metallization processes has advantages and disadvantages. Generally, selection is a matter of preference and application.

Package   We now have a power transistor chip constructed and ready to drop into a package. The package is probably every bit as important as the chip we have just built. It does much more than simply giving a means of attaching the transistor to a circuit. Three factors must be taken into consideration:

- Lead inductance must be kept to a minimum. Concentration must be on the common lead first. This may be the emitter or, more commonly, the base in power devices. Second priority for reduction in lead inductance is the input, and last is the output.

- Thermal dissipation of the case must be high. The most important criterion when operating power devices is to get the heat away from the chip and out to the chassis. This dissipation occurs when the package does its job.

- The package should be a reliable and easy-to-use device. Such factors as package type, shape, size, and material should all be considered.

As the frequency of operation of any component increases, you must become increasingly aware of the inductance involved with that component. The need for awareness becomes evident when examining the equation for inductive reactance ($X_L = 2\pi fL$). If this reactance is to remain low as the frequency increases, the inductance must be decreased or kept at a minimum. Fig. 5–22 shows the internal structure of a power transistor and shows how the inductance can be controlled within the device. Notice how every bonding wire coming from the chip is the same size and, more importantly, the same shape. This uniformity ensures that the lead inductance will be kept to a minimum.

As we have stated previously, the most important criterion when operating power devices is to get the heat away from the chip and out to the chassis. The thermal characteristics of the chip are largely dependent on the particular design. From heat transfer theory, a point heat source (the chip) dissipates heat through a solid in the shape of a cone. That is, the heat is concentrated at the origin and fans out as it travels from that point. This definition is similar to our definition of diffusion, with a high concentration moving to a lower concentration. Another way of looking at heat transfer is to consider a beam from a flashlight and how the light fans out and diminishes as you get farther away from the bulb (the source).

By understanding how the heat dissipates through a substance, you can see that if the chips (or sources) are too close together, there will be overlapping heat areas that produce

Figure 5–22.   Power transistor *(Courtesy of M/A Com).*

"hot spots." These hot spots are areas of high heat density that make it very difficult to get heat away from the chips at a sufficient rate to protect them.

To eliminate these hot spots, or interference areas, the cells of a chip can be spaced such that these areas exist in low heat density areas. A high-yield construction is used in this case, and a geometric symmetry arises from the use of cellular and interval matching techniques, virtually assuring thermal balance. As an example, thermal measurements were taken in a 32-cell, 125-mil $\times$ 75-mil chip and slow thermal gradients (temperature differences) of only 3% from cell to cell. Single-cell transistors often show temperature variations of 50°C from center to edge. Uniform heat distributions allow the junctions to operate at moderate temperatures, ensuring long life reliability of the device.

Mounting   We have now come to the point where the transistor is in its package. This package may be one of a variety available from different manufacturers. Fig. 5–23 shows one of these devices. The method of mounting the transistor should be carefully considered. Whether it is a flange-mounted device or studded device, there are specific ways of

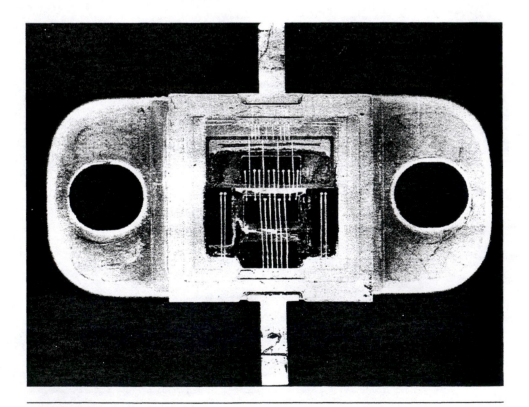

Figure 5–23.   Transistor package *(Courtesy of M/A Com)*.

properly mounting each. The following general rules should be followed for all devices. For specific requirements, consult the vendor data sheet.

- Be sure the metal area that the device is to be mounted to is flat and smooth.
- Remove all burrs from any mounting holes.
- Be sure the metal areas are clean—remove all oil and chips.
- Use thermal grease on the heat-sink area.
- Align the transistor properly; that is, be sure which lead is the base, emitter, and collector.
- Using the proper-sized screws for the flange-mount device and appropriate washers, mount the device to the heat sink.
- If a studded device is used, be sure to use the right torque when putting the nut on the stud. Too much torque will break the stud from the transistor. The following torques are recommended for various studs.

Stud	Torque
1/4 in	5 ± 1 in-lb
3/8 in	8 ± 1 in-lb
1/2 in	10 ± 1 in-lb

By using the proper mounting techniques, many potential problems will be eliminated both on initial turn-on and further down the road when the heat buildup destroys areas of the device.

5.2.8   Power Transistor Data Sheet

To finish up the topic of power transistors, we will look at a typical data sheet.

Power Output   The first item to naturally look at would be *output power*. This value is usually given as a power at a frequency—that is, 20 W at 2 GHz, for example. For this reason it is advisable to consult other pages of the data sheet to see what the device will do at other frequencies. Fig. 5–24 shows a power out versus frequency curve for a 20-W device. Notice the difference in the output levels with various input power levels. Obviously, this device was designed to operate at a 20-W level since that is the only place where the response is flat. All other input levels result in a bowed response at the output.

Gain   The second area to look at when deciding which device to use is the gain of the device. The device may not be much good if, for the 20-W output, 15 W must be the input. Some reasonable amount of gain must result from the device if it is going to be worthwhile. A gain figure of 6 dB is a typical number for a device such as the one we have been discussing (20 W at 2 GHz). This gain will mean an input of 5 W to get our 20 W out. These numbers correspond to the numbers in Fig. 5–24.

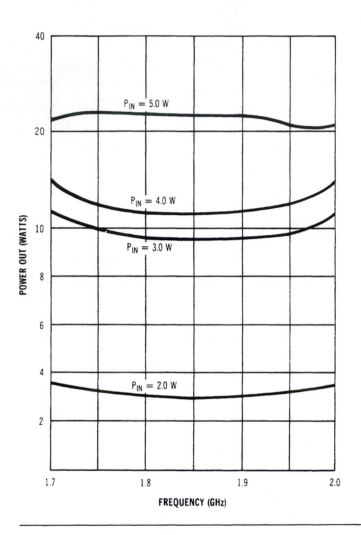

Figure 5–24.   Power out versus frequency.

Efficiency   Another area to be considered is the efficiency of the device. *Efficiency* is defined as the ratio of power out to power in. The power out is our 20 W. The power in is our 5-W input *plus* the *dc power* needed to operate the transistor. Typical efficiencies will run from 30 to 40%, which means that we must put in approximately 50 W to get 20 W out. So 5 W is input power, and 45 W is dc power. If we use a 28-V supply, we are allowed to draw 1.6 A of current (P = EI, 45 = 28 V and 1.6 A). These numbers will give the device an efficiency of 40%. Look very closely at efficiency numbers because as much power out as possible with an absolute minimum of power in is wanted.

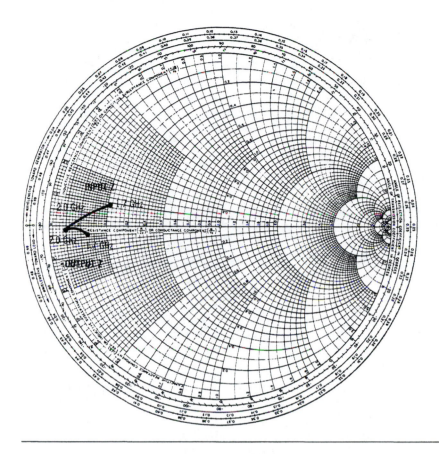

Figure 5–25.   Input/output impedances for a 20-W device.

Input and Output Impedance   A final item to consider when choosing a power transistor is the input and output impedances of the device. Fig. 5–25 shows the impedances for a 20-W device, and Fig. 5–26 shows them for a 2-W device. The two are shown to illustrate how the impedances can vary with the power level. One item to note is how low the impedances are that have to be matched to a 50-ohm source or load. This is why a good-sized microstrip pad is usually seen at the input and output of most of the higher power circuits to accomplish this match.

So, in review, consider the power output of a device, its gain, its efficiency, and its input and output impedances, and make provisions to get the heat out of the device and into the chassis. If all of these items are taken into consideration, a reliable operating power amplifier that will perform as expected will result.

5.3   Gallium Arsenide Transistors

In the mid 1970s the buzzword in the microwave industry was GaAs FETs (pronounced gasfets). That buzzword has evolved into a component that is now a necessity for high-performance, reliable microwave solid-state circuits.

Remember from Section 5.2 that the devices we covered were bipolar units. That is, two types of carriers cause the device to operate—electrons and holes (majority and minority carriers). The FET (field-effect transistor), on the other hand, is a unipolar device. Current in this type of device is carried by only one type of carrier, the majority carrier of the semiconductor.

The FET itself is not as new as some people are led to believe. The theory for FETs goes back to 1926, and William Shockley first proposed them in 1952. At that time, however, there were many technological and fabrication difficulties that kept the FET from arriving

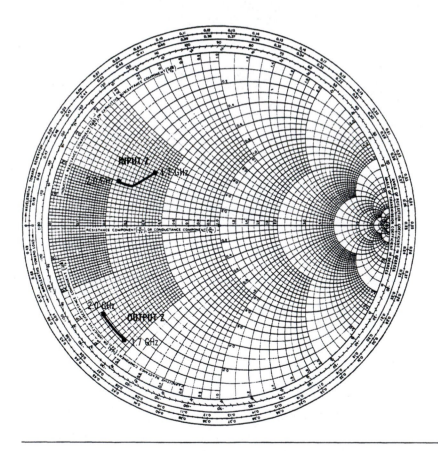

Figure 5–26.   Input/output impedances for a 2-W device.

on the scene until the early 1960s. At this time the silicon bipolar device was well defined and pushed the FET further out to the 1970s.

The earliest FET was the junction FET (JFET), which became available about the same time as the first microwave bipolar transistors. The JFET construction is shown in Fig. 5–27A. Advances in techniques and a need for low power resulted in the metal-oxide-semiconductor FET (MOSFET). This device also may be referred to as an insulated-gate FET (IGFET) because of the oxide layer between the gate and the substrate, as shown in Fig. 5–27B.

These FETs offered no competition to the well-established bipolar transistor in either of the forms shown in Fig. 5–27. However, when the metal-semiconductor FET (MESFET)

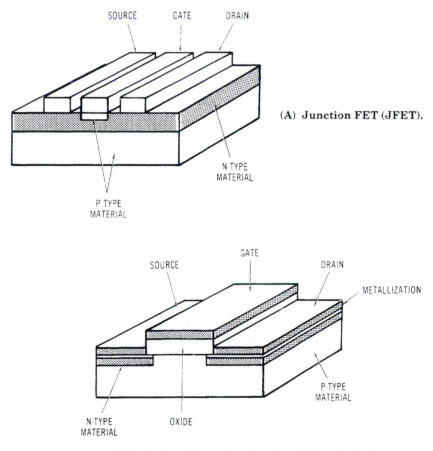

(A) Junction FET (JFET).

(B) Metal-oxide-semiconductor FET (MOSFET).

Figure 5–27. Field-effect transistors.

came on the scene, the bipolar began to shudder more than a little. This device used a Schottky barrier (described in Chapter 6) as the gate and is shown in Fig. 5–28. The *GaAs FET (gallium arsenide field-effect transistor)* is the type of device we will be discussing.

You may have noticed some strange designations in Figs. 5–27 and 5–28. These were *gate, drain,* and *source.* You have now been introduced to the terminology of the GaAs FET. They can be related to the bipolar devices as follows:

Bipolar	*GaAs FET*
Base	Gate
Emitter	Source
Collector	Drain

Think of the origin of the electrons as their *source,* their movement being controlled by a *gate,* and their release to the output by a *drain,* to be able to remember the terminology and make the transformation from bipolar devices. To aid in this change in terminology, Fig. 5–29 shows schematic representations of bipolar and GaAs FET devices with their terminal designations.

5.3.1   GaAs FET Construction

As in the case of the bipolar device, let us once again build a device to illustrate how different methods and substances are used. The first layer is a semi-insulating GaAs substrate. This layer consists of a doping of pure GaAs with chromium, which reduces input and output parasites within the transistor. This layer is the lowest one in Fig. 5–30, which illustrates the construction of a GaAs FET.

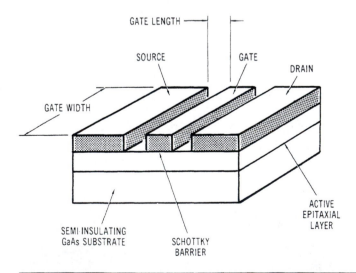

Figure 5–28.   Metal semiconductor field-effect transistor (MESFET).

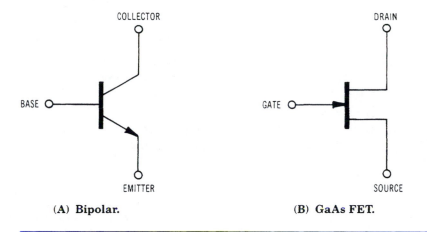

(A) **Bipolar.**   (B) **GaAs FET.**

Figure 5–29.   Bipolar and GaAs FET schematics.

Diffusion of impurities from the semi-insulating substrate during epitaxial growth can degrade the electrical performance of the conductive layer that supports the transistor element (source, gate, and drain). For this reason a very thin epitaxial film is grown to prevent such diffusion. Precise control of growth rate and doping concentration is critical to maintain a uniform thickness of the layer. By controlling the diffusion of impurities, the buffer layer reduces the transistor noise figure and prevents carrier trapping. This phenomenon will cause long- and short-term failures such as I_{DSS} instability. This instability is a fluctuation in drain current (I_{DSS}), which causes the entire device to be unstable. Other terms for this action are I_{DSS} looping or stepping.

The top layer of the device is the *active conductive epitaxial layer*. This layer can be formed in a variety of ways.

- VPE (vapor-phase epitaxy)—the conducting layer grown on the surface of the wafer uses gases that contain the required elements and compounds.

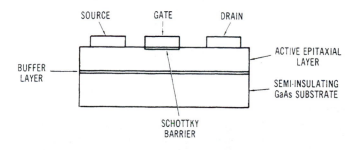

Figure 5–30.   GaAs FET transistor construction.

- LPE (liquid-phase epitaxy)—the conducting layer is grown on the surface from a liquid melt of GaAs and other elements.

- Implantation—the elements necessary to make the top layer conductive are "shot" (implanted) into the GaAs surface using a strong electric field (50,000 to 400,000 V).

- MBE (molecular-beam epitaxy)—an insulating wafer of GaAs is placed under very high vacuum near small furnaces called "effusers." The effusers melt gallium, arsenic, and other elements that condense on the surface of the wafer to form the active layer.

The source and drain of the transistor are produced by the same process and are treated the same in our discussion. These two elements are connected to each other by a conducting channel. The gate of the device modulates this channel thickness to produce the resulting power gains. The source and drain are deposited on the GaAs layer in a vacuum and then defined by a standard photolithographic process. The elements are made of alloys of gold and germanium (usually gold/indium/germanium) and are heated in a sintering process to produce low-resistance ohmic contacts.

The gate also uses standard photolithographic techniques to define its area. It is made of a layer of aluminum that forms a Schottky-barrier diode with the GaAs layer. This barrier enables the very short gate lengths to be achieved. Aluminum is used because of its low shunt resistance, purity, ease of deposition, and the elimination of the high-temperature gate diffusion process necessary with other materials. Other metals that could be used are tungsten, platinum, and molybdenum for forming the Schottky junction. These materials, however, require a layer of titanium put on the active layer first to act as a "glue."

The metallization techniques used for the aluminum take advantage of the good characteristics listed, but disallow the formation of any metallic compounds such as "white" or "purple" plaque. These plaques can result from bonding gold wires to aluminum pads.

Fig. 5–31A illustrates the construction of an actual GaAs power FET. The source, drain, and gate areas are further defined for this device in Fig. 5–31B. The pellet, as shown in Fig. 5–31A, is ready to be placed in a package and connected to external leads.

Fig. 5–32 shows a low-noise GaAs FET. Notice the difference in appearance between the low-noise device and the power FET in Fig. 5–31. Once again, source, drain, and gate areas are indicated in a drawing of the device, Fig. 5–32B. Notice the different types of structure used for different devices. Also, notice that the orientation of the elements (source, drain, and gate) is always the same regardless of what type of device is used. The gate is always between the source and drain so that it may perform its modulating function and produce gain through the device.

A point should be brought up before getting into the explanations of low-noise and power GaAs FETs. That point is the importance of the gate when specifying GaAs FETs. You will notice that with different manufacturers, GaAs FETs may be called 0.5 micron, or 1 micron, or some other value device. This number refers to the *gate length* and is the most important single FET element that determines high-

(A) Power GaAs FET internal construction.

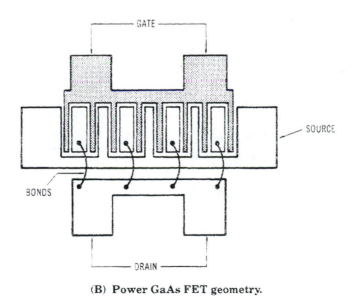

(B) Power GaAs FET geometry.

Figure 5–31.   Power GaAs FET internal construction and geometry *(Courtesy Microwave Semiconductor Corp.).*

235

(A) Low-noise GaAs FET internal construction.

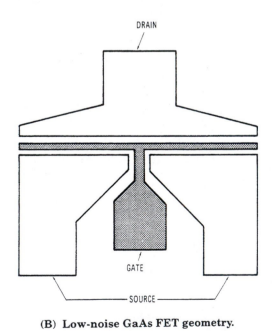

(B) Low-noise GaAs FET geometry.

Figure 5–32.   Low-noise GaAs FET with diagram *(Courtesy Varian Associates).*

frequency gain. Commercial, high-performance FETs have the 0.5-micron gate length referred to earlier and will operate to around 18 GHz. Lower-frequency FETs have 1.0-micron gate lengths and will operate in the range of 8 to 10 GHz. Some FETs have been built that have 0.2-micron lengths and will operate to somewhere around 30 GHz.

A characteristic of the gate that comes into play when discussing power FETs is the *width*. Small-signal FETs may have a gate width of 0.15 mm, and a power device might have a width of 26 mm. This difference is very evident when referring back to Figs. 5–31 and 5–32. Fig. 5–31 is an illustration of a power device, and Fig. 5–32 is an illustration of a low-noise device. Obviously the gate width of the power device is larger than the narrow pencil-thin width of the low-noise device.

To wrap up our building process for FETs, let us consider the package. Obviously, many problems would be eliminated if a FET chip could be used in your circuit. This use, however, is not always possible. For a variety of reasons, a packaged device must be used. The package, therefore, must be such that it will have a minimum effect on the overall device performance. Fig. 5–33 shows some cases used for GaAs FETs. Fig. 5–33A is a case used for low-noise devices. A typical case for a power FET is shown in Fig. 5–33B.

Fig. 5–33C is called a *chip carrier*. It is a case of sorts with a grounded source and 50-ohm input and output lines. This carrier has low thermal resistance and offers chip performance to those individuals who do not have chip handling capability. This carrier is also available with a bypassed source resistor preset for maximum linear power output.

5.3.2   Low-Noise FET Data Sheet

To wrap up the topic of GaAs FET devices, we will cover two devices—the low-noise FET and the power FET. Our coverage will involve a look at a typical data sheet for each and an explanation of each critical term.

Frequency   The first thing to look for when considering a low-noise GaAs FET is the *frequency* of operation. If possible, try to choose a device that operates at a higher frequency than needed so that the portion of the curve that is increasing in noise figure is avoided. For example, in Fig. 5–34, operating at 8 GHz or less would be needed if the device were used with the characteristics shown. Above that frequency the noise figure increases, and the associated gain decreases, which is not an ideal situation. So be sure your frequency of operation is below the maximum frequency of the device used. One note on Fig. 5–34 is that this curve holds for the condition given—$V_{DS} = 3$ V, and $I_{DS} = 10$ mA—that is, a drain-to-source voltage of 3 V and a drain current of 10 mA. If your FET is biased differently than these conditions, it will not perform as shown in Fig. 5–34.

Noise Figure   The *noise figure* on a data sheet can be termed *optimum noise figure* (NF_{OPT}), *minimum noise figure* (NF_{MIN}), or *spot noise figure*. Any one of these terms may be used, but they all say that this figure is the lowest noise figure when the conditions listed (V_{DS} and I_{DS}) are used and the corresponding S-parameters for these conditions are used.

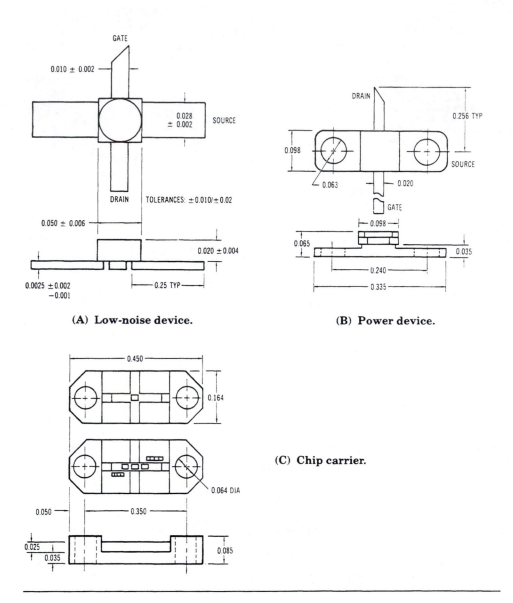

Figure 5–33.   GaAs FET packages. (All dimensions are in inches.)

Gain   *Gain* is a very important parameter for a low-noise circuit. There are actually two gain numbers involved when talking of low-noise GaAs FETs. One is the *maximum available gain* (G_{MAX} or MAG), and the other is the *gain at optimum noise figure* (G_{NF}).

The *maximum available gain* figure is not really concerned with low-noise performance. It is concerned with obtaining the most output available from a device. Conditions

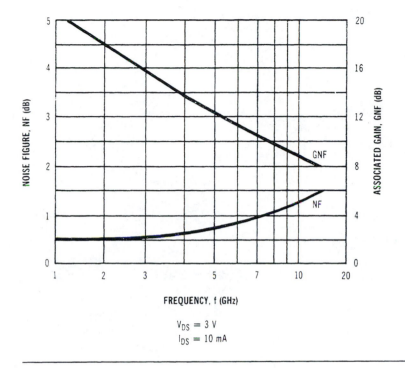

Figure 5–34. Noise figure and gain versus frequency.

are once again important when obtaining this gain figure. The conditions for the maximum gain are usually spelled out as $V_{DS} = 3$ V, $I_{DS} = 30$ mA, for example. Notice the greater amount of current necessary to obtain the increased gain.

The *gain at optimum noise figure* numbers are those obtained when the conditions are duplicated for the best noise figure obtained. That is, we have said that the best noise figure was obtained when $V_{DS} = 3$ V and $I_{DS} = 10$ mA. We, therefore, will duplicate the specified gain when these conditions are met. This gain, as might be imagined, is lower than the maximum available gain since we are operating with less drain current in order to get a low noise figure. Less drain current will automatically reduce the gain. The gain figure can be on the order of 2 to 3 dB difference from the maximum available gain figure. However, this difference can be made up in following stages if increased gain is really important. The prime function of the first stage is low noise.

Input and Output Match   An important part of obtaining low-noise performance is the input and output match of the device: S-parameters. Fig. 5–35 shows S_{11} and S_{22} (input and output reflection characteristics, respectively) for a chip device. Fig. 5–36 is the S_{11} and S_{22} values for a low-noise packaged GaAs FET device. Test conditions are slightly different, but notice how the packaged device spreads out S_{11}

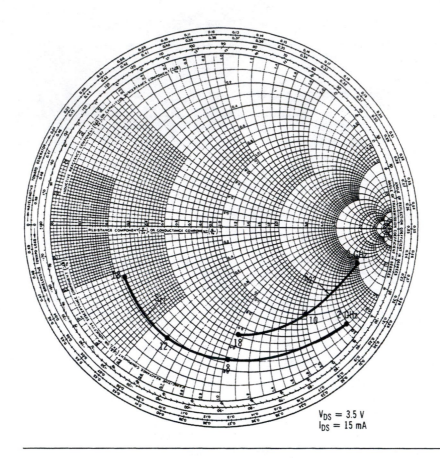

$V_{DS} = 3.5 \text{ V}$
$I_{DS} = 15 \text{ mA}$

Figure 5–35. S_{11} and S_{22} for GaAs FET chip.

and S_{22} considerably over the chip version. (S-parameters are covered in detail in Chapter 8.)

Other Terms Other parameters to be aware of on a low-noise GaAs FET data sheet are

I_{DSS}—Saturated drain current (maximum)

I_{GS}—Gate to source leakage current

I_{GD}—Gate to drain leakage current

I_{DS}—Drain current

V_{DS}—Drain to source voltage

V_{GS}—Gate to source voltage

g_m—Transconductance (ratio of change in drain current divided by the change in gate voltage)

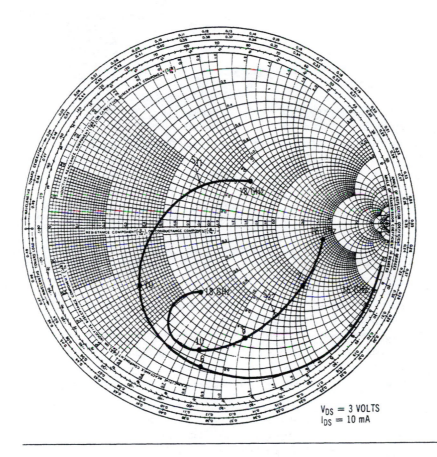

$V_{DS} = 3$ VOLTS
$I_{DS} = 10$ mA

Figure 5–36. S_{11} and S_{22} for GaAs FET packaged device.

5.3.3   Power GaAs FET Data Sheet

When choosing a power GaAs FET, specific parameters should also be considered.

Frequency—the first parameter to consider, of course, is the *frequency* of operation. If the short paragraph at the top of the sheet does not call out a specific frequency range, it will usually call out a band. If you are not familiar with the frequency bands, look farther down the data sheet and notice frequencies for specific parameters and determine the operating range of the device.

Output Power—the second parameter to be concerned with is the *output power,* which will be presented in two ways: power output and power output at the 1-dB compression point. The power output figure is the one that tells the *linear* power output—that is, the power output of the device when operating within a specified input power range. This range does not drive the device into compression. For this power output figure, there must be conditions set down that will result in this power output. These conditions are power input, V_{DS}, and either I_{DS} or V_{GS} (V_{DS} is drain to source voltage, I_{DS} is drain to source current, and V_{GS} is gate to source voltage).

Power output at the 1-dB compression point is found by increasing the power input in 10-dB steps and noting that the output also increases by 10 dB. When a 10-dB input results in a 9-dB increase in output, the 1-dB compression point has been reached. To increase the output further does not make any sense because you will get no more power at the output; the limit has been reached. Many times the 1-dB compression output power is used to keep the output of an amplifier constant over a wide range of environmental conditions or over a varying range of input powers.

Gain—gain, of course, is also an important parameter to consider in a power device. It would be foolish to supply dc and rf power to a device and produce only 1 or 2 dB of gain. Most power FETs will exhibit from 7 to 10 dB of gain up to around 8 GHz with very little effort. Be sure to check the particular device to be used to be sure to know what gain it should produce.

Temperature Range—the use of power devices brings to light a whole new set of parameters. Those parameters are thermal. First, check the operating temperature range of the device to be used. Most devices will operate from around −65°C to +175°C. These temperatures are somewhat less than the temperatures of their bipolar brothers, which will operate up to +200°C. When making the transition from bipolar to GaAs FETs, you should be aware of this temperature difference.

Thermal Resistance—the other thermal parameter, thermal resistance, is expressed in degrees Celsius per watt and tells how well the device gets the heat generated within the device out of that same device. This figure should be as small as possible since as little rise in temperature per watt of power as possible is wanted.

Impedance—the final parameter to be investigated is the impedance of the device to be used (S_{11} and S_{22}). Fig. 5–37 shows a typical plot of S_{11} and S_{22} for a power device that produces a 1-W output at 8 GHz. Notice how much lower the impedance is at the input than for the low-noise device. This lower impedance will make the input matching to a 50-ohm circuit a little more difficult and may require multisection matching if a broad band of frequencies is to be covered.

5.3.4   GaAs FET Handling Precautions

One area that is of prime importance, but is not covered on data sheets, is handling precautions for GaAs FETs. These precautions are for both low-noise devices and power FETs.

When handling the packaged unit, never pass the device from one person to the next by the leads, especially on a dry day, since static charges will destroy it. Although the MESFET is not as sensitive as a MOSFET, it is still quite delicate.

When soldering the device into a circuit, the soldering iron must be properly grounded. Soldering time should not exceed 20 seconds at 260°C. Never insert a FET into a

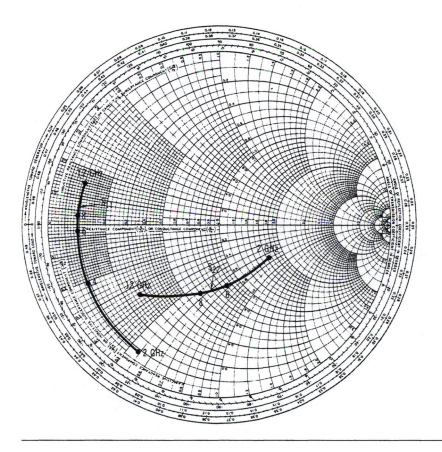

Figure 5–37. S_{11} and S_{22} for power FET.

prebiased circuit, particularly one where self-biasing is used; the bypass capacitor will charge through the forward-biased barrier and most likely will destroy the gate.

Once the device is soldered into the circuit, do not just "switch on" the gate and drain bias supplies. Many regulated power supplies feature large in-rush transients due to their stability-limited gain-bandwidth product. Adjust V_{GS} slowly to about -1.0 V, gradually increase V_{DS} from 0 V to the desired bias point, and, finally, readjust V_{GS} to give the desired drain current.

Remember, for a FET, $V_{DG} = V_{DS} - V_{GS}$, so place no more than 10 V between the drain and gate. If something is suspect, avoid the use of a digital multimeter to check the resistance because the voltage supply within the meter may destroy the gate.

We must also caution against the use of curve tracers. The high-voltage transformers used in most curve tracers may have high leakage currents that can, if not properly grounded, destroy the FET. If a curve tracer test is required, we suggest the following:

1. Ground the curve tracer to earth.
2. Set the gate-source voltage to zero voltage.
3. Increase the drain-source voltage slowly to the desired value ($\simeq 3$ V).
4. Adjust the gate-source voltage to the desired value.

When handling GaAs FET chips, the most important precaution to remember is that all operations must be performed in a clean environment by the fewest number of people possible. If incoming inspection is performed, leave the chips in the carrier and store in a dry environment, preferably in dry nitrogen. When removing the chip from the carrier, use a vacuum probe with a Teflon® tip. If tweezers are used, extreme caution is mandatory. Gallium arsenide is more brittle than silicon, and dust may break off and become lodged in the small channel.

The back of the chip is metallized with pure gold and can be die-attached with low-temperature epoxy, AuGe or AuSn 0.5-mil preforms. Whatever material is used, the maximum allowable exposure is 300°C for five minutes. The use of a die-attach machine is not recommended because the gate may be damaged.

If AuSn preforms are used, recommendations are heating the stage to 200°C and the probe to about 250°C. Probe pressure should be less than 22 grams. For specific chip handling procedures, consult the individual manufacturer for any peculiarities that may be present for their devices.

In general, three simple precautions will prevent the destruction of devices from transient breakdown:

1. Dc ground the device gate unless it is to be negatively biased.
2. Decouple the bias ports near the device with a low-inductance capacitor (1 μF tantalum) shunted with a zener diode (5.8 V, 1.3 W). In addition to limiting transients, the zener diode gives protection against over-voltage and reverse biasing.
3. Switch on power supplies and set their outputs to minimum before connecting the device.

5.4   High Electron Mobility Transistor

The High Electron Mobility Transistor (HEMT) is one that arrived on the microwave scene around 1978 to mixed reviews. For some, the HEMT did not really fit in or find any overwhelming application. Probably the reason for this is that the HEMT followed the GaAs FET too closely and most people were convinced that the GaAs FET was going to be the ultimate answer to all solid-state microwave problems. Also, the HEMT was devised to be a low-noise device only. It was not intended to be a power device. This caused limited consideration and, thus, use. The devices, however, have now found more applications, and a wide variety of literature can be found on areas where the HEMT is being used. This is good news for many circuit designers as well as semiconductor manufacturers.

If you look at the HEMT construction, you can see how designers might be turned off by it and simply say, "This is only a souped-up GaAs FET. It looks exactly like one." Well, both you and the designers would be exactly right. It does look like a GaAs FET, as shown in Figure 5.38. Figure 5.39 confirms this even more when we look at the HEMT geometry. As in most cases, looks can be deceiving. The actual difference between the GaAs FET and the HEMT is not in the structure of the device but in the semiconductor material it uses. These devices are based on the modulation-doped GaAs/AlGaAs heterostructure epitaxy in which the motion of the charge carriers is confined to the thin sheet within the GaAs buffer layer. These layers can be seen in Figure 5.38.

In the HEMT, a silicon-doped AlGaAs layer grown on top of an undoped GaAs layer brings about the formation of a two-dimensional electron gas on the GaAs side of the heterostructure. The gas forms because of the electrons' greater affinity to the GaAs. The two-dimensional gas layer is about 150 Å thick and forms the carrier channel that links the device source and drain.

When a Schottky barrier gate is placed on top of the AlGaAs layer, a depletion region forms beneath the gate. If the AlGaAs layer is sufficiently thick, the gate and the interface depletion region will not overlap, and the device will operate as a "normally on" transistor. In this *depletion mode* an application of negative bias to the gate will extend the gate depletion region to the heterojunction interface, thereby barring electron flow and pinching off the drain-source current. (This should be recognizable from introductory FET theory, where depletion mode MOSFETs are explained.)

Etching a recess to seat the gate deeper in the AlGaAs layer creates a device that operates in the *enhancement mode*. This type of device is "normally off." When a positive voltage greater than the threshold voltage is applied to this device, electrons will

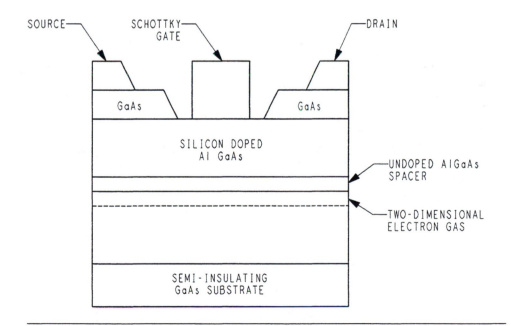

Figure 5–38. HEMT construction.

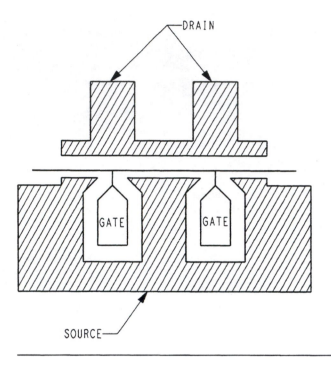

Figure 5–39.   HEMT geometry.

accumulate at the hetrojunction interface and form the two-dimensional gas, turning the device on. Thus, unlike the MESFET, the HEMT actually functions very much like the metal oxide semiconductor (MOSFET) that we referred to earlier. In the MOSFET the Schottky barrier gate controls the number of electrons in the two-dimensional electron gas by raising and lowering the interface barrier. The MESFET uses the bias applied to the gate to modulate the channel and thereby the number of electrons passing through it. With the HEMT, the channel thickness remains constant. The number of carriers is modulated.

In the HEMT, carrier transport in the two-dimensional electron gas is similar to movement in undoped GaAs. Unlike the highly doped channel of a conventional GaAs MESFET, there is little or no impurity scattering in the undoped GaAs in which the two-dimensional gas resides. As a result, electrons in HEMTs travel at twice the saturated velocity (about 2×10^7 cm/sec) at room temperature as those in GaAs FETs, with an electron mobility of nearly 8000 cm²/V-s (compared to 4000 cm²/V-s for the GaAs FET channel). At lower temperatures, 77° K, for example, the electron velocity increases to about 3×10^7 cm/sec while the mobility increases to approximately 80,000 cm²/V-s, depending on the epitaxial layer structure.

Thus, the HEMT joins its predecessors in the microwave field. It has had its triumphs to this point, and it has had its problems. These are the ups and downs that the bipolar and

GaAs FET have been through and, to some degree, are still going through. The HEMT has found many applications and will continue to find even more. (A variation of the conventional HEMT is the PHEMT, which is a pseudomorphic HEMT. These devices are primarily used in millimeter applications, above 20 GHz.)

5.5   Chapter Summary

We began this chapter by stating that the microwave transistor had probably accounted for some of the most significant advances in microwaves in recent years. You can now see how this statement is true. We covered bipolar transistors (low-noise, linear, and power) and showed applications where noise figures in the 1.0 dB range and powers of 20-30 watts/device were attainable. The GaAs FET (low-noise and power) expanded the frequency of operation up to the high microwave, and low millimeter, range. The High Electron Mobility Transistor (HEMT) has been the next step in the microwave solid-state progression.

The microwave transistor is a good example of the old saying: "Good things come in small packages."

5.6   References

1. Arden, J. A. "The Design, Performance, and Applications of the NEC V244 and V388 Gallium Arsenide field-Effect Transistor." *NEC*, June 1, 1976.

2. *FET Applications Note*. Plessey Optoelectronics and Microwaves. Publication #PSI843.

3. Fitchen, F. C. *Transistor Circuit Analysis and Design*, Second Edition. New York: Van Nostrand Reinhold, 1966.

4. Laverghetta, T. S. *Microwave Measurements and Techniques*. Dedham, MA: Artech House, 1976.

5. Riddle, R. L., and Ristenbatt, M. P. *Transistor Physics and Circuits*. Englewood Cliffs, NJ: Prentice-Hall, 1958.

6. *Transistor Designers' Guide*. Bulletin #5210, 1978.

7. Watson, H. A. *Microwave Semiconductor Devices and Their Circuit Applications*. New York: McGraw-Hill, 1968.

8. Laverghetta, T. S. "Modern Microwave Measurements and Techniques," Norwood, MA., Artech House, 1988.

Questions

5.1   Introduction

1. Describe a point-contact transistor.
2. What is different about an alloy-junction transistor as compared to a point-contact transistor?

3. Define *planar*.
4. What does *epitaxial* refer to?
5. Distinguish between a bipolar and a unipolar device.

5.2 Bipolar Transistors

6. Why is a device called *bipolar*?
7. Describe the planar construction of a bipolar transistor.
8. What area of a low-noise device is most critical?
9. Why is the interdigital construction used for low-noise transistors?
10. Define *diffusion*.
11. Name two other terms used to describe the noise figure of a transistor.
12. Describe *noise contours*.
13. Describe *gain contours*.
14. Why is the current kept low for low-noise devices?
15. What is a linear transistor?
16. Describe *compression*.
17. Describe *linearity* in a bipolar linear transistor.
18. Describe *oxidation* in a power transistor.
19. Describe *masking* in a power transistor.
20. Describe *metallization*.
21. What is the difference between gold and aluminum metallization in a power device?
22. Why is *efficiency* important in a power transistor?

5.3 Gallium Arsenide Transistors

23. Why is the GaAs FET a unipolar device?
24. Describe the MESFET.
25. What is the *schottky junction*?
26. Describe the difference in geometry between a low-noise FET and a low-noise bipolar.
27. List three precautions to be observed when working with FETs.

5.4 High Electron Mobility Transistors

28. How does a HEMT differ from a GaAs FET?
29. How can a HEMT be changed from *depletion mode* operation to *enhancement mode*?
30. What is a PHEMT, and how is it used?

6 Microwave Diodes

Objective

To introduce the reader to the world of microwave diodes. This chapter will dispel the myth that diodes can rectify only AC signals. The Schottky, PIN, Tunnel, Gunn, IMPATT, and TRAPATT diodes will be covered with detailed explanations and applications.

Key Terms

Carrier Lifetime

Delayed Domain

Drift Region

Epitaxial

Gunn Effect

Intrinsic

Junction Capacitance

Mesa

Negative Resistance

PIN

Planar

Resistive Cutoff Frequency

Schottky Junction

Thermal Resistance

Tunnel Diode

6.1   Introduction

There was a time when the term *diode* meant only one thing—rectification in a power supply. There were some diodes that were used as detectors in IF circuits of receivers, but generally their major application was in power supplies.

As the solid-state industry began to grow and new processes and techniques became available, the diode found new applications. This little two-element device (cathode and anode) could perform many other functions besides rectifying ac in a power supply. It could do many functions at microwave frequencies that would make the circuits smaller and more efficient. Besides rectifying ac or rf, it could do the same thing to microwaves to produce a dc voltage proportional to the microwave input (detector); it could mix microwave frequencies together to produce an IF output (mixer), attenuate microwave frequencies, switch microwave frequencies electronically, amplify microwave frequencies, and oscillate at microwave frequencies to become a source. All these functions for a diode were unheard of until solid-state electronics came of age and pointed out new and improved applications. Today, most of these applications are taken for granted. You would probably be hard pressed to name another single device that could perform all these functions. Yet, through all of this use, the diode still consists of two basic elements—a cathode and anode, as previously mentioned. Even though the structure may change, the device itself still contains only these two elements. To illustrate the device and its use in microwaves, we will cover six types of diodes: Schottky, pin, tunnel, Gunn, IMPATT, and TRAPATT.

6.2   Schottky Diodes

We referred to the Schottky barrier junction in the previous chapter when we were constructing the gate region of a GaAs FET. The Schottky barrier is set up by contact between a metal and a semiconductor and gets its name from early analysis in an area termed *majority-carrier rectification* by W. Schottky in 1938. The semiconductor material used is generally n-type silicon or n-type GaAs.

The properties of the operating Schottky diode are determined by majority carriers (electrons in this case) instead of by minority carriers as in the case of the normal pn diode junction. This results in a device that can be switched very rapidly from forward to reverse bias since there is no minority carrier storage effect to contend with. This effect is a time delay inherent in the pn junction device caused by the minority carriers being in a storage state rather than free to move about like the majority carriers of the Schottky junction. The energy, or Schottky, barrier exists at the metal to semiconductor interface because of the difference in the work functions of the two materials. The barrier is unaffected by a reverse bias, but is decreased by a forward bias—the familiar characteristic of a rectifying diode. Therefore, when the device is forward biased, the majority carriers (electrons) can be easily injected from the semiconductor material into the metal, where the energy level is now much higher. Once in the metal, the electrons give up this excess energy in a very short time, on the order of 10^{-15} seconds (or one femtosecond, which is one thousandth of a picosecond). After they give up this energy, they become a part of the free electrons of the metal. Of course, when reverse biased, the energy level of the barrier is too high for the electrons to overcome, and the device does not conduct. These characteristics indicate the potential of the Schottky barrier for being a very efficient high-frequency mixer and detector.

An equivalent circuit of a Schottky diode is shown in Fig. 6–1A. Fig. 6–1B shows where each of these terms comes from in the diode itself. The only term that is not shown in Fig. 6–1B is L_s, which is the inductance present in the wire that is bonded to the

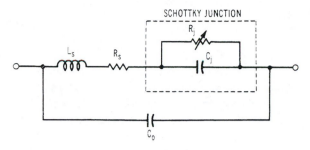

(A) Equivalent circuit.

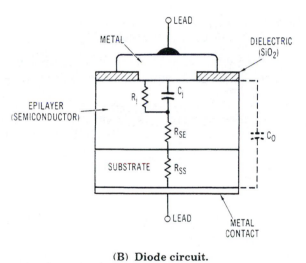

(B) Diode circuit.

Figure 6–1. Schottky diode.

metallized area over the junction. The remainder of the terms are an important part of the diode and are described in the following paragraphs.

Three resistances shown in Fig. 6–1B characterize the Schottky diode: R_j (junction resistance), R_{SE} (resistance of the epilayer semiconductor), and R_{SS} (resistance of the substrate). The term R_S in Fig. 6–1A is equal to R_{SE} plus R_{SS}. This term is the most important of the resistance terms when considering a diode for mixer or detector operations. If this resistance is high, power can be dissipated, and a maximum transfer of power within the device cannot take place. This situation can be improved by choosing a diode with a minimum value of R_S. A typical value for R_S is 4 to 6 ohms.

The capacitance terms associated with a Schottky diode are C_j and C_o. C_j is the capacitance at the Schottky barrier junction and is in shunt with the R_S value we just discussed. C_o is the overlay capacitance across the oxide layer of the device. Another capacitance not shown is C_p, which is the package capacitance of the diode. A total capacitance for the diode, C_T, would be C_j plus C_o plus C_p. The most important of these capacitances is the junction capacitance. This value should be kept small to ensure the high-frequency operation of the diode. Typical values for C_j are 0.3 to 0.5 pF.

There are three basic ways to make contact to the Schottky diode: point contact (whisker), C-spring contact, and thermo-compression bond. All three types are shown in Fig. 6–2. Fig. 6–2A shows the whisker contact, Fig. 6–2B shows the C-spring contact, and Fig. 6–2C shows the thermo-compression bond. The characteristics that determine the actual contact chosen for a particular Schottky device are the size of the junction and a particular package requirement. If we were to categorize each of these contacts, they would probably be something like this:

Whisker Contact—used for millimeter radiometry mixers since its construction minimizes parasitic capacitance. Also used in some glass packages.

C-Spring Contact—used in most glass packages and is the least costly to manufacture.

Thermo-Compression Bond—the most reliable contact mechanically.

Two common types of packages are shown in Fig. 6–3. The first (Fig. 6–3A) is the glass package. This diode is a relatively inexpensive package, and it can be seen that a whisker contact is used as mentioned earlier. The second case (Fig. 6–3B) is the pill package. The

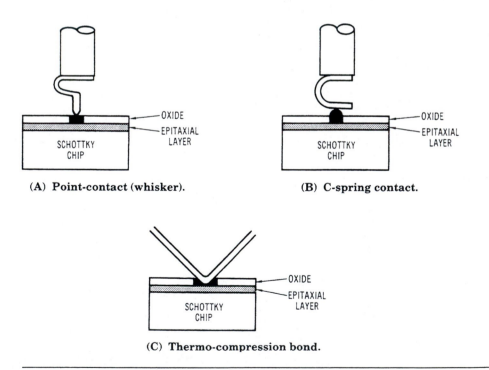

(A) Point-contact (whisker).   (B) C-spring contact.

(C) Thermo-compression bond.

Figure 6–2.   Schottky diode contacts.

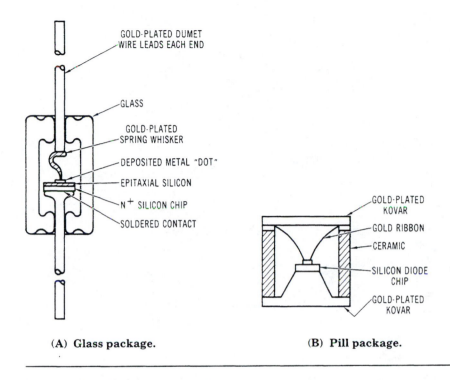

GOLD-PLATED DUMET
WIRE LEADS EACH END

GLASS

GOLD-PLATED
SPRING WHISKER

DEPOSITED METAL "DOT"

EPITAXIAL SILICON

N$^+$ SILICON CHIP

SOLDERED CONTACT

GOLD-PLATED
KOVAR

GOLD RIBBON

CERAMIC

SILICON DIODE
CHIP

GOLD-PLATED
KOVAR

(A) Glass package.   **(B) Pill package.**

Figure 6–3.   Schottky diode packages.

pill package is used in higher frequency miniature applications. Notice that the highly reliable thermo-compression bond is used for this package. It should be pointed out that these drawings of Schottky diode packages are not to scale. The glass package is about the size of a quarter watt carbon resistor while the pill package is very small, on the order of 0.020″ to 0.050″ square and only about 0.015″ high. These are the devices that are so small that you have to sweep under the bench when you drop one on the floor. They are not usually readily visible just by glancing down at the area where you dropped them. It usually requires a bit of house cleaning to recover them.

To wrap up our coverage of the Schottky diode, we will look at some of the terms on a typical data sheet. Remember from our previous discussions that three of the critical parameters for a Schottky diode are R_S (total resistance of the semiconductor and substrate), C_j (shunt junction capacitance), and L_S (inductance of the gold wire bonded to the metallized area over the junction; see Fig. 6–3A). All these terms will determine how well a specific diode will operate at microwave frequencies as either a detector or a mixer. The ranges to look for to ensure proper microwave operation are

- R_S 4 to 6 ohms
- C_j 0.3 to 0.5 pF
- L_S 0.4 to 0.9 nH

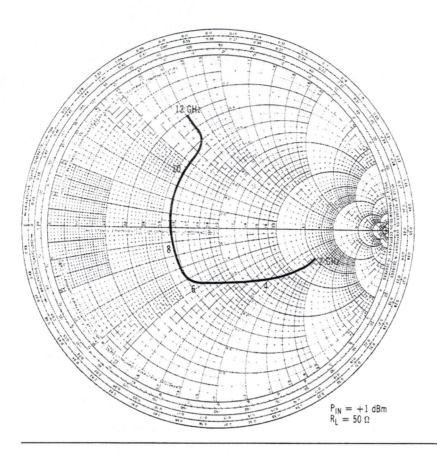

Figure 6–4.   Schottky diode impedance.

If the diode falls within these ranges, you will stand a good chance of obtaining properly operating microwave devices.

In the case of practically every component used in microwaves, impedance match is very important. This match is also important for the Schottky diode. Remember from Chapter 4 that an important block in the microwave detector is the input matching section. This matches the diode impedance, Z_D, to the characteristic impedance, Z_0, of the system. Similarly, when we discussed mixers, we discussed the input coupling network interfacing with the diodes. This interfacing can be done efficiently only if the impedance of the diode is matched to the impedance of the coupler. To ensure this match, the impedance of the diode for your frequency (or band of frequencies) of operation must be known. This impedance is an important part of the diode data sheet. A typical impedance plot for a Schottky diode is shown in Fig. 6–4. The figure shows the impedance of the diode from 2 to 12 GHz plotted on a Smith chart, with the reference being to a 50-ohm system. The Smith chart is covered in detail in Chapter 7. For now, we can say that the impedance of the diode can be read directly from the Smith chart since the chart displays every imaginable impedance as it appears in Fig. 6–4 in its "normalized" form (1.0 at the center). To obtain impedance directly in our 50-ohm system, multiply every number inside the chart by 50. That is, 1.0 becomes 50 ohms, 1.8 becomes 90 ohms, 0.4 becomes 20 ohms, and so on. Keeping in mind that each number must be multiplied by 50, we can read the impedance directly from the chart. If we choose, for example, the impedance at 4 GHz, we will read $1.4 - j0.95$ ohms. (The $-j$ is used since the portion below the horizontal axis is capacitive and all reactance values are $-j$. Those above the axis are $+j$, or inductive.) In a 50-ohm system, the impedance value read at 4 GHz would be $70 - j47.5$ ohms. In a similar manner, the impedance at 8 GHz would be $0.55 - j0.15$ ohms ($27.5 - j7.5$ ohms). Another parameter that must be given is the input power level to the diode when this impedance was read. Different power levels can result in different impedance readings. These particular readings were taken at $+1$ dBm. The input power to the diode when determining its input impedance is a very important parameter to know. This is because the impedance of the diode is a function of power level and may change drastically with various levels of power. So in order to accurately characterize a device you must know the operating conditions of that device, its input power.

The Schottky diode is a highly reliable, efficient means of detecting microwave energy and for mixing two microwave signals together to form a lower intermediate frequency for use in other electronic applications.

6.3   Pin Diodes

The normal diode junction consists of a p-type material brought together with an n-type material to form the familiar pn junction. The pin diode is distinguished from the normal pn junction type by an area called an *intrinsic region* sandwiched between the p+ doped and n+ doped silicon layers. (Doping is the term used to describe the addition of impurities to a semiconductor material to increase conduction.) This layer has almost no doping and, thus, has a very high resistance. When forward bias is applied across the diode, the conductivity of this "I" layer is increased, creating an electrically variable resistor at microwave frequencies. As the bias current is increased, the diode resistance decreases. This relation makes the pin diode a natural as a variable attenuator, among other applications.

6.3.1   Construction Techniques

A variety of construction techniques are used to obtain this variable-resistance characteristic in the pin diode. We will present four of these. The first two are standard techniques, and the second two are processes of Varian Associates and MA/COM, respectively. These techniques are *planar, mesa, Plesa*™, and *Cermachip*™.

Planar Processing   *Planar processing* is a well-established, high-yield method of producing reliable diodes, especially at low frequencies. Fig. 6–5 shows the planar construction. This type is made by diffusing boron through a window in the thermal oxidation passivation to form the junction. You will recall that a process similar to this was discussed in

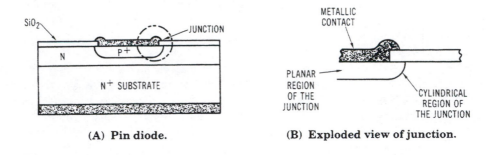

(A) **Pin diode.**   (B) **Exploded view of junction.**

Figure 6–5.   Planar pin diode construction.

Chapter 5 when we covered bipolar transistors and their construction. This type of fabrication is the most common when dealing with low-frequency diodes.

For microwave usage, planar devices have two drawbacks. First, the pn junction depletion region is shaped like a plane-paralleled plate capacitor in the center but is cylindrically shaped at the edges. This shape can be seen in Fig. 6–5B in the exploded view of the junction. This cylindrical edge reduces the voltage breakdown of the device considerably. Second, the interactive silicon surrounding the junction produces extra fringing capacitance. It also stores a charge that reduces the switching speed of the device.

The planar device does, however, have some good characteristics. The series resistance (R_S) versus current characteristics are very linear when plotted on log-log axes. They retain this linearity over a wide dynamic range of forward-bias currents. This characteristic, as well as the fact that they are inexpensive to fabricate, makes planar pin diodes very useful for attenuator applications.

Mesa Processing   Normally, to make a microwave pn junction, a slice of n or n+ epitaxial silicon of the proper doping density and thickness is selected. Then boron is diffused to produce a p-type layer on the front surface. This is shown in Fig. 6–6A. The next step is to metallize the front and back of the slice to produce low-resistance ohmic contacts to the silicon. Fig. 6–6B illustrates the device at this point. Circular areas of the proper diameter are masked on the p+ side, using photolithography, and mesas are etched on the slice, as shown in Fig. 6–6C. After mesa etching the individual devices must be broken apart and mounted in packages. An individual mesa is shown in Fig. 6–6D. This type of construction is very common for pin diode fabrication. The structure allows the active intrinsic region to very closely resemble a parallel-plate capacitor with minimum fringing capacitance on the substrate, contrary to the planar device, which had an excess of fringing capacitance. A thermal oxide (SiO_2) is used as the surface passivation layer in this device, which ensures maximum reliability.

Plesa™ Processing   The *Plesa* fabrication process is a hybrid of the classical mesa and planar processes, extracting the advantages from each. To produce a pn junction epitaxial

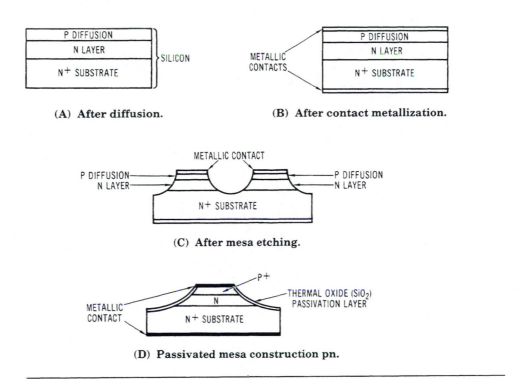

(A) **After diffusion.**

(B) **After contact metallization.**

(C) **After mesa etching.**

(D) **Passivated mesa construction pn.**

Figure 6–6. Mesa fabrication.

diode, an n or n+ epitaxial silicon slice is selected as with mesa or planar technologies. Prior to any diffusions, silicon mesas are formed. Masks are formed for etching by first pyrolytically depositing a silicon nitride (Si_3N_4) layer and an SiO_2 layer on the n-layer surface. Suitable etches result in an array of masking dots on the epitaxial silicon, as shown in Fig. 6–7. The mesas are then etched in the silicon slice, and a very dense thermal oxide is grown on the sides. An important consideration is that the pn junctions have not yet been formed, allowing high temperature passivation without unwanted contaminants. At this point in the Plesa processing, the die shown in Fig. 6–7B is passivated with a window prepared for the diffusion. This die is the same as the planar form, except for the geometry of the epitaxial silicon.

Conventional diffusions through the windows produce junctions under the dense passivation, as in the planar method. Because of the unique processing, the junction is in a mesa geometry, resulting in a near parallel-plate capacitance and excellent electrical characteristics. Final metallization for ohmic contacts completes the devices.

Plesa devices have an integrity of passivation not achieved in any other type of mesa diode. The low-reverse-bias leakage currents and exceptional leakage current stability under time-temperature stress are evidence of the reliability achieved with this process. As we have previously mentioned, Plesa is a construction process used by Varian Associates.

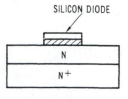

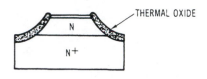

(A) Silicon mesas formed.   **(B) Passivated with a window.**

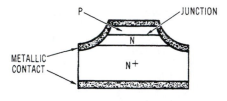

(C) After contact metallization.

Figure 6–7.   Pin diode Plesa™ construction.

Cermachip™   The fourth type of pin diode construction, Cermachip, is shown in Fig. 6–8. This construction is used by MA/COM. The basic geometry of the chip is the same as for the passivated mesa. The major difference is in the passivation. In the Cermachip a thick hard glass passivation is used instead of the thermal oxide of the passivated mesa construction. This glass passivation allows the device to replace a fully packaged hermetically sealed diode chip in many applications. Because of process limitations, however, this type of device cannot be used for low-level pin applications. Only chips with large "I" regions (2 mils or greater) and active region diameters greater than 6 mils can be made with this process. Thus, the Cermachip process is used primarily for high-power applications such as high-power switches and phase shifters. An example of such applications would be in a phase array antenna system for radar usages.

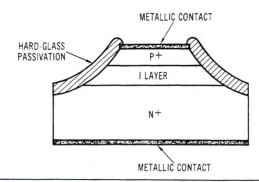

Figure 6–8.   Cermachip™ construction.

6.3.2 Applications

We have mentioned some general applications of pin diodes throughout our discussion so far. To appreciate how the device works in microwaves, a few specific applications will be covered. Remember that, basically, the pin diode can only switch and attenuate. With these two characteristics, a wide variety of components can be fabricated.

Since the pin diode is classified as a microwave variable resistor, the first application must logically be that of a variable attenuator. This attenuator may be either a step or the continuously variable type. Two basic attenuators are shown in Fig. 6–9. They are the series and shunt attenuators.

Attenuation in the series pin circuit (Fig. 6–9A) is decreased (more power appears at the output) as the rf resistance of the diode is reduced. This resistance is reduced by increasing forward bias on the diode. Conversely, attenuation in the shunt circuit (Fig. 6–9B) is decreased when the rf resistance of the diode increases because less power is absorbed in the diode and more appears at the output. Forward bias must be decreased at the diode to accomplish this decrease in attenuation. If we assume that the diode is a pure resistance, the attenuation for the series and shunt circuit can be calculated as

$$\alpha_{(series)} = 20 \log \left(1 + \frac{R_D}{2Z_0} \right) \tag{6.1}$$

and

$$\alpha_{(shunt)} = 20 \log \left(1 + \frac{Z_0}{2R_D} \right) \tag{6.2}$$

where, Z_0 is the characteristic impedance,
      R_D is the resistance of the diode at a specified bias current.

One point to notice in these equations is that the attenuation is not a function of frequency but only a ratio of circuit and diode resistances. As the bias on the diode is varied, the load resistance experienced by the source also varies. Attenuation, therefore, is achieved primarily by reflection.

These two pin circuits operate on the principle of reflections. Fig. 6–10 shows three types that operate on the principle of absorption of energy. All three are shown as variable-resistance devices on the left and the pin diode counterpart on the right. Remember the first two (π and T) from our discussions on fixed attenuators in Chapter 4. The third type (bridged T) is a modification of the basic T circuit. All three will absorb the undesired energy (or the amount wanted attenuated), as previously mentioned. This absorption is controlled by the bias applied to the diodes. Insertion losses of less than 1 dB are realizable with these devices, with attenuation figures on the order of 20 dB. Of the three circuits, the π configuration is usually considered to be superior to the T or bridged T, having greater dynamic range and linearity.

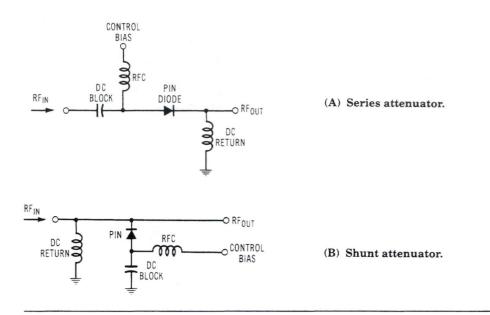

Figure 6–9.   Basic pin attenuators.

   Before proceeding any further, it would pay to note that all these previous circuits have a zero-bias condition and a maximum-bias condition, and in between are various values of attenuation. With these conditions and values in mind, consider the result of starting at zero bias and going rapidly to maximum bias. Going from a low loss to a high loss quickly would create a new component—a switch. All these attenuators are, thus, potential switches if the need arises.
   The circuits discussed so far are generally used at the low end of the microwave spectrum. The component that moves the pin attenuator (or switch) well up into the microwave region is the quadrature hybrid. From our discussions in Chapter 4, remember that we defined a quadrature hybrid as a directional coupler whose two outputs are equal in amplitude and separated from one another by a constant 90° phase. These characteristics provide an attenuator with a constant impedance because of the phase relationship and balanced levels of each of the diodes involved. Fig. 6–11 shows two pin attenuators using quadrature hybrids in their construction. A single-hybrid coupled pin attenuator is shown in Fig. 6–11A, and a double-hybrid device is shown in Fig. 6–11B.
   In the single-hybrid attenuator, the power at port 1 of the hybrid is divided equally between ports 2 and 3. Port 4 is an isolated port as a result of the conjugate action of the hybrid. This relationship is explained in the quadrature hybrid discussion in Chapter 4. The mismatch produced by the pin diode resistance in parallel with the load resistance (Z_O) at ports 2 and 3 reflects part of the power back. If the diode is forward biased and the resistance is low, a high power level is reflected back, and the power at port 4 (RF_{out}) is lower from the input only by the insertion loss of the device. Increasing the diode resistance

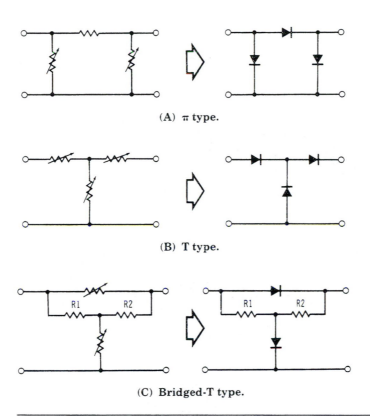

(A) π type.

(B) T type.

(C) Bridged-T type.

Figure 6–10.   Pin diode attenuators.

causes less and less power to be reflected to the output. Less power reflected to the output causes the attenuator action of the device. The maximum attenuation that can be achieved with this type of attenuator depends on the directivity of the coupler and the quality of the terminations at ports 2 and 3 when the diodes are unbiased. The VSWR at port 1 will depend on the power split and how equal a split can be attained. Equal reflection coefficients will also determine if a good input VSWR is obtainable. The equality of reflection coefficients can be ensured by having matched pin diodes in the circuits as well as having them equally spaced from ports 2 and 3. The insertion loss will depend on the losses and the equality of the power division in the hybrid coupler and the minimum resistance of the pin diode when forward biased.

Circuits of the single-hybrid type are easily and economically constructed and are useful over a wide frequency range. However, due to the hybrid directivity, there will be a considerable ripple in the high-attenuation state. This ripple characteristic can be eliminated by using two identical hybrids, as shown in Fig. 6–11B. This circuit provides the designer with flatter attenuator responses with frequency and better overall VSWR characteristics. The biasing scheme is improved with the two-hybrid scheme also. Instead of using ceramic rf bypass capacitors, as in the single-hybrid case, the diodes make direct

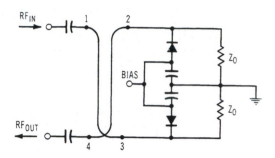

(A) Single-hybrid attenuator.

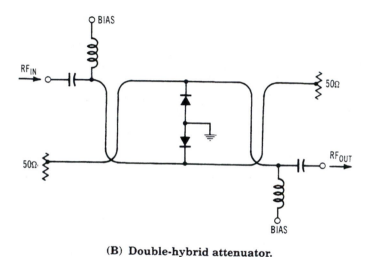

(B) Double-hybrid attenuator.

Figure 6–11.   Hybrid-coupled pin attenuators.

contact with the rf ground planes, and bias is applied through inductors that appear as a high impedance of the rf signal. Just as with the single-hybrid device, the two-hybrid scheme requires matched devices and equal spacing for optimum performance.

Just as with the π, T, and bridged-T methods of attenuation, the pin diodes can be used as switches, and so can the single- and double-hybrid schemes. The bias must be turned from max to min instead of some value in between, creating the necessary switching action.

We stated earlier that the pin diode, basically, can only attenuate and switch. These limitations are obvious from other applications. The pin diode, for example, can be used as a modulator. This use is a very popular application for a test setup where pulse and am circuits must be evaluated. The device still, however, involves only attenuation for amplitude modulation and switching techniques for pulse operation.

Another widely used application of pin diodes is for phase shifters for phased-array radar systems. These devices operate by switching line lengths and loading susceptances or balanced-phase bits into the system as needed. Many of these systems are very complex arrays that use precise phase increments. Still, the basic task of the pin diode is to switch these phase increments in or out as required.

6.3.3 Data Sheets

To wrap up our discussion of the pin diode, we will present and explain the important terms found on a data sheet. The primary terms that will characterize the pin diode are

V_B—Breakdown voltage (volts)

$C_j(v)$—Junction capacitance (pF)

$R_S(i)$—Series resistance (ohms)

τ_L—Carrier lifetime (nsec)

θ_j—Thermal resistance (°C/W)

Breakdown Voltage   The breakdown voltage of the pin diode is controlled by the width of the intrinsic (I) layer of the device. This number limits the rf voltage swing that may be applied to the diode. If this voltage is exceeded, you will be operating basically with a pn junction diode.

Junction Capacitance $C_j(v)$ is the junction capacitance of the diode when it is in a fully depleted state. The character v in this term is an important part of the parameter. It tells what voltage must be applied to result in the junction capacitance shown on the sheet. For example, $C_j(-50)$ gives the junction capacitance of a device with 50 V applied to deplete the I layer. This voltage is usually more than enough to deplete the layer. This voltage may vary with different manufacturers and, of course, will never exceed the V_B value for the device. Be sure to check the capacitance value for $C_j(v)$ as well as the value of voltage used.

Series Resistance $R_S(i)$ is the total rf resistance of the diode when a certain value of forward current is flowing through it. Unless otherwise specified, the value is measured at 1 GHz. Just as in the case of $C_j(v)$, $R_S(i)$ may have different values of current depending on the device used. For the values of forward current given, the resistance of the I layer is usually very small. The value of $R_S(i)$ is, therefore, very close to the minimum series resistance of the diode.

Carrier Lifetime   *Carrier lifetime* (τ_L) is a measure of the ability of a pin diode to store charge. A pure silicon crystal has a theoretical lifetime of several milliseconds. However, impure doping quickly reduces this lifetime to microseconds and nanoseconds. The actual measurement of the average carrier lifetime is accomplished by injecting a known amount of charge into the I-region of the diode and measuring the amount of time to get it back using reverse bias current. The pulse used to perform this test should have a fast rise time, and the pulse width should be several times that of the expected lifetime of the diode being tested. Characterization of the carrier lifetime of a pin diode is of utmost importance when designing switching circuitry.

Thermal Resistance   The *thermal resistance* (θ_j) of a pin diode is a measure of the ability of the diode to withstand heating effects due to both rf and dc power dissipation. It is defined as the ratio of steady-state temperature rise (°C) of the junction per watt of steady-state power dissipated within it. Thus, the unit of θ_j is °C/watt. This parameter is of primary importance when working with high-power switching applications.

Package   Typical packages for pin diodes are shown in Fig. 6–12. Each of these diode packages introduces two parameters that must be considered when completing your final design. These are package inductance (L_p) and package capacitance (C_p). In some applications these parameters can be tuned out or even put to use in the circuit. They do, however, limit the bandwidth over which you can operate. Once again, the packages shown in Fig. 6.12 are not to scale. The largest package would be the double stud, while the smallest would be the LID, with the pill package close behind.

In conclusion, we can say that the pin diode is a very versatile component that has many uses. Summarized in one statement, the pin diode is a voltage- (or current-) controlled

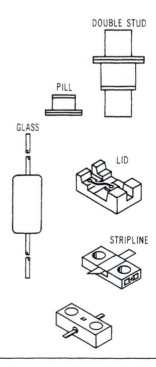

Figure 6–12.   Pin diode packages.

microwave resistor that can be either gradually varied or switched rapidly. It is much more than the classic idea of a diode or rectifier.

6.4   Tunnel Diodes

You may have heard of *tunnel diodes* before, or you may recognize such terms as *TDAs (tunnel diode amplifiers)* or *TDOs (tunnel diode oscillators)*. But why would a device have such a name as *tunnel*?

The tunnel diode gets its name from the *tunnel effect*. Tunnel effect is a process whereby a particle, obeying all the laws of quantum theory, can virtually disappear from one side of a potential barrier and appear instantaneously on the other side. This transfer occurs even though the particle does not have enough energy to surmount the barrier—it is as though the particle can "tunnel" under the barrier.

In the case of the tunnel diode, the *barrier* is the space charge depletion region of a pn junction. This same barrier prevents current from flowing in the reverse direction in the case of the ordinary rectifier diode. In the tunnel diode, this barrier is made very thin (less than 0.000001 inch). It is so thin, in fact, that penetration by means of the tunnel effect becomes possible. This penetration results in additional current in the diode at very small forward bias that disappears when the bias is increased. This additional current produces the negative resistance in the tunnel diode.

The tunnel diode differs from the pn junction primarily in the doping level. The impurity level in microwave pn junction diodes is in the range of 10^{16} to 10^{18} atoms/cubic centimeter, but values of 10^{19} atoms/cubic centimeter are quite common for tunnel diodes.

The tunnel diode is characterized as a negative resistance device. This does not mean that an ohmmeter would read negatively; it is a dynamic AC resistance rather than a DC resistance, which would be measured with an ohmmeter. That is, Ohm's Law says that $E = IR$. Any increase in the voltage has a corresponding increase in current. A decrease in voltage will similarly decrease the current. In the case of the tunnel diode there is a region where an increase in voltage results in a decrease in current. So it is rationalized that in this area $E = I(-R)$, representing a negative resistance device.

Two curves characterize the tunnel diode. They are shown in Fig. 6–13. The first curve (Fig. 6–13A) is the I-V (current versus voltage) curve, and the second (Fig. 6–13B) is the R-V (resistance versus voltage) curve that emphasizes the negative resistance characteristics of the device. Two very important points are on these curves. The first is R_m *(minimum negative resistance)* and is the smallest value of negative resistance exhibited by the diode. When this parameter is measured, it is measured at a low frequency, so the reactive components present can be neglected.

The second term is R_n *(negative resistance at a minimum K)*. The value K *(noise constant)* is a term dependent on temperature, diode current, junction resistance, and diode voltage at which the measurement is made. This value K is also referred to as *shot noise*. Remember from Chapter 5 that we defined *shot noise* as noise generated due to the random passage of current carriers across a barrier or semiconductor junction. This definition, of course, applies to the tunnel diode as well as to the transistors we were discussing earlier in Chapter 5. The normal operating point for tunnel diode amplifiers is in the region of R_n and

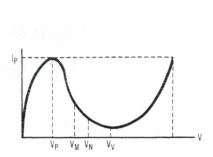

(A) Current versus voltage curve.

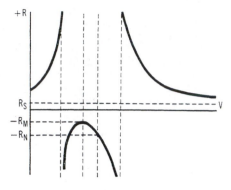

(B) Resistance versus voltage curve.

Figure 6–13. Tunnel-diode curves.

is defined as the value of negative resistance at minimum K. The noise constant does not vary rapidly with junction resistance in this region, so the exact operating point is usually determined by linearity and gain requirements. Maximum linear dynamic range is ensured when operating exactly at R_m, and the lowest noise figure is obtained when the device is operated at R_n. There must, therefore, be a trade-off to determine operating point versus requirements. Typical values of R_m vary from 35 to 70 ohms depending on I_p. Values of R_n will range from a low of 40 ohms up to around a 120-ohm maximum. The value of R_n will also depend on the value of I_p.

6.4.1 Construction

The construction of a typical tunnel diode is shown in Fig. 6–14. Fig. 6–14A is the diode in its package, and Fig. 6–14B is an expanded view of the semiconductor chip area showing the junction. There are several techniques of forming very narrow pn junctions for use in microwave tunnel diodes. The most popular are (1) ball-alloy, (2) electronic-pulse, and (3) planar processes. For microwave diodes the most common technique is the ball-alloy process shown in Fig. 6–14. Electronic pulsing has been used to produce the highest frequency devices, and the planar technique results in very consistent geometry and electrical characteristics.

Ball-Alloy Technique   As mentioned, the tunnel diode shown in Fig. 6–14 uses the ball-alloy technique. In this process, a small sphere (ball) of metal containing a heavy concentration of impurities (dopants) is alloyed to an oppositely doped semiconductor. An example would be a ball composed of a solid solution of indium and gallium alloyed into n-type arsenic-doped germanium. An electrical connection is made from the top of the dot (alloyed melted ball) to the top of the package (as shown in Fig. 6–14), and the semiconductor material under the dot is etched to reduce the active junction area and obtain a

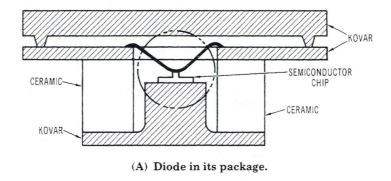

(A) Diode in its package.

(B) Expanded view of semiconductor chip.

Figure 6–14.   Tunnel-diode construction.

predetermined peak current. The semiconductor material that remains to support the dot is referred to as the *neck*. This neck area is important in considerations of series resistance and inductance, which will be covered later in this section.

Ball-alloy junctions have found wide applications in the microwave diode area because they are relatively easy to make and require the precise control of only one variable (peak current). This variable is usually attained with the reduction of junction diameter by a simple etching process. There is, however, an extra series resistance component with this technique caused by the neck formed during etching.

Electronic-Pulse Technique   The electronic-pulse technique can be used for tunnel diodes with many different combinations of material. Tunnel diodes have been made using the electro-pulse process on gallium antimony (GaSb), germanium (Ge), gallium arsenide (GaAs), and silicon (Si). Some of the dopants used in this construction are technetium (Tc), zinc (Zn), gallium (Ga), arsenic (As), selenium (Se), boron (B), and phosphorus (P). So you can see that a variety of combinations of materials is truly possible when using this process. In this process the heavily doped semiconductor wafer is first brought in contact with a pointed wire or tape of the carrier alloy. One or more low-voltage pulses of short duration (2 V at 8 μsec) are then applied between the wafer and the carrier alloy. These pulses lead to a localized alloying between the carrier alloy and the semiconductor.

Between pulses, the characteristics of the diode are measured, and the process is continued until suitable characteristics are obtained. Of course, if this process is allowed to go too far, the entire device will have to be rejected and started over. Electronic pulsing has been used in device construction for all useful peak-current ranges and well into the millimeter range. However, the small market and great difficulties involved in this high-quality construction have limited the extended use of this technique.

Planar Processing   The *planar* process for tunnel-diode fabrication combines planar technology (covered previously) with various alloy techniques to define many similar alloy tunnel diodes on one slice.

One planar technique is the *solution regrowth process,* in which a heavily doped germanium slice is first dipped into a solution of oppositely doped germanium. The solution is then cooled slightly so that a very thin epitaxial growth takes place, and the slice is withdrawn. Individual junctions are defined by etching and photoresist techniques. In another planar process, the doped metal is evaporated onto and alloyed with the slice, and the junction areas are defined by etching. Alternatively, the metal may be evaporated through holes in an etched oxide and alloyed. The technique of evaporating through holes in an etched oxide has been used to fabricate planar tunnel diodes with quite uniform electrical characteristics as a consequence of identical processing on all junctions within the slice. The planar technique also provides improved mechanical stability since the devices contain neither the fragile neck area of the ball-alloy devices nor the point contact of the electrically pulsed diodes.

6.4.2   Component Definitions

The equivalent circuit of a tunnel diode is shown in Fig. 6–15. The solid components are those associated directly with the diode, and the dashed line represents induced capacitance due to the package the diode is placed in. We will explain each of these terms so that you might have a better understanding of the makeup of a tunnel diode.

Series Inductance (L_S)   The effect of *series inductance* is somewhat dependent on diode mounting in the circuit. Generally, the diode is mounted in some coaxial configuration. In

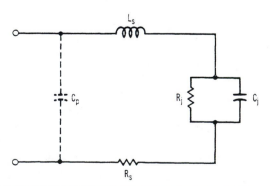

Figure 6–15.   Tunnel-diode equivalent circuit.

this type of application, the series inductance is determined primarily by the linear distance from the top of the package to the bottom. Thus, the parameter arises in part from the tape lead that joins the top of the encapsulation and in part from the inductance associated with the neck of a ball-alloyed device. Generally, the series inductance is considered to be a function of package design and the method of forming the junction. It should be kept at a minimum for best circuit operation and is measured at the valley bias point (V_v) on the I-V curve.

Junction Resistance (R_j)   *Junction resistance* is the negative resistance component of the device and is a dynamic (ac) resistance of the diode junction. The value of R_j varies from almost zero when the diode is heavily backbiased, to infinity at the peak current point, then to a minimum negative value, and back to infinity at the valley current point. Values of negative resistance for commercially available diodes range from 40 ohms to 120 ohms at approximately one half the peak current (I_p) of the diode. (When a value of R_j is given, a corresponding value of current is also given.)

Series Resistance (R_S)   *Series resistance* of a tunnel diode is a combination of the resistance of the ohmic contact, the spreading resistance in the wafer, and the resistance in the neck in the vicinity of the junction. Contact resistance in properly constructed diodes is sufficiently low to be neglected. Spreading resistance is a function of the resistivity of the semiconductor material and the diameter of the neck at the semiconductor surface. The resistance in the neck is a function of the effective length and diameter of the neck as well as the resistivity of the material. For the ball-alloy technique, the neck resistance frequently dominates, but in some of the other techniques the spreading resistance becomes a greater factor. The resistance (R_S) is measured by pulsing the diode far into the forward region (where R_j is low), then measuring the incremental resistance with a small sampling pulse on top of the biasing pulse.

Junction Capacitance (C_j)   Tunnel diode *junction capacitance* is basically the capacitance of a step junction. This capacitance corresponds to a space-charge depletion width and varies with applied bias. It is determined by measuring the total capacitance with the diode biased at the valley voltage (V_v) and subtracting the package capacitance. The operating point junction capacitance is defined to be 75% of the valley point junction capacitance.

Package Capacitance (C_p)   The *package capacitance* is a function of the package style. In high-frequency diodes, the junction capacitance (C_j) is often much less than package capacitance (C_p). This term must, therefore, be considered in the design phase since it is such a large part of the overall diode capacitance.

Resistive Cutoff Frequency   One additional parameter for tunnel diodes should be mentioned before we get into tunnel diode applications. This parameter is *resistive cutoff frequency* (f_{ro}). This is the frequency where the resistive part of the parallel combination of C_j

and R_j equals R_S. This frequency is specified at R_m and the relationship $|R_j| = |R_m| + |R_S|$. The resistive cutoff frequency can be calculated by

$$f_{ro} = \frac{1}{2\pi |R_j| C_j} \sqrt{\frac{R_j}{R_S} - 1}$$ (6.3)

6.4.3 Applications

With the tunnel diode presented and defined, the next step is to put it to work. For many years the primary application of the tunnel diode was as a tunnel-diode amplifier (TDA). This is a lightweight, inexpensive, low-noise component that is useful at the front end of microwave receivers. This device improved the sensitivity of communications systems, satellite systems, radar systems, and countermeasure systems. The TDA exhibits extremely low power consumption and also is capable of wide bandwidth operation.

We said that the TDA was the primary applications for tunnel diodes. This, unfortunately (or fortunately depending on how you look at it, especially if you were a TDA designer), is no longer the case. The GaAs FET replaced the TDA for many of the applications listed above and has found even more applications in this rapidly changing microwave world, where commercial applications are so important. The main applications now are tunnel diode oscillators and detectors (the circuit shown in Fig. 4.12B would be used with tunnel diodes substituted into the circuit instead of Schottky diodes).

In order to understand the tunnel diode oscillator (TDO), it will be necessary to explain how a tunnel diode amplifier works so that when the appropriate feedback is incorporated a tunnel diode oscillator will result. That is our approach in this text.

Two methods of using the tunnel diode in an amplifier application are shown in Fig. 6–16. (Notice the different symbol used for the tunnel diode as compared to other diodes.) Fig. 6–16A uses a circulator to isolate the incident signal from the amplified wave of the tunnel diode. A circulator is probably the most common type of circuit in use when TDAs are in operation. Fig. 6–16B uses a quadrature hybrid for this isolation. Notice the similarity between this circuit (a reflection-type tunnel diode amplifier) and the pin

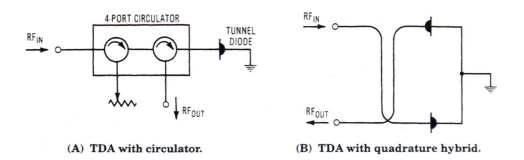

(A) TDA with circulator.   (B) TDA with quadrature hybrid.

Figure 6–16.   Tunnel-diode amplifiers.

attenuator shown in the previous section (Fig. 6–11A), which is a reflection-type attenuator. A design guide for adequate TDA gain and low noise is

$$f_{ro} \simeq 4f_o \qquad (6.4)$$

where, f_{ro} is the resistive cutoff frequency of the tunnel diode as described earlier,

f_o is the amplifier operating frequency.

Since tunnel diodes in tunnel-diode amplifiers are operated in the negative resistance region, adequate consideration must be given to the prevention of instabilities. Switching may occur unless the biasing circuit is well regulated and is of low impedance. Likewise, oscillations may occur if off-band frequencies are not terminated properly or if the self-resonant frequency is less than the diode resistive cutoff frequency. (This is one method of making a TDO.)

Two requirements must be met to obtain a low-noise figure-tunnel-diode circuit. First, the tunnel diode must itself exhibit low noise; secondly, the circuit must be optimized for low-noise performance. The basic performance of the tunnel diode can be described by a noise figure of merit that is a function only of the diode parameters, its bias point, and the operating frequency.

When the diode is used as an amplifier, the circuit noise figure can be made to asymptotically approach the noise figure of the diode. The latter is normally the minimum limit. The noise figure of the amplifier is made to approach the figure of merit of the diode by suitable adjustment of circuit parameters. The dynamic range of any amplifier is the input signal variation over which the amplifier functions effectively. The lower limit is determined by the noise figure, the upper limit by saturation. To compare the effect of semiconductor materials on dynamic range, consider gallium antimonide, germanium, and gallium arsenide tunnel diodes with the same values of R_m. The peak current values for equal R_m are in the respective ratios of 1 to 1.5 to 2.5. The level at which saturation occurs is roughly proportional to $10 \log I_p^2$. Thus, for a 1 to 2.5 variation of I_p, the saturation level changes by about 8 dB. Fortunately, this variation is considerably larger

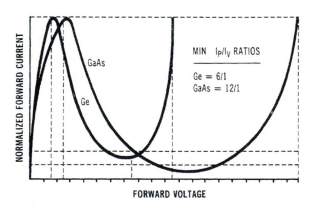

Figure 6–17. Tunnel-diode characteristics.

than are the noise-figure variations for each material; a net result of about 5 dB in dynamic range is effected by using a gallium arsenide diode rather than one of gallium antimonide.

An increased dynamic range and a low-noise figure are possible by constructing dual-stage amplifiers, using a gallium antimonide unit in the first stage and a gallium arsenide device in the second.

Since microwave tunnel diodes are free from transit-time effects, they are a logical choice for use as high-frequency microwave oscillators. Such devices tend to be simple, compact, rugged, and reliable. Gallium arsenide offers the best choice for oscillator applications because of a wide dynamic range in the I-V characteristic. As evidenced in Fig. 6–17, GaAs has about 9% greater current swing and twice the voltage range obtainable with germanium diodes. The power output from a tunnel-diode microwave oscillator may be calculated from

$$P_{out} \simeq \frac{3}{16} \Delta I \, \Delta V \left(1 - \frac{f_o^2}{f_{ro}^2} \right) \tag{6.5}$$

where, $\Delta I \; = I_p - I_v,$
$\qquad \Delta V = V_v - V_p,$
$\qquad f_o$ is the operating frequency,
$\qquad f_{ro}$ is the resistive cutoff frequency.

Below self-resonance, the diode reactance is capacitive; therefore, in the lower-frequency range, the load at the desired frequency must be inductive. Conversely, above self-resonance, the diode reactance is inductive; therefore, the load at the desired frequency must be capacitive.

The tunnel diode is a versatile device that finds many uses in microwaves. Even though the GaAs FET has eliminated the primary use of tunnel diodes, there still are others available, as we have shown.

6.5   Gunn Diodes

The Gunn diode, like the tunnel diode, is a negative resistance device. Its origin goes back to 1963, when J. B. Gunn discovered that when a dc voltage is applied to the ohmic contacts on the ends of a bar of n-type GaAs (gallium arsenide) or InP (indium phosphide), the current first rises linearly as the voltage increases from zero and then begins to oscillate when a certain threshold value is exceeded. The period of the oscillations was very close to the transit time of the carriers between the contacts. Since Gunn's initial discovery, the Gunn (or bulk-effect) devices have progressed to obtain cw signals of high-spectral purity upwards to 100 GHz, with outputs near the 100-mW level. Indium phosphide (InP) material recently has extended oscillator applications up to 140 GHz. The oscillator is the primary application of Gunn diodes in the microwave and millimeter areas.

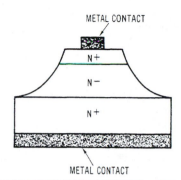

Figure 6–18. Gunn-diode constuction.

6.5.1 Construction

Construction of the Gunn diode is shown in Figure 6.18. It is of Mesa construction and consists of layers of GaAs (Gallium Arsenide) or InP (Indium Phosphide). You will notice that we call the device a *Gunn diode*. You will also notice that there is no junction in the device. We refer to the Gunn diode as a "diode" simply because it is a two-element device and not because it exhibits the typical junction diode properties. The I-V curve for a Gunn device is shown in Fig. 6–19 and indicates a decrease in current as bias voltage is increased. The drop in current is not as pronounced as in the case of the tunnel diode but does indicate a reversal, indicating a negative component.

When the Gunn diode is biased in the negative-resistance region, the field distribution within the sample rearranges itself to form a thin, high-field domain, with the rest of the sample experiencing a subthreshold field. The domain drifts through the sample at the saturated drift velocity and collapses when it reaches the anode. The bias current then momentarily increases until a new domain (or bunch of current carriers) forms at the cathode. This periodic current increase results in high-frequency power being made available to the external circuit.

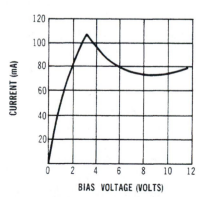

Figure 6–19. Gunn-diode I-V curve.

The output frequency, thus, is determined basically by transit time effect (the time it takes for one bunch to get from the cathode to the anode), but in practice to vary the frequency over about one octave is possible by tuning the external resonant circuit (because of modulation of the bias voltage by the rf voltage developed in the resonant circuit, which can cause the instantaneous diode voltage to fall below the diode threshold). Formation of subsequent domains is then delayed, giving rise to the name *delayed domain* for the mode of operation.

Limited space-charge accumulation (LSA) operation in Gunn diodes is obtained by tuning the diode in a high-impedance circuit at a frequency such that a full domain cannot form during the period of an oscillation. The rf swing necessary is such that the pulsed operation is normally used. The principal advantage of this mode is that since the material is displaying bulk negative resistance, the dimensions of the diode are no longer limited by transit-time considerations, meaning that millimeter-wave frequencies can be achieved without the use of impossibly small active regions. A disadvantage of LSA operation is that if domains do start to form, the rf voltages involved are so high that the domain fields may cause an avalanche breakdown, leading to the destruction of the device. Since domains can be created by even the slightest disruption in the consistency of the material, the requirements of the material for LSA operations are extremely strict. Also, it is possible to use Gunn diodes in other modes (quenched domain, multiple domain, and hybrid, for example), but these suffer from some or all of the disadvantages of being little understood, being difficult to control, or requiring diodes of special design.

6.5.2 Precautions

Certain precautions should be observed when operating Gunn diodes. They should be operated from a constant-voltage power supply. The required operating voltage range is determined by the threshold field, power dissipation, and breakdown considerations. The data sheets should be very carefully checked for maximum voltage ratings. Although in some circumstances a greater output can be obtained by operating at higher voltages (pulsed conditions), no guarantee of reliability is given. The quoted ratings are maximum values, and the power supply should be free of any overvoltage transients (which are most likely to occur during switch-on). Bias polarity must, of course, also be heeded very carefully. Since the frequency-pushing characteristics of these devices can be as high as 30 MHz/V, noise and ripple on the power supply can produce a significant amount of fm noise on the oscillator output. A well-stabilized power supply is, therefore, necessary to reduce frequency-noise modulation.

Bias-circuit oscillations may occur in certain circumstances due to a slightly negative resistance in the diode dc characteristics just above threshold, and they are found generally in the range of 1 to 100 MHz. Although they are not damaging to the diode, they can result in a poor output spectrum. These oscillations can be suppressed by connecting a 0.01 μF to a 1.0 μF capacitor across the oscillator bias terminals.

6.5.3   Applications

Gunn diodes are now widely used as power sources for pumps, sweepers, intrusion alarms, and fuses. They also are used as transmitters in radars, beacons, transponders, speed sensors, radio, and data links, and they serve as local oscillators and amplifiers for communications equipment. Recent advances in process technology are extending the power levels into some medium-power applications.

An expanding use of Gunn diodes is in the field of wideband, tuned oscillators, such as the yttrium-iron garnet (YIG) type for sweepers and electronic countermeasure (ECM) receivers. The low-noise characteristic and wide bandwidths of Gunn diodes yield full-band tuned oscillators for wideband receivers.

Fixed or trimmable low-power, low-noise local oscillators have been one of the first uses for Gunn diodes and continue to replace reflex klystrons in most new equipment. Retrofit programs are continuing to replace klystrons where noise and reliability of the tubes have not been satisfactory. Excellent work has been done in this area to provide a superior local oscillator at reduced costs, using Gunn diodes.

The InP (indium phosphide) Gunn device is being developed specifically for low-noise amplifiers from 26 GHz to 100 GHz and for medium-power low-noise oscillators from 40 GHz to 140 GHz. Typical oscillator powers so far obtained are shown here.

Frequency (GHz)	Power Output (mW)
56	200
90	100
95	68
100	44
55 (pulsed 10% duty)	860

Indium phosphide (InP) has advantages over GaAs (gallium arsenide) for high-frequency millimeter-wave oscillator applications. Two of these advantages are

- Higher peak-to-valley ratio,

- Reduced scattering time due to a high-threshold field.

Peak-to-valley ratio is illustrated in Fig. 6–20. Fig. 6–20 is a plot of particle velocity versus electric field for n-type GaAs and InP. InP has a peak (maximum)-to-valley (minimum) ratio of approximately 3, and the GaAs shows less than 2.4. This ratio must be high because the basic efficiency of a transferred-electron oscillator (Gunn device) is strongly influenced by this ratio of the material used. The InP ratio has been shown to be higher than that of GaAs, but even better is the fact that when the materials are heated, the ratio in GaAs is significantly reduced, and the InP reduces both peak-and-valley velocities nearly the same amount, preserving the high resultant ratio.

The scattering process defines the movement of charges (or particles) from one point to another. Each of the scattering processes within the Gunn device takes a certain amount of time to move between these points. (The points referred to are termed the central valley, which is the main valley of the curves, and the satellite valley, which is a secondary valley.) Times considered are those required to move from the central valley to the satellite

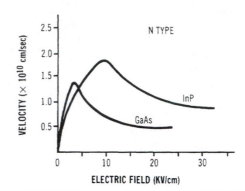

Figure 6–20. Peak-to-valley
comparison.

valley and the reverse time. These times have been shown to be a factor of two smaller in InP than in GaAs. Thus, if we have a GaAs Gunn device operating at 90 GHz, we can expect similar performance of InP at 180 GHz, which is a sizable improvement in overall performance.

A third characteristic that makes InP more desirable at millimeter frequencies should also be mentioned. That characteristic is the higher impedance of InP. The design of millimeter-wave oscillators is significantly simplified when a large device of negative resistance and small reactance is present, which is the case for InP Gunn devices. The obvious higher-peak velocity of InP (shown in Fig. 6–20) permits an increased active layer length, maintaining a transit-time relationship with a low bias-to-threshold ratio.

As a final note, it should be pointed out that the wideband negative resistance characteristics of an InP Gunn device, coupled with its low-noise properties, make the device an excellent choice as a reflection amplifier also. The applications are suitable for millimeter-wave narrow and broadband circuits. To sum up the Gunn device applications, we can say

- Microwave application—GaAs,
- Millimeter application—InP.

6.6   IMPATT Diodes

The IMPATT (*IMP*act *A*valanche and *T*ransit *T*ime) diode is actually a diode in the sense we have been discussing (pn junction) and has other properties that, as the name implies, deal with the transit time within the device. In our previous discussions we have said that the Gunn diode was a negative resistance device. It was said to have a dynamic negative resistance. This meant that, over a certain range current decreased with an increase in applied voltage, and vice versa. This particular point was pursued no further, it being taken for granted that any device which exhibits a dynamic negative resistance for dc (which was used previously) will also exhibit it for ac. That is, if an alternating voltage is applied, current will rise when voltage falls at the same ac rate. We may thus now redefine negative resistance as the property of a device that causes the current through it to be 180 degrees out of phase with the voltage across it.

This point is important because this is the only kind of negative resistance exhibited by the IMPATT diode.

A combination of delay involved in generating avalanche current multiplication and delay due to transit time through a drift space provides the necessary phase shift between the applied voltage and the resulting current in the diode. A drawing of the IMPATT structure is shown in Fig. 6–21 (notice the pn junction in the device).

An extremely high voltage gradient is applied to the IMPATT diode, in the order of 400 kV/cm, which results in a very high current. A normal diode would very quickly break down in this condition, but the IMPATT is constructed so that it is able to withstand these conditions with no damage or breakdown occurring. With this high potential gradient, back-biasing the diode causes a flow of minority carriers across the junction. If it is now assumed that oscillations exist, we may consider the effect of a positive swing of the RF voltage superimposed on top of the high dc voltage. Electron and hole velocity has now become so high that these carriers form additional holes and electrons by knocking them out of the crystal structure. This is called *impact ionization*. These additional carriers continue the process at the junction, and it now snowballs into a full-blown avalanche effect. If the original dc field is just at the threshold of allowing this situation to exist, this voltage will be exceeded during the entire RF positive cycle, and avalanche current multiplication will be taking place during the entire time. However, since it is a multiplication process, avalanche is not instantaneous. Fig. 6–22 shows that the process takes time such that the current pulse maximum, at the junction of the device, occurs at the instant when the RF voltage across the diode is zero and going negative. A 90-degree phase difference between voltage and current is obtained.

As described, the current pulse in the IMPATT diode is situated at the junction. It does not, however, stay there. Because of the reverse bias, the current pulse flows to the cathode at a drift velocity dependent on the presence of the high dc field. The time taken by the pulse to reach the cathode depends on this velocity and, of course, the thickness of the highly doped (n) layer. Also, the thickness of the drift region is selected so that the time

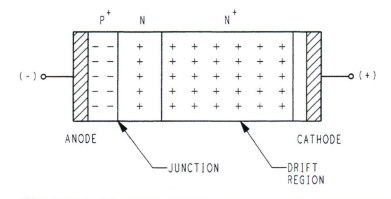

Figure 6–21.   IMPATT schematic.

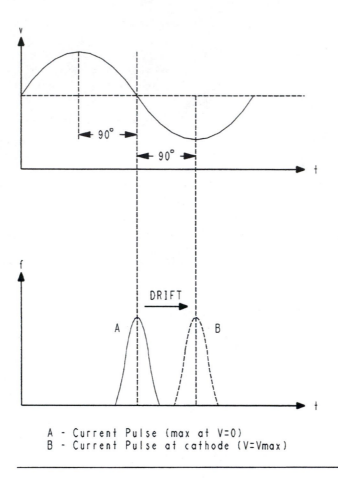

A - Current Pulse (max at V=0)
B - Current Pulse at cathode (V=Vmax)

Figure 6–22. IMPATT behavior.

taken for the current pulse to arrive at the cathode corresponds to a further 90-degree phase difference. So, as can be seen in Fig. 6–22, when the current pulse actually arrives at the cathode, the RF voltage is at its negative peak. Thus, the voltage and current in the IMPATT diode are 180 degrees out of phase, and a dynamic RF negative resistance has been shown to exist.

Commercial IMPATT diodes have been available for some time. They are made of either silicon or gallium arsenide, sometimes epitaxial, and usually of mesa construction, with some having Schottky junctions. When gallium arsenide (GaAs) is used, it gives lower noise, higher efficiencies, and higher maximum operating frequencies.

It should be noted that practical IMPATT diodes were unlike Read's original proposal, which called for a double drift region. We have shown in Figures 6–21 and 6–22 that there was a single drift region. The reason for this departure from original theory is that this type of structure was very difficult to fabricate for some time. This is not as much of a problem anymore, and GaAs RIMPATT (read-IMPATT) devices are now available that show higher efficiencies.

6.7   **TRAPATT Diodes**

The TRAPATT (*TRA*pped *P*lasma *A*valanche *T*ransit *T*ime) diode is derived from the IMPATT previously covered.

Consider an IMPATT diode mounted in a coaxial cavity so arranged that there is a short circuit one-half wavelength away from the diode at the device operating frequency. When oscillations begin, most of the power will be reflected across the diode, so the RF field across it will be many times the normal value for IMPATT operation. This will rapidly cause the total voltage across the diode to rise well above the breakdown threshold value. As avalanche now takes place, a plasma of generated electrons and holes is generated, placing a large potential across the junction. This opposes the applied dc voltage. The total voltage is thereby reduced, and the current pulse is trapped behind it, so to speak. When this pulse travels across the n-drift region of the semiconductor chip, the voltage across it is thus much lower than in IMPATT operation. This has two effects. The first is a much slower drift velocity and, thus, longer transit time so that for a given thickness the operating frequency is several times lower than for corresponding IMPATT operation. The second point is that when the current pulse does arrive at the cathode, the diode voltage is much lower than in an IMPATT diode. Thus, dissipation is also much lower, and the efficiency is much higher. This is very similar to Class C operation of transistor circuits, and if we look close we will see that the TRAPATT lends itself to pulsed operation rather than CW situations.

Commercial TRAPATT diodes are constructed of silicon and GaAs, with structures corresponding to those of the IMPATT diode. The main difference is that there is gradual rather than abrupt changes in doping levels between the junction and the anode. The TRAPATT structure is shown in Fig. 6–23.

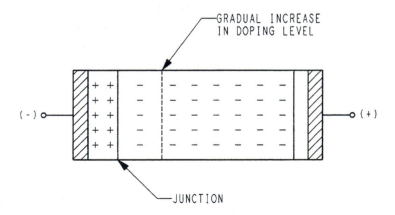

Figure 6–23.   TRAPATT schematic.

Because the drift velocity of a TRAPATT diode is much less than in an IMPATT, either operating frequencies must be lower or the active regions must be made thinner. Actually, both these considerations are incorporated. Most good TRAPATT results have been obtained at frequencies under 8 GHz. It has also been found that by the time 5 GHz is reached, the width of the depletion layer is only about 2 μm, which is the lower practical limit of the devices. TRAPATT pulses are rich in harmonics. This can be taken advantage of and is, by tuning amplifiers and oscillators to these harmonics and making very efficient circuits.

6.8   Chapter Summary

We have covered six types of diodes used in microwave and millimeter applications. The Schottky diode found application in detectors and mixers; the pin was used as a switch or attenuator; the tunnel and Gunn were negative resistance devices used for reflection amplifier and oscillator applications; and the IMPATT and TRAPATT are oscillators. Each of these diodes has specific areas of use in microwaves, and they exhibit characteristics that single them out over all other devices. One thing they all have in common is that none of them falls into the old diode stereotype of rectifiers in a power supply.

6.9   References

1. *Applications of PIN Diodes.* Hewlett-Packard Co., Applications Note 922, Palo Alto, CA.
2. Brehm, G. E., and Mao, S. "Varactor-Tuned Integrated Gunn Oscillators." *IEEE Journal of Solid-State Circuits,* Vol. SCS, September 3, 1968.
3. *Diode and Transistor Designers' Catalog.* Hewlett-Packard Co., 1980.
4. Fank, F. B., DiCrowley, J., and Berenz, J. J. "InP Material and Device Development for Millimeter Waves." *Microwave Journal,* June, 1979.
5. Fank, F. B. "InP Emerges as Near-Ideal Material for Prototype Millimeter-Wave Devices." *MSN,* February, 1982.
6. Laverghetta, T. S. *Microwave Measurements and Techniques.* Dedham, MA: Artech House, 1976.
7. *PIN Diode Designer's Guide.* Microwave Associates, 1980.
8. *Receiving Diode Handbook.* Microwave Associates, Bulletin 4006, 1980.
9. *Reference Data for Radio Engineers,* Sixth Edition. Indianapolis: Howard W. Sams & Co., 1975.
10. Watson, H. A. *Microwave Semiconductor Devices and Their Circuit Applications.* New York: McGraw-Hill, 1968.

Questions

6.2 Schottky Diodes

1. Describe a Schottky barrier junction.
2. Is a Schottky diode a bipolar or unipolar device? Explain.
3. Describe the resistances associated with a Schottky diode.
4. Describe the capacitances associated with a Schottky diode.
5. Which Schottky diode bonding method is considered to be the best mechanically?

6.3 Pin Diodes

6. Define *intrinsic*.
7. Describe Mesa processing.
8. Which type pin attenuator is a reflective device?
9. Which type pin attenuator is an absorptive device?
10. How can pin attenuator be changed to a pin switch?
11. Describe *carrier lifetime*.

6.4 Tunnel Diodes

12. Describe *tunnel effect*.
13. How does the tunnel junction vary from the conventional pn junction?
14. Describe *negative resistance*.
15. What is *minimum negative resistance*?
16. Describe the *negative resistance at a minimum k* point on the tunnel diode R-V curve.
17. Describe the resistive cutoff frequency of a tunnel diode.
18. List and describe two applications of tunnel diodes.

6.5 Gunn Diodes

19. What is the *Gunn effect*?
20. Is a Gunn diode a diode in the traditional sense? Explain.
21. What type Gunn diode is used at microwave frequencies?
22. What type Gunn diode is used at millimeter frequencies?

6.6 IMPATT Diodes

23. What two phenomena determine the operation of an IMPATT diode?
24. Describe IMPATT behavior.
25. How does the 180-degree phase shift occur in the IMPATT diode?
26. What is an application of an IMPATT diode?

6.7 TRAPATT Diodes

27. What is the difference between the IMPATT and the TRAPATT diode?
28. Name an application of a TRAPATT diode.

Problems

6.3   Pin Diodes

1.   We need a pin diode series reflective attenuator with a loss ranging from 2dB to 20 dB (Z_o = 50Ω). What diode resistance range is needed for this requirement?

2.   Find the values of diode resistance if a shunt reflective attenuator was used in problem 1.

6.4   Tunnel Diodes

3.   We have a tunnel diode with a junction resistance of 52Ω, a series resistance of 6Ω, and a junction capacitance of 0.5 pf. What is the resistive cutoff frequency?

4.   A tunnel diode is to be used as a TDO. The cutoff frequency is 14.8 GHz, we want to operate at 7.6 GHz, current change is 4 ma., and voltage change is 0.5 volts. What is the expected power output?

7 The Smith Chart

Objective

To provide a basic introduction to the Smith chart. This will be accomplished by constructing the chart from scratch in order to demystify it. Following this construction all parts will be described, and examples of its use will be presented.

Key Terms

Imaginary Impedance	Reflection Coefficient
Mismatch Loss	Return Loss
Normalized Chart	Transmission Line
Real Impedance	Wavelength
Reflection Angle	

7.1   Introduction

We have mentioned so far many names of individuals who have contributed to the field of microwaves: Schottky and Gunn for diodes, and Brattain, Bardeen, and Shockley for transistors, to name a few. All these people provided valuable contributions to microwaves in their own particular area. But the name that probably has contributed the most to the overall field is that of Phillip H. Smith.

In the January 1939 issue of *Electronics,* Phillip H. Smith published an article titled "Transmission Line Calculator" in which he presented a chart that could be used to analyze transmission lines without the use of long, involved equations. This chart, of course, became known as the *Smith chart* and is one of the most useful tools a microwave engineer can have.

The Smith chart is based on two sets of orthogonal circles that represent microwave impedances. These sets are needed because every microwave impedance is expressed as a combination of two components:

$$Z = R \pm jX \qquad (7.1)$$

where, R is the resistive, or real, component,

X is the reactive, or imaginary, component.

Both components must be spelled out to completely characterize the impedance, even if one of them is equal to zero. A 50-ohm system, for example, is called just that, 50 ohms. Actually, the system is a $50 + j0$ ohms, but most of the time only the 50 ohms is shown, and the reactive component is assumed to be zero. The same is true with pure reactance: j20 ohms, for example. The $+j$ indicates an inductance and implies that there is no real part to the impedance at all ($0 + j20$ ohms would express both components accurately). Most of the time, however, zero components do not need to be worried about because the impedance to be dealt with consists of two distinct components. When two components need to be dealt with, the Smith chart with the two sets of circles is very useful.

The most useful type of chart to use is a *normalized* chart. That is, the center of the chart is 1.0 and all the impedances are divided by the system characteristic impedance (Z_0), both real and imaginary. The resistive (real) part of the impedance can be found along one of the complete circles of the chart (R/Z_0). The reactive (imaginary) part of the impedance, (jX/Z_0), is found along one of the lines that show up as arcs but are actually large circles, as will be shown later. We will completely construct a Smith chart in the next section to clarify the statements made above.

7.2   Smith Chart Construction

Fig. 7–1 shows the Smith chart. At first glance it looks a little confusing, with numbers and lines and circles literally covering the whole area. Actually, once you understand it, the chart is relatively simple. To aid in this understanding, we will "build" a Smith chart and explain each component along the way. The building process will take place in three steps: the resistive (or real) components, the reactive (or imaginary) components, and the reflection angles and portions-of-a-wavelength markings shown around the outside edge of the chart.

7.2.1   Resistive Components

The first step is to look at the resistive portion of the chart. Fig. 7–2A shows its *pure resistance* line, which goes through the center of the chart. Any points on this line are *resistive only* and contain no reactive components; along this line is where such values as $50 + j0$,

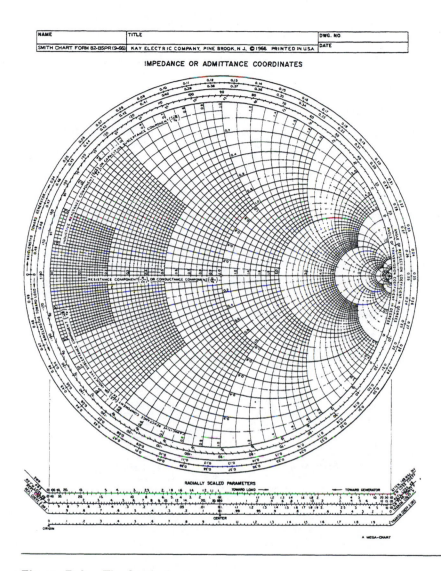

Figure 7–1.   The Smith chart.

27 + j0, and 100 + j0 would be plotted. Note the values all along its line. The range is from zero ohms on its left to infinity on the right. In between these two extremes are various points all referenced to Z_0 (characteristic impedance). Note that the center of the line states only Z_0 and not any specific number, which enables this chart to be used for any impedance instead of being limited to a single-impedance system. This center line is the beginning of a *normalized* Smith chart. Fig. 7–2B shows where this line is on the chart.

The six points shown on the resistive axis are just that, single points. Each of these points, as well as many other values, can be displayed over one half-wavelength by what

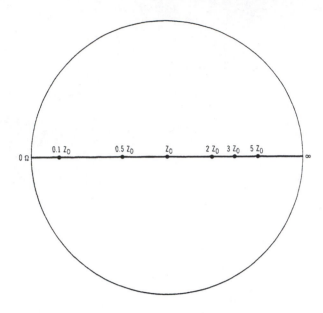

(A) Pure resistance line on plain background.

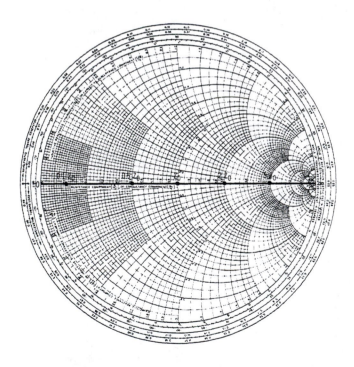

(B) Pure resistance line on the Smith chart.

Figure 7–2. Plotting pure resistance lines.

are known as *constant resistance* circles. The radius for these circles can be determined by using this relationship:

$$\text{radius} = \frac{1}{1 + \dfrac{R}{Z_0}} \tag{7.2}$$

where, R is the resistance point required,
　　　　Z_0 is the system characteristic impedance (usually 50 ohms).

The reference used for these radii is the radius of the outside circle in Fig. 7–2. To show that this reference is true, we will plug 0 ohms into the equation. You can see that the radius now becomes 1/[1 + (0/50)] (for a 50-ohm system). The radius very quickly becomes 1 and is the reference for all other resistances. To further check this relationship we can use Z_0 as another example. Note that this point is in the exact center of the chart. This location means the radius for a circle going through this point should be exactly one-half that for the 0-ohm case.

EXAMPLE 7.1

Verify the radius for $Z_0 = 50\ \Omega$ and $R = 50\ \Omega$.

$$\text{radius} = \frac{1}{1 + \dfrac{R}{Z_0}}$$

$$= \frac{1}{1 + \dfrac{50}{50}}$$

$$= \frac{1}{1 + 1}$$

$$= \frac{1}{2}$$

So it can be seen that the R = ½ is verified.　　　　　　　　　　　　　　　　　　　◆

This equation will work for any value of resistance and for any characteristic impedance. Be sure, however, that the characteristic impedance chosen remains the same for all calculations. Fig. 7–3 shows the constant resistance circles for the six resistance points shown in Fig. 7–2. You can see where these and other resistance circles fall on the actual Smith chart in Fig. 7–3B. As a check of the radius relationship, you might want to measure the total radius of the chart and check the measured versus the calculated radii for the six points we have emphasized.

And so the resistance (or real) part of the Smith chart has been constructed. Notice on the completed Smith chart that not all the resistance circles are complete. The circles from $R = 2Z_0$ down to $R = 0$ do not close at the infinity end of the chart in order to avoid all the

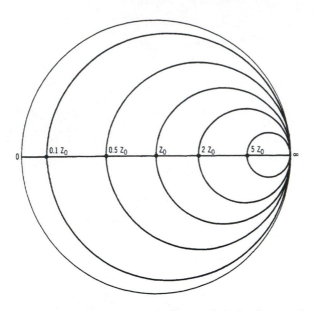

(A) Constant resistance circles on plain background.

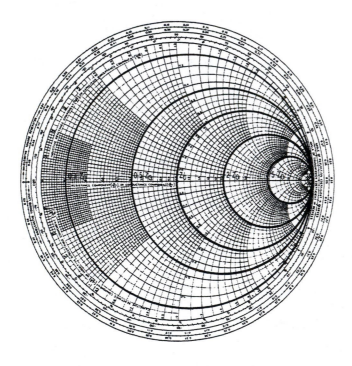

(B) Constant resistance circles on the Smith chart.

Figure 7–3. Plotting constant resistance circles.

clutter that would result if all the circles converged on that single area. To pick out any one particular circle with all of them ending at the same location would be very difficult. Therefore, the chart leaves this area open, and where the circles will go must either be estimated or a compass used to complete the circles.

7.2.2   Reactive Components

The second portion of the microwave impedance (R $\pm$ jX) is the reactive, or imaginary, component. This component may be either an inductive reactance or capacitive reactance and is distinguished by where it appears on the Smith chart.

The *pure reactance* (that is, no resistive component) values are displayed around the outside circumference of the chart. These values are called constant reactance points, and some of them are shown in Fig. 7–4A. Once again, they are a constant times Z_0 ($+j3Z_0$, for example), just as in the case of the constant resistance points we discussed before. If the measurements from a capacitor or inductor impedance were plotted, they would theoretically appear around the outside edge of the Smith chart. We say *theoretically* because the component will not be a pure reactance. Since nothing can be perfect, a certain amount of resistance will be associated with each component. The impedance, therefore, will be complex (R + jX) rather than only reactive. For many analysis situations, however, the idea of pure reactance can be used, and these values are on the outer edge of the chart. Fig. 7–4B shows where the reactive parts appear on the actual Smith chart.

We previously mentioned that the reactive component may be inductive or capacitive, depending on where it appears on the Smith chart. If the reactance appears above the pure resistance line, it is an inductive reactance. If it is below the line, it is a capacitive reactance.

Fig. 7–5 shows six cases where the component is an inductive reactance. Notice once again that the values are a function of Z_0 and not specific numbers. This increases the versatility of the chart and makes it a *normalized* Smith chart.

The reactance circles are determined by the ratio of Z_0/X. This may be $-jX$ or, as in the case of our circles in Fig. 7–5, $+jX$. This ratio is the radius of the circle. To illustrate how this process works, we will do some examples. Consider Fig. 7–6, which shows the derivation of three circles whose curves are shown in Fig. 7–5. These are $+j3Z_0$, $+jZ_0$, and $+j0.5Z_0$.

We will use point A as a starting point for each of the radii. If we consider the first line to be $+jZ_0$ and put this into our equation, we will find that Z_0/jX equals $Z_0/Z_0 = 1$. The radius for this circle is, therefore, 1, which means that the radius of the circle is equal to the radius of the entire circle that has the reactance points on the outside edge. This radius is measured from the center of the chart to the outside (point A). If we construct a line perpendicular to point A, called a *radius line*, we can measure all reactance circle radii along it. With a compass point at point A, we measure the reference radius (Z_0 to A) and swing up to the radius line to point B. The compass point is then placed at point B, and the $+jZ_0$ circle is drawn.

The $+j0.5Z_0$ is constructed by first finding the radius value (Z_0/jX), which is $Z_0/0.5Z_0$, which is 2. This value means that the radius to be used is equal to the outside diameter of

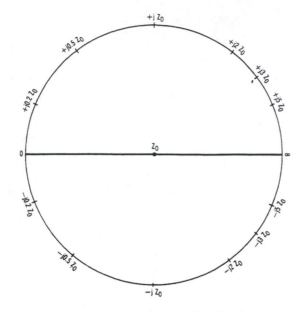

(A) Constant reactance points on plain background.

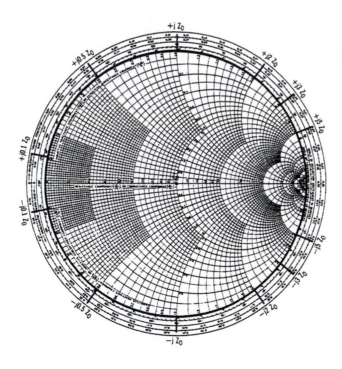

(B) Constant reactance points on the Smith chart.

Figure 7–4. Plotting constant reactance points.

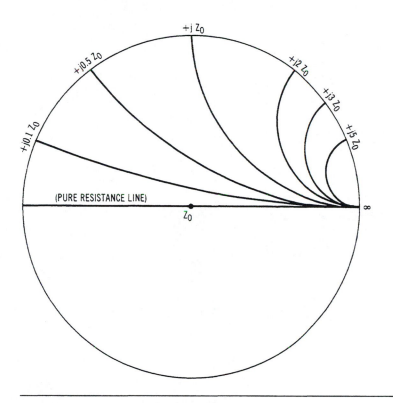

Figure 7–5.   Inductive reactance circles.

the chart circle. By using the same procedure as discussed previously, we come up with point C as a starting point. From this point the $+j0.5$ circle is drawn.

In the opposite direction, the $+j3Z_0$ radius is equal to 0.33. Point D is the starting point for the circle. From this point the $+j3Z_0$ circle can be drawn. From these discussions and figures, it can be seen that the reactive lines on the inside of the chart appear as arcs, or portions of circles, as we have previously said, while in reality they are complete circles. Most of the circle does not appear within the confines of the chart.

Fig. 7–7 shows the three circles we have just constructed as well as three additional circles we initially showed in Fig. 7–5. From this chart, notice how each of the circles can be constructed and how the radius of the circle gets larger as the reactance gets smaller and how the circle gets smaller for larger reactances. One additional point should be made at this time. Notice in Figs. 7–5 and 7–6 that we have made complete circles for each reactance; that is, we started out on infinity (Fig. 7–5) or point A (Fig. 7–6) and extended the radius until it reached the other side of the circle. Now notice these same lines in Fig. 7–7. None of them go continuously from the reactance point to the infinity portion of the curve because the area to the right side of the chart would become very congested and would be impossible to read. This situation is the same as we referred to in our discussion of

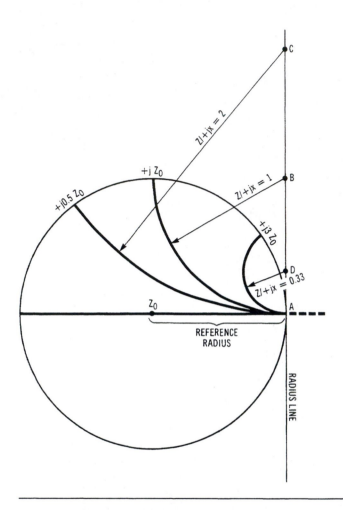

Figure 7–6. Construction of inductive reactance lines.

resistive circles. All those circles were not complete, in order to prevent congestion, and the reactance circles do their part in reducing clutter by not coming completely from one edge of the chart to the other.

Since the top of the Smith chart is inductive reactance, the lower portion is, logically, capacitive reactance. Fig. 7–8 shows six capacitive reactance circles. Notice, once again, that the reactance is expressed as a function of characteristic impedance (Z_0), which supports the idea of a *normalized* Smith chart. Fig. 7–9 shows the construction of these circles. They are the same as Fig. 7–6 for the inductive reactance except they are a negative value ($-jX$) instead of positive as before. Once again, we chose $j0.5Z_0$, jZ_0, and $j3Z_0$. The only difference is that the capacitive reactance has a $-j$ and the inductive reactance has a $+j$.

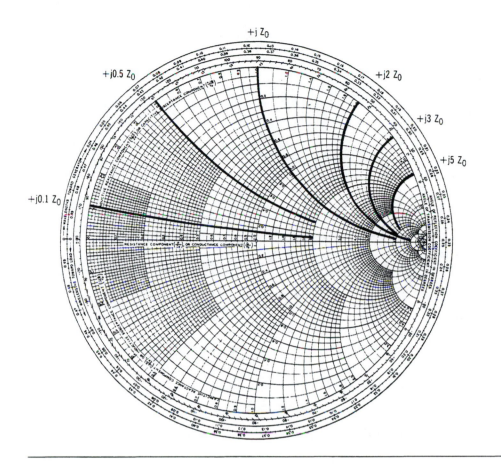

Figure 7–7.   Inductive reactance circles on the Smith chart.

The radii used are the same as the inductive case, with the exception of the direction used for the radii A to B, A to C, and A to D.

   Fig. 7–10 shows the three curves we constructed and the three additional curves from Fig. 7–8 and where they appear on the Smith chart. Notice once again how none of the curves completely finish in order to reduce the congestion on the high-impedance side of the chart.

7.2.3   Reflection Angles and Wavelengths

With the components of a complex microwave impedance (R $\pm$ jX) explained and their portion of the chart constructed, we can now get into some of the chart's additional features that make the impedances needed much easier to find. One such feature is the *angle of reflection coefficient*. Fig. 7–11 shows this section of the chart. Note that the chart starts

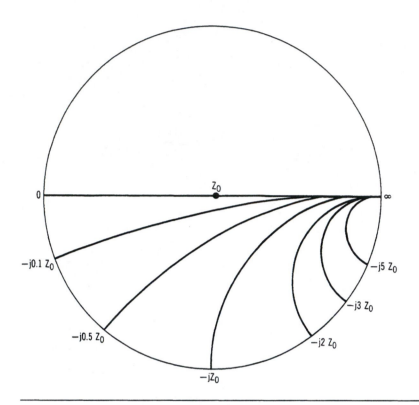

Figure 7–8. Capacitive reactance circles.

on the right side at $0°$ and goes counterclockwise to $180°$. It also starts at $0°$ and goes clockwise to $-180°$. This figure verifies the statement made in Chapter 3 that defined conditions, first of all, at a short circuit having an angle of reflection coefficient of $180°$. This is shown at the left of the chart as 0 ohms. Likewise, an open circuit has an angle of reflection coefficient of $0°$. This is at the right side, where the resistance line shows infinity. These angles are relative phase angles that range over $\pm 180°$. They are the phase angles of the reflected signal coming from a mismatched load with respect to the phase of the incident (or input) signal. They are relative angles because they compare one phase to another.

Between the two extremes ($0°$ and $180°$), there are various angles depending on the amplitude and characteristic (inductive or capacitive) of the load being considered. These angles are very important when using S-parameters for circuit designs. Each S-parameter has an amplitude and angle, and both are important in characterizing a parameter device. S-parameters will be covered in detail in Chapter 8.

One more area should be covered on the chart itself. This area consists of the two extreme outside scales labeled *wavelengths toward generator* and *wavelengths toward load*. These scales are shown in Fig. 7–12. Remember from Chapter 1 how we discussed wavelength and related it to frequency and the dielectric constant of the medium through which

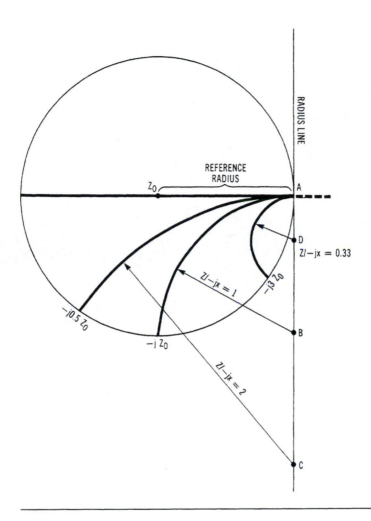

Figure 7–9. Construction of capacitive reactance lines.

it traveled. This calculation gave us an absolute number for wavelength. The numbers we refer to here on the Smith chart are those found one step after we calculate the absolute wavelength.

To illustrate this point, consider the following example.

EXAMPLE 7.2

We have a coaxial line with a short circuit at the end. The line has a Teflon® dielectric ($\epsilon = 2.1$) and a 5-GHz signal at the input. A stub is located 1.3 cm (0.511 inch) from the load (short circuit). We would like to know what fraction of a wavelength the stub is away from its load.

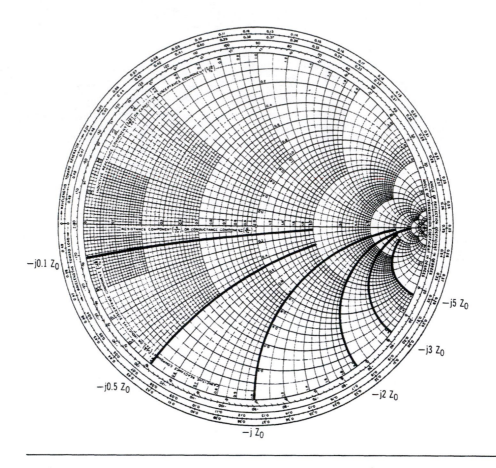

Figure 7–10.   Capacitive reactance curves on the Smith chart.

The first item we need to know is how long one wavelength is for our example. By using the wavelength (λ) formula we find the following:

$$\lambda = \frac{C}{f\sqrt{\epsilon\lambda}}$$

$$= \frac{3 \times 10^{10}\ \text{cm/sec}}{5 \times 10^{9}\ \sqrt{2.1}}$$

$$= \frac{3 \times 10^{10}}{5 \times 10^{9}\ (1.45)}$$

$$= \frac{3 \times 10^{10}}{7.24 \times 10^{9}}$$

$$= 4.14\ \text{cm (1.63 inches)}$$

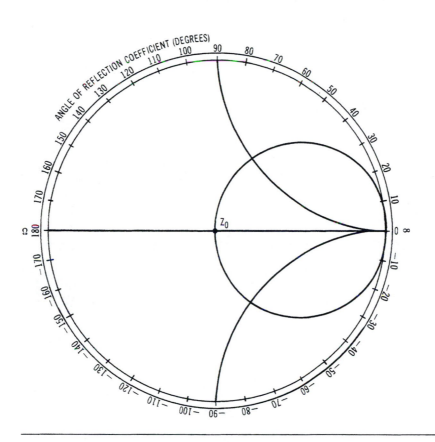

Figure 7–11.   Angle of reflection coefficient in degrees.

The stub in our example is 1.3 cm (0.511 inch) back from the load. To find how many wavelengths that is, we divide 1.3/4.14 and get 0.314λ distance. With this information, we can now use the chart. First, we must have a starting point. That point is the load and is a short circuit in our case. Normally, this load will be some value of complex impedance, but since our explanation is only on the wavelength section of the chart and no resistive or reactive lines are involved, we will use the convenient short. Examples will be presented later in this chapter that will show complex impedances and how the wavelength portion of the chart is used with them.

Since our starting point is a short (0 ohms), we begin at point A in Fig. 7–12. This, you will recall, is the 0 ohms point on the resistive line of the Smith chart. The stub we have on our line is 1.3 cm from the load, or toward the generator. We, therefore, use the outside scales and move clockwise around the chart. We move around until we find 0.314 on the scale, which is point B in the figure and indicates what the impedance at the stub would be if we had a full Smith chart presented.   ♦

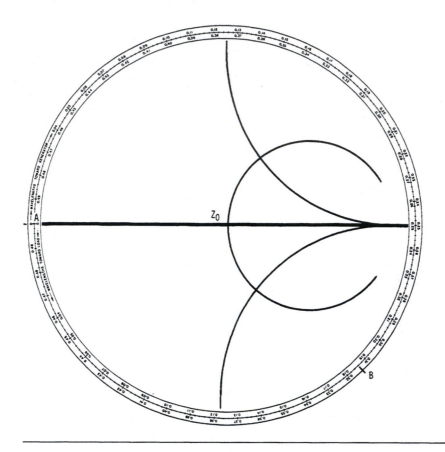

Figure 7–12.   Wavelength charts.

Notice from these scales how easy it will be to have an impedance at one point and be able to find out what another point is doing some distance away. This distance may be either toward the generator or the load. The wavelength portion of the Smith chart is really the most representative of what line lengths mean to microwave circuits and systems.

7.3   **Smith Chart Evolution**

We have now "built" the basic Smith chart as you see it today, with the exception of the radially scaled parameters that will be covered later. You should be starting to comprehend the wealth of information contained in the chart. The original chart developed by Phillip Smith in 1939 was not as complete as the one we presented in Fig. 7–1. This original chart is shown in Fig. 7–13.

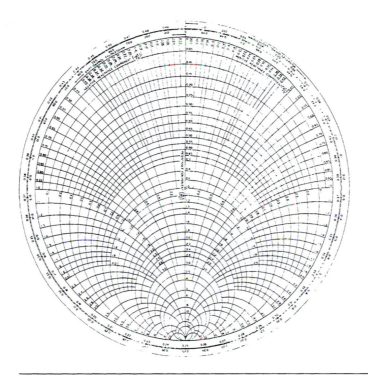

Figure 7–13.   Original Smith chart (1939).

There are certain differences between the original Smith chart (called a *transmission line calculator*) in 1939 and the one we are familiar with today:

- The original chart was rotated 90° clockwise from the chart of today, which puts the real resistance axis vertical as opposed to horizontal.
- No angle of reflection coefficient was shown on the first chart.
- Some of the resistance circles on the original chart were complete, and some were not.
- Both the resistance and reactance circles went only to a value of $20Z_0$.
- Many more points on both the resistive and reactive curves were marked (0.5, 0.55, 0.6, etc.).

In January, 1944, *Electronics* published an article titled "An Improved Transmission Line Calculator." The chart presented at that time resembles very closely the chart we are familiar with today. This chart is shown in Fig. 7–14. Some obvious changes on this chart are

- The angle of the reflection coefficient has been added.
- The excess numbers have been removed to reduce clutter.
- The resistance and reactance circles are continuous, as they are in the present chart.

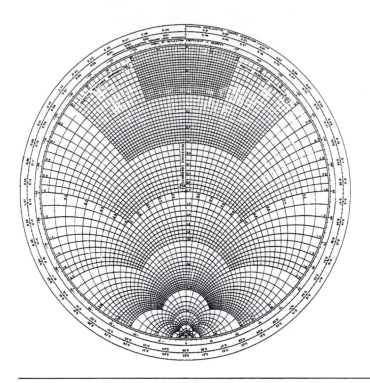

Figure 7–14.   Improved Smith chart (1944).

(You will note, however, that the chart is still oriented as the original one was—with the real resistance axis being vertical. It was not until the late 1940s that the chart was turned and appears as it does today.)

7.3.1 Radially Scaled Parameters

One area on the Smith chart that is often overlooked is at the bottom of the page and is termed *radially scaled parameters*. Actually, the Smith chart (or transmission line calculator as it was called) has had radially scaled parameters since it first appeared in 1939. There were not anywhere near the number that we presently have, but there was still a scale that had a set of numbers for a term ρ that was representative of 1/VSWR on it. This scale was referred to as an arm, pivoted at the center of the chart, and had a transparent slide that could be moved to the proper values.

7.3.2 Arm and Scales

Fig. 7–15 shows this "arm" as it appeared in the original article in 1939. The lower scale is the ρ scale and, as mentioned, represented 1/VSWR. The arm was attached to the center of

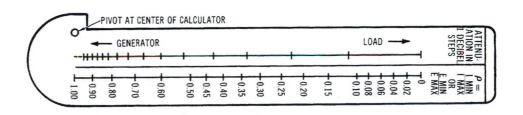

Figure 7–15.   Parameter arm for original Smith chart.

the chart, rotated to the known impedance, and a value of ρ determined from the scale. The transparent slide was placed at the value of ρ. The arm was then moved to a new area of the chart, depending on the fraction of a wavelength away from the impedance started with. The new impedance was then read off the chart, determined by the slide marker set previously. The inside scale on the arm was used for transmission lines with losses up to 15 dB. This scale was used by setting up an impedance at one end of a line and setting its transparent slide line on that impedance. This point becomes 0-dB attenuation. The arm was then moved to its impedance at the other end of the line and the slide moved to this setting. The difference was read off the scale to tell how many decibels of loss were in the transmission line. No numbers are on this scale. The readings are all relative. You can use any mark as a starting point, and the scale is set up in 1-dB steps.

The 1944 chart not only improved the original chart but also its arm. This arm is shown in Fig. 7–16. Notice that there are many more scales on this arm than the original arm of 1939. The two scales carried over from the original arm are the attenuation in 1-dB steps and the SWR loss coefficient. Additional scales are voltage (or current) ratios and decibels, voltage or current minimum and maximum values, and reflection characteristics that are the reflection loss in decibels and the value of the reflection coefficient. The greater capability of this arrangement as opposed to the original setup is noticeable. This arm was, once again, attached to the center of the chart and rotated as before.

The one problem with this arrangement was that the arm was attached to the chart and would sometimes get in the way. The solution to this problem is what we see today at the bottom of our present Smith charts. Basically, today's arm is the same arm used in the 1944 charts unfolded; that is, the four scales on the left were rotated until all eight scales were in a line. Then some changes were made in the scales, and the lower portion of Fig. 7–17 was a reality. A wealth of information is available on these scales. They are out of the way enough to allow use of the chart separately but are readily accessible when needed for use. Fig. 7–18 shows the parameter scales by themselves. To understand these scales we will discuss the headings and then present a representative example to show how each parameter is read from the scales. We will cover the following terms:

- Reflection coefficient (voltage and power)
- Return loss
- Transmission loss

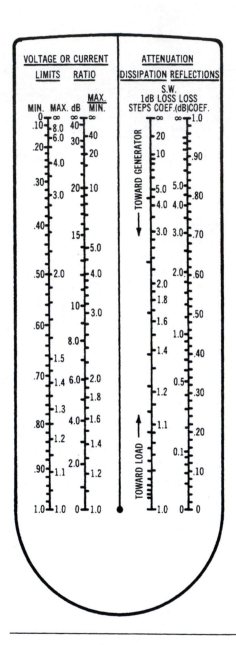

Figure 7–16. Improved parameter arm.

302

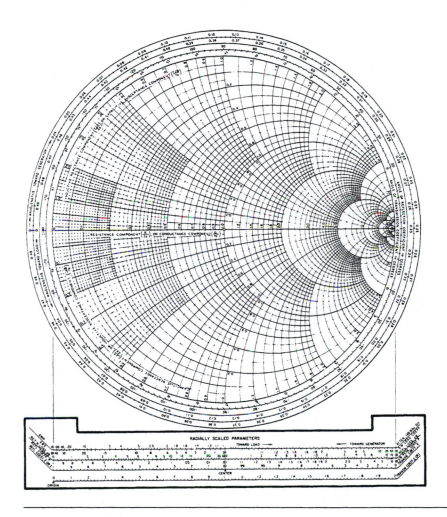

Figure 7–17.   Radially scaled parameters on the Smith chart.

- Standing wave ratio
- Mismatch loss

Reflection Coefficient   The reflection coefficient, read on the fifth scale down on the left in Fig. 7–18, is a ratio of reflected signal to incident signal—that is, E_r/E_i for voltage and P_r/P_i for power. The scales on the radially scaled parameters indicate both coefficients for the voltage and power. Consider point A in Fig. 7–18; on the voltage scale we read 0.5, which is the value of ρ. The power reflection coefficient, r^2, should be 0.25 and is read as that value on the scale above. Any value of voltage reflection coefficient can be converted to a power by simply reading the scale above.

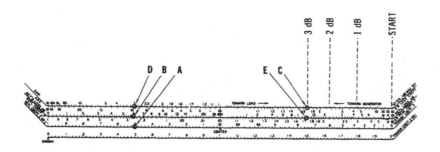

Figure 7–18.   Radially scaled parameters.

Return Loss   The return loss scale is also on the left side of Fig. 7–18. Return loss is the ratio of incident power (P_i) to reflected power (P_r) at a point on a transmission line and is expressed in decibels. The return loss is actually a measure of how well the transmission line is matched to the load it is driving. With a good match, the majority of the power will be absorbed into the load, and very little will be reflected back (or returned). Our point A on the reflection coefficient scale translates up two scales and becomes 6 dB of return loss. This amount can be confirmed by substituting into the following equation:

$$\text{Return loss} = 20 \log \rho \qquad (7.3)$$
$$= 20 \log (0.5)$$
$$= 20 \, (-3.01)$$
$$= -6.02 \text{ dB}$$

(The negative sign indicates a loss.) This 6-dB figure is shown at point B.

Transmission Loss   Transmission loss is the equivalent *one-way loss* in a signal on a transmission line. Transmission loss is in contrast to the return loss, which is a *two-way loss*. If, for example, a 3-dB attenuator were put in a line, a 6-dB difference in return loss and only 3-dB difference in transmission loss would be noted. This idea of the transmission loss being one half the return loss can be verified by using dividers and marking off return loss on the appropriate scale, point B, and then measuring off the same distance on the *transmission loss scale* in the upper right portion of the parameter scales (point C). Notice that 6 dB is read at point B, and 3 dB is read on the 1-dB *steps* scale. The amount, 3 dB, is found by starting at the right end of the scale and counting marks on the scale while moving to the left *(toward generator)* as shown in Fig. 7–18. The *transmission loss coefficient* is calculated by using the *power reflection coefficient* ($\rho^2 = 0.25$, in our case) and using this relationship:

$$\text{Transmission coefficient} = \frac{1 + \rho^2}{1 - \rho^2} \qquad (7.4)$$

$$= \frac{1 + 0.25}{1 - 0.25}$$

$$= \frac{1.25}{0.75}$$

$$= 1.667$$

This number is verified in Fig. 7–18 on the loss coefficient scale at point C on the top right scale.

Standing-Wave Ratio   The standing-wave ratio (SWR) (in decibels) is found by the relationship

$$20 \log \text{VSWR} \tag{7.5}$$

where, VSWR is the voltage standing-wave ratio.

The VSWR is found by taking dividers and placing one end at the center and the other end at the impedance being measured. Then place the dividers on the *standing-wave chart* in the upper left portion of Fig. 7–18. Our 3:1 VSWR, corresponding to a $\rho = 0.5$ at point A, is shown at point D. You can see that the SWR in decibels is 9.5 dB on the dBs scale. This number is verified by substituting 3 into the SWR equation.

$$\begin{aligned} \text{SWR (dB)} &= 20 \log \text{VSWR} \\ &= 20 \log 3 \\ &= 20\,(0.477) \\ &= 9.54 \text{ dB} \end{aligned}$$

Mismatch Loss   The mismatch loss is a measure of the loss caused by reflection and is the ratio of incident power to the difference between incident and reflected power, expressed in decibels as follows:

$$\text{Mismatch loss (dB)} = 10 \log \frac{P_i}{P_i - P_r} \tag{7.6}$$

$$= 10 \log \frac{E_i^2}{E_i^2 - E_r^2} \tag{7.7}$$

where, P_i is the incident power,
$\quad\quad\quad$ P_r is the reflected power,
$\quad\quad\quad$ E_i is the incident voltage,
$\quad\quad\quad$ E_r is the reflected voltage.

As an example, a VSWR of 3.0 represents a mismatch loss of 1.25 dB. This is shown at point E in Fig. 7–18 (on the reflection loss in dB scale on the right).

7.4   Smith Chart Applications

At this point the entire Smith chart has been built and explained. The next step will be to illustrate examples of how to use it.

7.4.1   Expanded-Scale Smith Chart

Before we get into examples, however, we should mention one special-case chart—that is, the *expanded-scale Smith chart*. There are times when fine-grain low VSWR readings must be taken. You can readily see that at the center of the standard chart it would be difficult to distinguish the difference between a 1.15 and a 1.17 VSWR, for example. For this reason, the center of the chart is expanded. Fig. 7–19A shows the most common expanded chart in relation to the full Smith chart. Fig. 7–19B shows the chart itself. This chart is one with 1.59:1 full scale. In other words, the center circle and its associated radially scaled

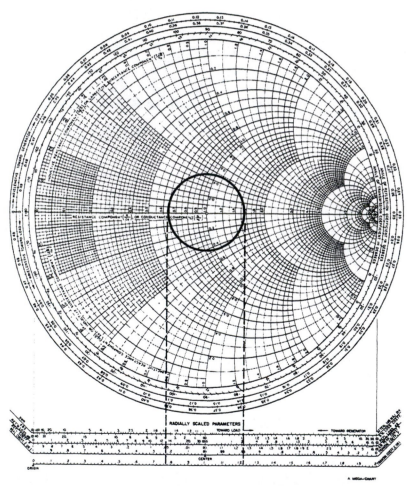

(A) Full-sized Smith chart showing section to be expanded.

Figure 7–19.   Expanded chart in relation to the full Smith chart.

parameters are displayed in the same area as a full chart normally is. The expanded-scale chart is like putting the center of the chart under a microscope and blowing it up. The one section of the chart that always remains the same is the outside circles— that is, wavelength toward the generator, wavelength toward the load, and angle of reflection coefficient in degrees. These sections remain the same no matter how far the center of the chart is expanded.

For even finer grain measurements, an expanded chart is available that has an outside VSWR of 1.12:1, which is about the limit for the expanded-scale Smith chart.

The best way to illustrate the value of the Smith chart is to show some of its uses. Now that we have covered the chart and its variations, we will present applications.

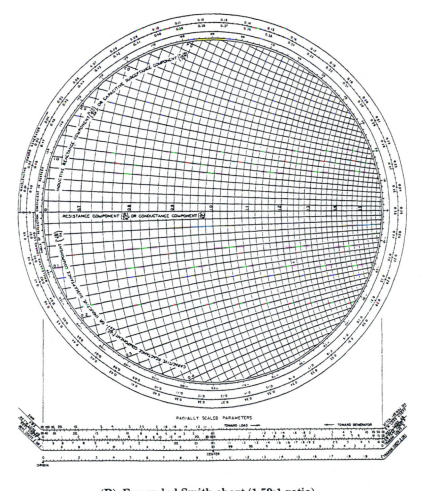

(B) Expanded Smith chart (1.59:1 ratio).

Figure 7–19. *(continued.)*

7.4.2   Applications Example I

In this example, we will prove a statement made at the beginning of this chapter. We said that a wealth of information was available from the Smith chart. To show that this statement is true, we will give a single complex impedance, and from that we will find eight different parameters that characterize the transmission line we are using. (All values are normalized.)

Given,

$$Z_L = 0.8 + j0.85 \; \Omega$$

Find,

VSWR

VSWR (dB)

Angle of reflection coefficient

Reflection coefficient

Reference position

Return loss

Mismatch loss

Admittance

Fig. 7–20 shows the solution to this problem. First, the impedance (Z_L) is plotted on the chart. To plot Z_L, first find the 0.8-ohm resistance circle on the chart. Follow the circle in a clockwise direction, since the reactance is +j, until the 0.85 reactance circle is intersected. This intersection is shown in Fig. 7–20 and designated as Z_L.

With the impedance plotted, our first parameter is VSWR. This value can be found in one of two ways. First, take a compass, put the point at the center of the chart, put the pencil point at Z_L, and draw a circle as shown in Fig. 7–20. This circle is called a *VSWR circle*. When the VSWR circle intersects the real resistance circle on the right side (point A), the VSWR can be read directly. The VSWR is read as 2.5 in our example. Also, the VSWR may be found by using the radially scaled parameters. On the upper-left scale, put the point of the compass on the center of the scale and mark off the same radius as used for the VSWR circle. The VSWR is read directly off the scale (point B = 2.5), and the VSWR in decibels can be read at the same time (8.0 dB).

The reflection coefficient and angle of reflection coefficient are the next to be found. The reflection coefficient is read on the fifth scale down on the left of the radially scaled parameters. It is found by measuring from the center of the chart to Z_L and marking off that distance on the scale. The reflection coefficient is 0.43, as indicated by point C in Fig. 7–20. The angle of this reflection coefficient is read directly off the chart as +78° (point D).

The reference position of Z_L is shown at point E in Fig. 7–20 and is just that: a reference to be used if any further information is needed for future calculations at a different spot in the transmission line. The reference position is read as 0.142 λ.

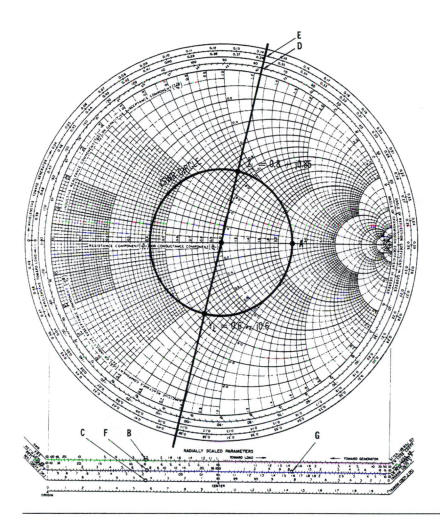

Figure 7–20. Applications example 1.

Return loss and *mismatch loss* are shown at points F and G in Fig. 7–20 on the radially scaled parameters. The return loss is read as 7.5 dB (which corresponds to a 2.5:1 VSWR), and the mismatch loss is 0.9 dB.

The final parameter to be found in our example is *admittance*. Remember from your earlier learning of electronics theory that admittance is found by taking the reciprocal of the impedance. Nowhere is this more graphically represented than on the Smith chart. To find the admittance, simply draw a line from the impedance, Z_L, through the center of the chart and right on through until the VSWR circle is crossed on the other side. Where the line intersects the VSWR circle is where the corresponding admittance is. In our case it is $0.6 - j0.6$ mhos.

7.4.3   Applications Example 2

Given,

$$Z_L = 1.6 + j1.2 \ \Omega$$

Find,

Input impedance (Z_{in}) of a line 4 cm long, the VSWR, and the output admittance (Y_L). Frequency = 2 GHz; dielectric constant of the line (ϵ) = 2.1.

The first order of business is to plot Z_L and obtain a starting reference point. Fig. 7–21 shows the point. The 1.6-ohm real-resistance circle is followed until it intersects the

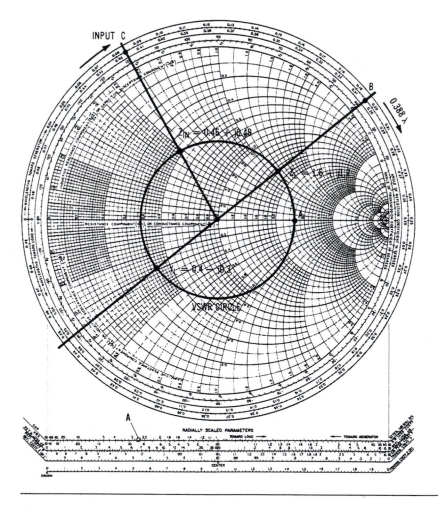

Figure 7–21.   Applications example 2.

1.2-ohm reactance scale shown as Z_L in Fig. 7–21. Now draw a line using the center of the chart and the point Z_L to locate a reference wavelength point. This line is drawn completely through the chart, as shown in the figure. The next step is to draw the VSWR circle. Once again, place the point of the compass at the chart center and the pencil point at Z_L. Draw the complete circle on the chart. At the same time use the VSWR radial chart and mark off the radius. The VSWR can now be read both on the radial chart and the main Smith chart as 2.75:1 (point A in both cases).

We now need to find out how far back we have to move to get to the input. We have already established our starting point (Z_L) by extending the line through to the wavelength scale. This point reads 0.196 λ (point B). To find how far we must move, we need the wavelength figure for 2 GHz with a dielectric constant (ε) of 2.1.

This equation is as follows:

$$\lambda = \frac{C}{f\sqrt{\epsilon}}$$

$$= \frac{3 \times 10^{10} \text{ cm/sec}}{2 \times 10^9 \text{ Hz } \sqrt{2.1}}$$

$$= \frac{3 \times 10^{10}}{2 \times 10^9 \ (1.45)}$$

$$= \frac{3 \times 10^{10}}{2.9 \times 10^9}$$

$$= 10.3 \text{ cm}$$

This figure is one full wavelength, and our input is 4 cm of this total wavelength. We, therefore, must divide the 4 by 10.3, and the result is that the input is 0.388 λ back from the load (toward the generator). We must now move toward the generator (clockwise direction) 0.388 λ. We are starting at 0.196 λ, so our final destination is 0.388 + 0.196, or 0.584 λ on the wavelength scale. Since the numbers stop at 0.5 λ on the scale, we need to move an additional 0.084 λ toward the generator. This point is at point C and represents the input 0.388 λ from the load. We can now read Z_{in} input as 0.46 + j0.49 ohms directly off the chart.

The load admittance is obtained exactly as before, by moving directly across the circle and reading it off the chart. The value is 0.4 − j0.3 mhos.

7.4.4   Applications Example 3

Given,

$$Z_L = 70 + j40 \ \Omega \text{ (50-ohm system)}$$

Find,

Matching circuit to match Z_L to 50 ohms.

The first step in this example is to normalize the load impedance (Z_L). Since this example is a 50-ohm system, we divide each component by 50 and obtain $Z_L = 1.4 + j0.8$ ohms, which is now plotted in Fig. 7–22. The line from the center of the chart through Z_L is extended completely through the chart to be used as a reference point. The ultimate goal of a matching network is to eliminate any of the reactive component of an impedance. With this thought in mind, we could add $-j0.8$ (40 ohms of capacitive reactance) to the load. This addition would result in Z_L moving down the 1.4-ohm real-resistance circle to point A. This can be done since we are changing only the reactance component and not doing anything to the resistive part. Thus, we move around on the resistance circle only.

When we get to point A, we will read an impedance of $1.4 + j0$, or 70 ohms. This has accomplished the goal of making the impedance purely resistive but has not matched the load to 50 ohms. To match the load to 50 ohms, we must end up with an impedance of $1 + j0$ ohms. The reference point referred to previously now comes into play. This point is point B

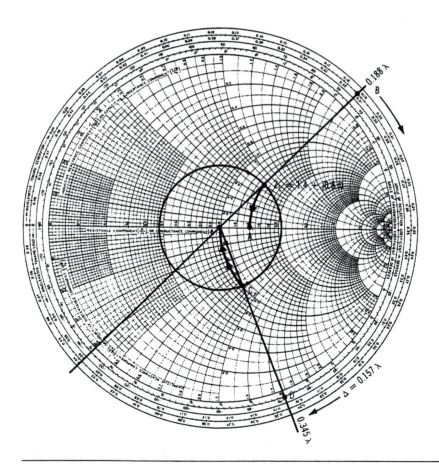

Figure 7–22.   Applications example 3.

in Fig. 7–22 and is 0.188 λ. By moving clockwise (toward the generator), the R = 1.0 (50-ohm) circle will be intersected, which is shown as point C in Fig. 7–22 and indicates an impedance of 1.0 − j0.80 ohms. The location of point C is found by subtracting 0.188 λ (point B) from 0.345 λ (point D). The resultant is 0.157 λ. Thus, we are saying that if a series reactance of +j0.8 (+40 ohms) is added, 0.157 λ from the load, a perfect match of the load and transmission line (1 + j0) will be the result. Once again, we accomplished this by moving directly from point C along the R = 1.0 circle because we are operating only with the reactance and leaving the resistance portion of the circle untouched.

7.4.5 Applications Example 4

Given,

$$Z_L = 60 + j50 \ \Omega \ \text{(50-ohm system)}$$

Find,

Impedance-matching circuit using a single stub transformer (short circuit). A single stub transformer is a specific length of transmission line placed in parallel with the existing transmission line to provide impedance matching. The stub can be either short circuited or open circuited on the opposite end. This is an excellent method to use to match a complex impedance over a narrow band of frequencies.

To solve this problem we must once again convert Z_L to a normalized impedance by dividing each component by 50. The result is that $Z_L = 1.2 + j1.0$ ohm, which is shown in Fig. 7–23. To operate easily with shunt stubs we must use admittance rather than impedance. With this approach we are able to add susceptances directly to a line rather than using complex parallel impedance relationships. This change from impedance to admittance, remember, can be accomplished simply by moving to the other side of the VSWR circle on the Smith chart. This point is shown as point A in Fig. 7–23 ($Y_L = 0.48 − j0.4$). For a match we still are shooting for a 1 + j0 impedance, which is also a 1 + j0 admittance.

We must move from point A (Y_L) to the unity circle to accomplish our first operation in stub matching. The starting point is where Y_L intersects the wavelength chart at 0.426 λ. We will move from the load admittance (Y_L) point clockwise (toward the generator) until we intersect the unity circle, which is point C. This point reads 0.159 λ, and the total distance moved is 0.233 λ. This distance, d_1, is shown at the bottom of Fig. 7–23. Two distances need to be calculated: the distance of the stub from the load (d_1) and the length of the shorted stub (d_2). We have just found the distance (d_1), away from the load where the shorted stub will be attached.

The stub length must now be determined. Our starting point for this operation is point C. Remember that we went to the unity circle in order to have an admittance of 1. What the stub must do is cancel out the +j0.95 susceptance so that our admittance will end up as 1 + j0. If we have +j0.95 susceptance, we need a stub that exhibits a −j0.95 value. To achieve this stub value, we move point C around the chart until we reach the inductive susceptance point on the unity circle corresponding to point C, which is shown to be point D. To obtain a pure susceptance, we move down to the susceptance line, as shown, to

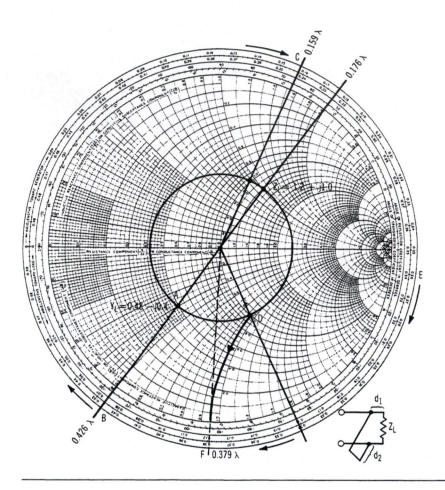

Figure 7–23.   Applications example 4.

point F. This point, 0.379 λ, is where only pure susceptance (−j0.95) will be realized to determine the length of the stub. To accomplish this we must start at the short-circuit end. A short circuit in impedance is 0 ohms; a short in admittance is an open circuit. We, therefore, start from infinity, point E, and move clockwise to point F for a −j0.95 value. This distance is found as 0.379 λ − 0.250 λ = 0.129 λ.

So we have now created a stub transformer that will match a 60 + j50-ohm load impedance to 50 ohms. This transformer is a short-circuit stub that is 0.129 λ long and placed 0.233 λ from the load.

The previous four examples illustrate the Smith chart for a variety of applications. From simple impedance relationships, to return loss, to VSWR, to complex matching networks, all can be found if you remember the basic relationships of the chart and what each circle and line does.

7.5   Chapter Summary

We have stated more than once in this chapter that there is a wealth of information to be obtained from the Smith chart. By now we hope you are convinced that this is a true statement.

Also, you may realize now that the Smith chart is not the complex black magic instrument that it may have seemed to be in the beginning of the chapter. Our "building" process was designed to tear away any of that black-magic implication and show how very basic the chart can be.

The examples chosen to illustrate the applications are typical problems you may encounter in labs and be called upon to solve. With some careful bookkeeping and not losing track of where the impedances and admittances are, you should be able to solve any problem you encounter with microwave transmission lines.

In conclusion, we can say that Mr. Phillip H. Smith supplied us with a very valuable tool of the microwave trade. The microwave industry lost this gentleman in 1987. Even though he is no longer with us, every person who works in microwaves should thank him for his tremendous contribution.

7.6   References

1. Brown, R. G., Sharpe, R. A., and Hughes, W. *Lines, Waves, and Antennas*. New York: The Ronald Press, 1961.

2. Lance, A. L. *Introduction to Microwave Theory and Measurements*. New York: McGraw-Hill, 1964.

3. Laverghetta, T. S. *Microwave Measurements and Techniques*. Dedham, MA: Artech House, 1976.

4. Ragan, G. L. *Microwave Transmission Circuits*. New York: Dover Publications, Inc., 1965.

5. Smith, P. H. "Transmission Line Calculator." *Electronics*, January, 1939.

6. Smith, P. H. "An Improved Transmission Line Calculator." *Electronics*, January, 1944.

7. Smith, P. H. *Electronic Applications of the Smith Chart*. New York: McGraw-Hill, 1969.

Questions

7.2   Smith Chart Construction

1. What is the line across the center of the Smith chart called?
2. What is a normalized chart?
3. Where are pure inductance components indicated on the Smith chart?
4. Where are pure capacitance components indicated on the Smith chart?
5. Define *angle of reflection coefficient.*

6. Where is the angle of reflection coefficient indicated on the Smith chart?
7. Which direction are the *wavelength toward the generator* terms moved?
8. Which direction is the *wavelength toward the load* moved?

7.3   Smith Chart Evolution

9. What was the original Smith chart called in 1939?
10. What was the second version of the Smith chart called in 1944?
11. What are two differences between the original chart in 1939 and the present chart?
12. What are *radically scaled parameters*?
13. What is the difference between the voltage reflection coefficient and the power reflection coefficient?
14. Define *mismatch loss*.

7.4   Smith Chart Applications

15. Why are expanded-scale Smith charts used?
16. What is the outside VSWR of the highest expanded-scale Smith chart?
17. How are impedances normalized?
18. How is admittance found on a Smith chart?
19. Can an impedance be matched with a single transmission line? Explain.
20. Describe an open-circuit single-stub transformer.

Problems

7.2   Smith Chart Construction

1. Normalize and plot the following impedances on a Smith Chart ($Z_o = 60\Omega$):

    a. $55 + j33\Omega$ c. $5 + j126\Omega$
    b. $16 - j85\Omega$ d. $162 - j31\Omega$

2. For the following impedances, find the corresponding admittances ($Z_o = 50\Omega$):

    a. $62 - j62\Omega$ c. $4.2 - j6\Omega$
    b. $196 + j30\Omega$ d. $35 - j35\Omega$

3. Plot the following reflection coefficients:

    a. $0.3 < -51$ c. $0.15 < 43$
    b. $0.9 < 21$ d. $0.5 < 45$

7.3   Smith Chart Evolution

4. We have an impedance of $Z = 45 - j61$ ($Z_o = 50\Omega$). Using the radially scaled parameters on the Smith chart, find the VSWR, return loss, voltage reflection coefficient, and mismatch loss.
5. We have a reflection coefficient of 0.35 at an angle of +78 degrees. Plot this point on a Smith chart and use the radially scaled parameters to find the VSWR, return loss, and power reflection coefficient.

7.4   Smith Chart Applications

6. A load impedance is $Z_L = 75 + j60 \ \Omega$, with $Z_o = 62 \ \Omega$. Find Z_{IN} for a line 6.2 cm long when the frequency of operation is 4.2 GHz and $\epsilon_r = 2.32$.

7. For the transmission line in problem 6, find the load and input admittances.

8. For a certain transmission line ($Z_o = 50 \ \Omega$), $Z_L = 122 + j76 \ \Omega$. Any reactive load on this line will cause great problems. How far down the transmission line, in cm, must we move to have a real impedance available? What is this impedance value?

9. A reflection coefficient of 0.67 at an angle of 22 degrees is read at the end of a transmission line. What is the VSWR, return loss, and impedance 1.8 inches from the load? (frequency = 1.7 GHz, $\epsilon_r = 2.1$)

10. A load impedance of $Z_L = 80 + j71 \ \Omega$ is on a transmission line with $Z_o = 35 \ \Omega$. Find the dimensions (in inches and wavelengths) of a single open-circuit stub-matching circuit.

S-Parameters

Objective

To introduce S-parameters for use in microwave systems. Measurement of the parameters along with a design example where S-parameters are used is also presented.

Key Terms

Gain Bandwidth Product   S-Parameters
H-Parameters   Unity Gain Crossover
Incident Signal   Y-Parameters
Parameters   Z-Parameters
Reflected Signal

8.1   Introduction

To describe an item to someone requires that the same language and terminology be understood so that the characteristics of the device can be well defined. For example, such terms as engine size, horsepower, and miles per gallon would be used to explain a certain type of automobile; weight and composition would be used to define a bowling ball. Most items have particular terms to describe them, and these terms vary from item to item. These terms are called *parameters*. In electronics, parameters are used to describe specific

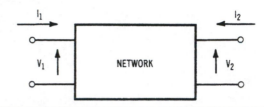

Figure 8–1. Two-terminal (two-port) pair device.

components. Components can be reduced to what are called two-terminal (or two-port) pair devices or, basically, the input and output of the component.

Fig. 8–1 is a two-terminal pair device showing input and output voltages and currents. This type of diagram is often referred to as a "black box" representation. Within this black box is a linear, active, bilateral network— a transistor, for example.

Because the external conditions (V_1, I_1, V_2, I_2) are measurable, the device in the box can be characterized. The most common parameters used for characterization are Z, Y, and h. The *Z-parameters* are called impedance parameters, the *Y-parameters* are admittance parameters, and the *h-parameters* are called hybrid. The parameters come from coefficients in the simultaneous equations that can be written relating all the measurable voltages and currents.

The box in Fig. 8–1 can be described, for example, by these equations:

$$V_1 = Z_{11} I_1 + Z_{12} I_2 \tag{8.1}$$

$$V_2 = Z_{21} I_1 + Z_{22} I_2 \tag{8.2}$$

These are the Z-parameter relationships. Similarly, the Y and h parameters can be represented:

$$I_1 = Y_{11} V_1 + Y_{12} V_2 \tag{8.3}$$

$$I_2 = Y_{21} V_1 + Y_{22} V_2 \tag{8.4}$$

and

$$V_1 = h_{11} I_1 + h_{12} V_2 \tag{8.5}$$

$$I_2 = h_{21} I_1 + h_{22} V_2 \tag{8.6}$$

For each of the equations, it is possible to draw an equivalent electrical circuit that can be considered to be the contents of the black box. An equivalent circuit using Z-parameters for Fig. 8–1 is shown in Fig. 8–2.

Each of the Z-parameters in Fig. 8–2 has the dimensions of impedances (ohms), thus justifying the name of *impedance parameter*. This justification may be difficult to see from the figure when noting that $Z_{12} I_2$ and $Z_{21} I_1$ are dependent voltage sources. If, however, I_2 were equal to zero by open circuiting the output terminals, then Z_{11} can be referred to as the input impedance with the output open circuited, or

$$Z_{11} = \frac{V_1}{I_1} \bigg| I_2 = 0 \tag{8.7}$$

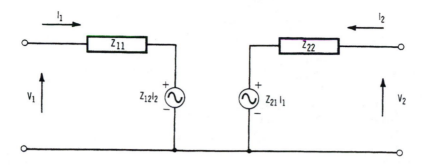

Figure 8–2.   Z-parameter equivalent circuit.

similarly for the output impedance with an open circuit input,

$$Z_{22} = \frac{V_2}{I_2} \Big| I_1 = 0 \tag{8.8}$$

also for the reverse transfer impedance with an open circuit input,

$$Z_{12} = \frac{V_1}{I_2} \Big| I_1 = 0 \tag{8.9}$$

and for the forward transfer impedance with an open circuit output,

$$Z_{21} = \frac{V_2}{I_1} \Big| I_2 = 0 \tag{8.10}$$

The Y-parameters, as previously mentioned, are admittance parameters and, as such, are dimensioned in siemens. The parameters are defined as follows for the input admittance with a short-circuit output:

$$Y_{11} = \frac{I_1}{V_1} \Big| V_2 = 0 \tag{8.11}$$

similarly for the forward transfer admittance with a short-circuit output,

$$Y_{21} = \frac{I_2}{V_1} \Big| V_2 = 0 \tag{8.12}$$

also for the reverse transfer admittance with a short-circuit input,

$$Y_{12} = \frac{I_1}{V_2} \Big| V_1 = 0 \tag{8.13}$$

and for the output admittance with a short-circuit input,

$$Y_{22} = \frac{I_2}{V_2} \Big| V_1 = 0 \tag{8.14}$$

The h-parameters are primarily used for low-frequency transistors and are dimensioned as impedance and ratios. The parameter h_{11} is the input impedance (ohms) with the output short circuited, h_{22} is the output admittance (siemens) with the input open circuited, h_{12} is the voltage feedback ratio with the input open circuited (dimensionless), and h_{21} is the current amplification with the output short circuited (dimensionless).

This combination of dimensions of the parameters (ohms, mhos, and dimensionless) justifies the name *hybrid* for these parameters.

The parameters are presented as follows with dimensions:

$$h_{11} = \frac{V_1}{I_1} \,\bigg|\, V_2 = 0 \qquad \text{(ohms)} \tag{8.15}$$

$$h_{12} = \frac{V_1}{V_2} \,\bigg|\, I_1 = 0 \qquad \text{(dimensionless)} \tag{8.16}$$

$$h_{21} = \frac{I_2}{I_1} \,\bigg|\, V_2 = 0 \qquad \text{(dimensionless)} \tag{8.17}$$

$$h_{22} = \frac{I_2}{V_2} \,\bigg|\, I_1 = 0 \qquad \text{(mhos)} \tag{8.18}$$

For a common-emitter configuration, the parameter h_{21} is termed β (beta). For a common-base mode, it is termed α (alpha). Both α and β are defined for short-circuit loading conditions.

The h-parameters were defined earlier to provide a basis for parameters used to characterize transistors. As previously mentioned, the parameters are used primarily for low-frequency transistors. By having values for h_{11}, h_{12}, h_{21}, and h_{22}, an engineer is able to have a transistor completely characterized and with this information can design an amplifier or oscillator circuit for a specific application.

All the parameters described are used in low-frequency or rf applications. They are used in this range because all of the parameters can be readily measured and characterized. When microwave frequencies are involved, the Z, Y, or h parameters cannot be used for a variety of reasons:

- Equipment is not readily available to measure voltages and currents at the input and output ports. Also, to obtain a specific voltage or current is difficult since the amplitudes are varying at different points in the line. Any variation in measuring point will cause a different reading.

- Short and open circuits are difficult to achieve over a broad band of frequencies. To characterize an open circuit is difficult in particular since capacitance between the center conductor and outer conductor of an open circuit must be accounted for. If this value has not been measured and printed on the open circuit, measurements will not be accurate.

- Active devices, such as transistors and tunnel diodes, are very often unstable when an open or short circuit is applied to them.

8.2   S-Parameter Definition

To have a set of parameters to be used at microwave frequencies, a unit must be chosen that is readily measurable. That unit is power. Whereas the inputs and outputs using Y, Z, and h parameters are expressed in voltage and current, the microwave parameters used have inputs and outputs expressed in power. These microwave parameters are called *S-parameters*.

S-parameters are transmission and reflection coefficients. The transmission coefficients are commonly called gains or attenuation, and the reflection coefficients are directly related to VSWRs and impedances. Using the convention that "a" is a signal into a port and "b" is a signal out, a relationship similar to those previously presented for Y, Z, and h parameters can be set up. Fig. 8–3 shows this relationship. It can be seen that "a_1" is the input signal, "b_1" is the signal reflected at the input, "b_2" is the output signal, and "a_2" is the output reflection. These designations are *incident* and *reflected signals* (*a* is incident; *b* is reflected) at a port. The diagram in Fig. 8–3 is not a case where *a* is on the center conductor and *b* is on the return; both signals are on the main transmission line. The only difference is that one is the input signal to a network (incident) and the other is the reflected portion of that signal. Thus, when we match an input or output port of a network to its characteristic impedance, the result is the one desired. To characterize the network's S-parameters, we set up zero reflections (b_1 or b_2 equal zero). If we express the input/output relations as we did previously, we would have

$$b_1 = S_{11} a_1 + S_{12} a_2 \qquad (8.19)$$

$$b_2 = S_{21} a_1 + S_{22} a_2 \qquad (8.20)$$

To solve for individual parameters, a_1 or a_2 must be made equal to zero; that is, the input signal terminated (a_1) or the output signal terminated (a_2 reflection = 0). With these conditions satisfied, the parameters are as follows:

$$S_{11} = \frac{b_1}{a_1} \bigg|\ a_2 = 0 \text{ (output terminated in } Z_0) \qquad (8.21)$$

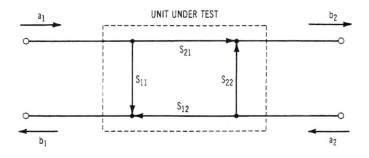

Figure 8–3.   S-parameter relationship.

$$S_{21} = \frac{b_2}{a_1} \bigg| \; a_2 = 0 \text{ (output terminated in } Z_0\text{)} \qquad (8.22)$$

$$S_{12} = \frac{b_1}{a_2} \bigg| \; a_1 = 0 \text{ (input terminated in } Z_0\text{)} \qquad (8.23)$$

$$S_{22} = \frac{b_2}{a_2} \bigg| \; a_1 = 0 \text{ (input terminated in } Z_0\text{)} \qquad (8.24)$$

From this relationship, we can place an identification on each parameter.

S_{11} is the input reflection coefficient with the output matched to Z_0,

S_{21} is the forward transmission gain or loss,

S_{12} is the reverse transmission (or isolation),

S_{22} is the output reflection coefficient with the input matched to Z_0.

To illustrate how these parameters are obtained, we can refer to Fig. 8–4 for S_{11} and S_{21} and Fig. 8–5 for S_{12} and S_{22}. Remember from Chapter 3 that when the load impedance (Z_L) equals the characteristic impedance of the transmission line (Z_0), the VSWR is 1.0:1 and the reflection coefficient (ρ) equals zero. This matched condition is one where there,

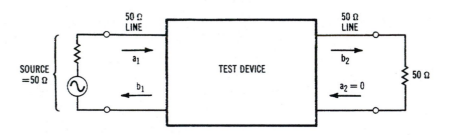

Figure 8–4. S_{11} and S_{21} test conditions.

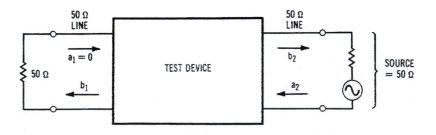

Figure 8–5. S_{12} and S_{22} test conditions.

theoretically, are no reflections. Therefore, by placing a termination equal to Z_0 at the output for S_{11} and S_{21} testing and at the input to get S_{12} and S_{22}, we have fulfilled the requirements that will make a_1 or a_2 zero. Note from the diagrams how the termination will result in the corresponding $a_2 = 0$ and $a_1 = 0$ to produce the needed parameter. In both cases, we have a 50-ohm transmission line terminated in 50 ohms. We, therefore, have our theoretically perfectly matched system we referred to earlier. Thus, a_2 will be zero in Fig. 8–4, and a_1 will be zero in Fig. 8–5.

There are many advantages to using S-parameters at microwave frequencies. The root of these advantages lies in the previously mentioned termination of the system in its characteristic impedance (Z_0). Some of these advantages are as follows:

- The termination is accurate at microwave frequencies. Remember that the Y, Z, and h parameters required opens and shorts. The lead inductance and capacitance of these components make them undesirable at microwave frequencies. The 50-ohm loads are very accurate at microwave frequencies.

- No tuning is required with a termination. An open or short requires an adjustment of electrical lengths to put it in its proper position. When a termination (Z_0) is placed at the end of a line, the device being tested will see the characteristic impedance, regardless of the length of the line.

- Broad-band swept-frequency measurements are possible since the device under test will remain terminated accurately over a broad frequency range. If an open or short is used, it will have to be retuned at every frequency or after a narrow band of frequencies is covered.

- When the device being tested is terminated in its characteristic impedance, it ensures that the device will remain stable. Some negative resistance devices (tunnel diodes, for example) have a tendency to oscillate under certain loading conditions. When they are terminated in Z_0, they will remain stable during the measurements.

8.3   S-Parameter Uses

With S-parameters defined and their advantages explained, let us now investigate where they are used. We will concentrate on one of the most widely used applications in microwaves—that of the microwave transistors.

The S-parameters of a microwave transistor can be measured either while the transistor is in its package or while it is a chip before being put into a package. Obviously, differences in the readings will occur because the package will have additional inductance and capacitance that the chip will not have.

To illustrate these differences, let us look at a transistor and its S-parameters both in chip and package form. The configuration used will be a common emitter, and all four parameters, S_{11}, S_{12}, S_{21}, and S_{22}, will be covered. Bias conditions for the device will be $V_{CB} = 15$ V and $I_C = 15$ mA (that is, a collector-to-base voltage of 15 V and a collector current of 15 mA). The bias conditions used are very important when measuring S-parameters. We will see later that the S-parameters can vary greatly with bias. Therefore, be sure that the S-parameters being used reflect the bias conditions that the device will experience in actual operation.

The first order of business, as with any test setup, is to have a test fixture for both the chip and packaged device. Be sure that the fixtures used are the proper ones for the device being tested. It is very important that the device fits correctly and that when the measurements are made, they are only of the transistor and not any of the lines in the fixture. Fig. 8–6 shows the S_{11} parameters for both the chip and packaged device.

Notice from the plot how the chip basically follows a constant resistance circle on the Smith chart. At lower frequencies (1 GHz) it displays capacitive reactance. As the frequency increases, it exhibits somewhat less resistance and begins to show inductive reactance as 10 GHz is approached. An equivalent circuit is shown in Fig. 8–7A. The resistance shown comes from the bulk resistivities of the transistor base regions plus any contact resistance present when contact is made between the device and the outside world.

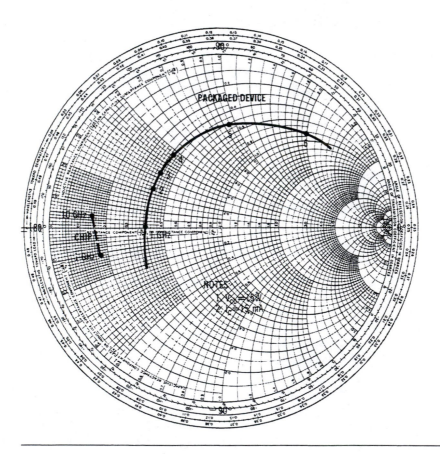

Figure 8–6. S_{11} for chip and package.

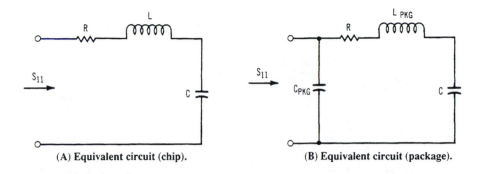

(A) Equivalent circuit (chip).   (B) Equivalent circuit (package).

Figure 8–7.   Equivalent input circuits for chip and packaged transistor.

The capacitance is due primarily to the base-emitter junction, and the inductance is the result of the emitter resistance being referred back to the input by a very complex β (transistor current gain) at high frequencies.

If we now look at the packaged device, we will note that it also starts out capacitive at the lower frequencies and moves very rapidly to the inductive region as the frequencies increase. Notice the change in curve shape as well as the spread-out curve appearance as compared to the chip.

The equivalent circuit of the packaged device is shown in Fig. 8–7B with the addition of package capacitance and inductance. These components are the cause of the curve in Fig. 8–6 moving inward toward the center of the chart and exhibiting more inductive reactance as compared to the chip. They also are the cause of the display being spread out more than when we looked at the chip.

Fig. 8–8 shows the S_{22} characteristics of the same chip and packaged device. Notice that the curves are entirely in the capacitive reactance region. If an admittance Smith chart were overlayed on the chip parameters, the chip parameters would follow very closely a constant conductance circle. The span of the response is determined by the capacitive elements, and the distance from the center of the chart is a function of the real elements R_s and R_o. These parameters are shown in Fig. 8–9A, which is an equivalent circuit of the chip.

The output reflection coefficient of the packaged device is shifted in, and the angle covered is extended. The equivalent circuit of Fig. 8–9B once again shows an increase in capacitance and the addition of inductance because of the package. The added inductance causes the S_{22} parameter to shift away from the previously followed constant conductance circle when the chip was used.

The forward transmission coefficient, S_{21}, is a gain in the case of our transistor but could also be a loss if a passive device were being used. Fig. 8–10 shows the values of S_{21} for both the chip and the packaged device. The values displayed are voltage gains. The packaged device exhibits slightly less gain than the chip, and the *unity gain crossover,* called f_s (to be defined later), is around 4 GHz as opposed to approximately 4.5 GHz for the chip device. An equivalent circuit representing S_{21} is shown in Fig. 8–11.

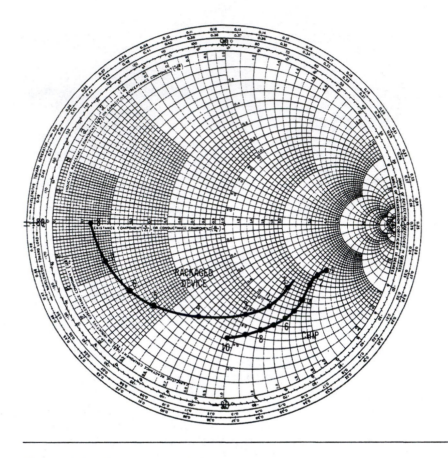

Figure 8–8. S_{22} characteristics.

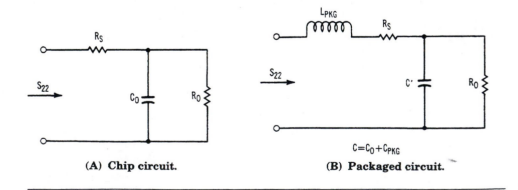

(A) Chip circuit.

(B) Packaged circuit.

$$C = C_0 + C_{PKG}$$

Figure 8–9. S_{22} equivalent circuit for chip and packages.

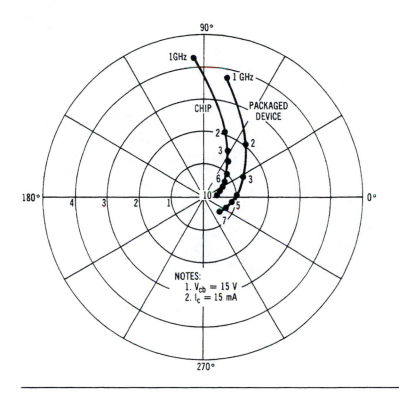

Figure 8–10. S_{21} parameters.

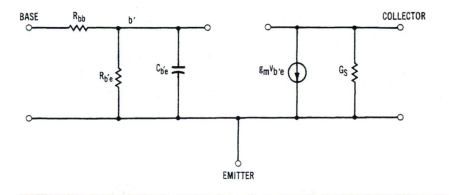

Figure 8–11. S_{21} equivalent circuit.

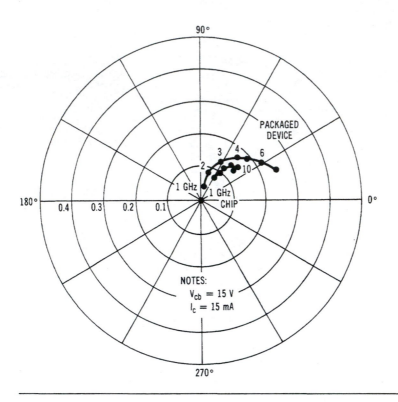

Figure 8–12. S_{12} diagram.

Fig. 8–12 shows the plots of the reverse transmission characteristic, S_{12}, which is, in effect, the reverse isolation figure for the transistor. Since a transistor is not a unilateral device—that is, it will not be an ideal conductor in one direction and provide complete isolation in the other direction—S_{12} will have some finite value. In Fig. 8–12 the S_{12} characteristics approximate a circular path. Also, notice how small and bunched the values of S_{12} are for the chip device. The circle opens up somewhat with the packaged device since the parameters of the package will cause the isolation to be decreased. Obviously, if at all possible, chips should be used in your circuits. They will improve overall performance greatly.

8.4   S-Parameters and the Transistor Data Sheet

Having covered S-parameters in an introductory way and shown plots of each of them, it is now appropriate to present a data sheet where they find wide applications—that is, the microwave transistor. The data sheet we will refer to is one of a commercially available device used in many microwave circuits. The numbers and plots shown are typical of transistors that are used throughout the microwave industry.

8.4.1   Additional Parameters

Before getting into the actual data sheet, some additional parameters are associated with the S-parameters that should be explained. These are f_t, f_s, and f_{max}.

- f_t is the frequency at which the short-circuit current gain of the device is equal to one. This term is also called the *gain bandwidth product*—that is, the product of the maximum gain of the device over the bandwidth it will operate (gain times operating bandwidth).

- f_s is the frequency where $|S_{21}|$ is equal to one or the power gain of the device; $|S_{21}|^2$ expressed in decibels is zero. Most of the time $|S_{21}|^2$ will appear on data sheets because this power gain figure is used by most designers for their amplifiers or oscillators.

- f_{max} is the frequency where the maximum available power gain, G_{max}, of the device is equal to one. F_{max} is also referred to as the maximum frequency of oscillation.

A chart showing the relationship of these three terms is shown in Fig. 8–13. The f_s of a transistor-connected common emitter is found by driving the base with a 50-ohm voltage source and terminating the collector at 50 ohms. A gain versus frequency plot is obtained that decays at the rate of approximately 6 dB/octave. Where the gain reaches 0 dB is f_s.

You can see from Fig. 8–13 that the curve that determines f_t is obtained by plotting $|h_{fe}|^2$. The term h_{fe} is referred to on microwave transistor data sheets as the forward

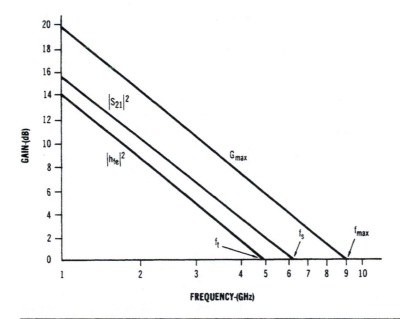

Figure 8–13.   Relationship of f_t, f_s, and f_{max}.

current transfer ratio and is applicable only to bipolar transistors. In the low-frequency applications where h-parameters are used, it is the short-circuit current amplification factor and is sometimes referred to as β. Regardless of its designation, it is always found by short circuiting the load of the device. At low frequencies this short circuiting is not much of a job. However, as we have said before, it is very difficult to obtain a true short at microwave frequencies. Remember that this difficulty was one of the reasons for going to S-parameters. Since $|h_{fe}|$ is determined by using a short circuit, we cannot receive direct data as with low-frequency devices. It can, however, be derived from measured S-parameter data, and in this example, it follows the $|S_{21}|^2$ curve very closely. This close correlation between the two is generally the case and results in f_t being slightly less than f_s.

To determine f_{max}, it is necessary to conjugately match the device being tested. This produces the maximum available power gain as a function of frequency. Note that this curve (f_{max}) is higher than the curve labeled $|S_{21}|^2$ because for G_{max}, we have perfectly matched the transistor; for the curve labeled $|S_{21}|^2$, the device is matched only for what is termed insertion power gain. Under matched conditions, the transistor is usable above f_s. In Fig. 8–13 it is usable, theoretically, at 9 GHz as opposed to an f_s of 6 GHz. In actuality, the device is most likely usable between 7.5 and 8 GHz.

8.4.2 Data Sheet Example I

Now let us consider an actual example. The data shown in Fig. 8–14 are for an actual commercially available bipolar transistor. You can see that the S-parameters are shown in different ways. They are listed in tabular form and also presented as plots on a Smith chart and polar display. Also, note that the parameters are given for different bias conditions. They range from a V_{CE} (collector-to-emitter voltage) of 6 V at a collector current (I_C) of 4 mA to a V_{CE} of 6 V and an I_C of 10 mA. Notice that the forward transmission gain (S_{21}) increases as the value of I_C increases. The reverse transmission (S_{12}), or isolation, also increases with an increase in I_C.

The input reflection coefficient (S_{11}) and output reflection coefficient (S_{22}) are both improved with the higher value of I_C. Note that the Smith chart plot and the polar plot are both for the $V_{CE} = 6$ V and $I_C = 10$ mA conditions. These conditions result in a display of the best input and output reflection coefficients and greatest gains and reverse isolation. Using conditions that result in the best data has become a common practice with transistor vendors.

Many times, however, the best input and output coefficients are not necessary, and the maximum gain is not needed. At times like these, a different bias condition is needed. Fig. 8–15 shows another section of the data sheet. This section is for the same device and shows many parameters listed for conditions of less collector current. Note in particular that the minimum noise figure is obtained with lower collector current. This can be understood when we consider that the majority of the noise generated within the transistor is due to current flowing through it, as we stated previously in Chapter 5. It, therefore, stands to reason that the less current we can draw from the collector supply, the lower will be the noise figure of the transistor.

Conversely, note that when f_t (the gain bandwidth product discussed earlier) and the maximum available gain, G_{max}, are involved, a larger amount of collector

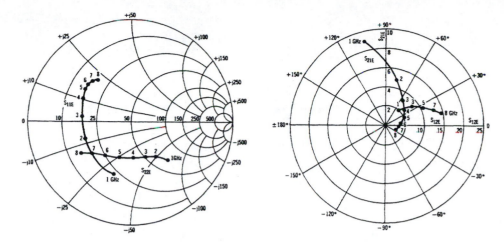

(A) Smith chart display of data.	(B) Polar display of data.

S-MAGN AND ANGLES:
V_{CE}=6 V, I_C=4 mA

FREQUENCY (MHz)	S11		S21		S12		S22	
1000	0.69	−83	7.03	115	0.07	45	0.74	−35
2000	0.51	−131	4.14	82	0.08	31	0.57	−35
3000	0.45	−165	2.97	60	0.10	26	0.50	−65
4000	0.47	166	2.26	40	0.11	18	0.45	−83
5000	0.52	150	1.80	21	0.12	13	0.44	−103
6000	0.53	143	1.48	6	0.12	9	0.48	−123
7000	0.53	138	1.23	−8	0.13	6	0.56	−139
8000	0.50	135	1.07	−19	0.14	4	0.62	−147

V_{CE}=6 V, I_C=6 mA

	S11		S21		S12		S22	
1000	0.61	−97	8.25	109	0.06	42	0.66	−40
2000	0.48	−144	4.68	79	0.08	32	0.49	−40
3000	0.45	−176	3.25	58	0.09	28	0.42	−67
4000	0.48	159	2.44	40	0.10	23	0.40	−84
5000	0.52	145	1.94	21	0.11	19	0.40	−104
6000	0.53	139	1.61	7	0.12	15	0.44	−124
7000	0.53	134	1.34	−7	0.13	11	0.52	−139
8000	0.49	131	1.17	−18	0.14	7	0.58	−147

V_{CE}=6 V, I_C=10 mA

	S11		S21		S12		S22	
1000	0.54	−113	9.34	103	0.05	41	0.58	−42
2000	0.46	−157	5.09	75	0.07	37	0.44	−53
3000	0.45	174	3.48	56	0.08	35	0.38	−66
4000	0.48	152	2.59	38	0.10	30	0.36	−83
5000	0.53	140	2.06	21	0.11	25	0.37	−104
6000	0.54	134	1.71	7	0.12	20	0.41	−124
7000	0.53	130	1.43	−6	0.13	15	0.49	−140
8000	0.49	127	1.25	−18	0.15	11	0.56	−147

(C) Chart with data.

Figure 8–14. S-parameter data.

SYMBOL	PARAMETERS AND CONDITIONS	UNITS	MIN	TYP	MAX
f_T	Gain Bandwidth Product at $V_{CE} = 6$ V. $I_C = 10$ mA	GHz		10	
$|S_{21E}|^2$	Insertion Power Gain at $V_{CE} = 6$ V. $I_C = 5$ mA. $f = 1.0$ GHz $f = 2.0$ GHz $f = 4.0$ GHz	dB dB dB		19 14 8.2	
NF_{min}	Minimum Noise Figure[1] at $V_{CE} = 6$ V. $I_C = 5$ mA. $f = 2.0$ GHz $f = 4.0$ GHz	dB dB		1.7 2.7	3.0
$M+1$	Noise Measurement[2] at $V_{CE} = 6$ V. $I_C = 5$ mA. $f = 4.0$ Ghz	dB			3.4
GNF	Power Gain at Optimum NF at $V_{CE} = 6$ V. $I_C = 5$ mA. $f = 2.0$ GHz $f = 4.0$ GHz	dB dB	8.0	13.0 8.5	
G_{MAX}	Maximum Available Gain[2] at $V_{CE} = 6$ V. $I_C = 10$ mA. $f = 2.0$ GHz $f = 4.0$ GHz $f = 6.0$ GHz	dB dB dB		17 12 9.0	
h_{FE}	Forward Current Gain at $V_{CE} = 6$ V. $I_C = 5$ mA		50	100	250

NOTES:

1. Input and output are tuned for optimum noise figures.
2. Maximum available gain (G_{MAX}) is calculated from the device S-Parameters using the equation.

$$G_{MAX} = |S_{21E}|^2 \times \frac{1}{|1 - S_{11}|^2} \times \frac{1}{|1 - S_{22}|^2}$$

$$(8.25)$$

Figure 8–15. Transistor parameters.

current is being drawn. This statement is not to say the absolute maximum current for the device is being used, but rather a much higher amount than would be used for low-noise characterization. The absolute maximum value of collector current is shown so that you will be aware of this value and stay well below it. In the data sheet shown, a 10-mA value is used as the highest value. This value is the highest at which the S-parameters are measured in this case. This value, however, does not mean that S-parameters could not be found for higher current values. However, to get too close to the maximum value of current is not a good idea because the device would be overstressed and the reliability of the device would be decreased. For this reason the 10-mA reading is a good rating to use as a maximum value for efficient, reliable operation.

The values of G_{max} given are for 2, 4, and 6 GHz. If a value is needed other than these, it can be calculated as indicated by note 2 in Fig. 8–15. The three S-parameters S_{21}, S_{11}, and S_{22} can be taken directly off the data sheet and plugged into the formula. To obtain the answer in decibels, remember to take the log and multiply by 10.

In Fig. 8–13 we showed a theoretical relationship of f_t, f_s, and f_{max}. To show that these are valid, we will construct the curves for the device just shown. Fig. 8–16 shows these curves. As indicated on the data sheet, f_t is at 10 GHz. The values for f_s are found by plotting $|S_{21}|^2$, which is slightly above f_t at 10.5 GHz. The parameter f_{max} is found by plotting G_{max} and ends up to be in excess of 17 GHz. The device, however, would never be used in this region because it would be useless to have an active device with 0-dB gain. It does, however, provide a point that characterizes the device and tells if any given problems could occur at the high frequencies that could cause oscillations. The plots of G_{max} and $|S_{21}|^2$ were made with points from the actual data sheet, as shown by the points in Fig. 8–16.

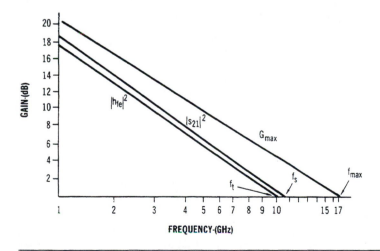

Figure 8–16.   The f_t, f_s, and f_{max} for a commercially available device.

Typical S- Parameters V_{CE} = 10 V, I_C = 3 mA

Freq. (MHz)	S_{11}		S_{21}			S_{12}			S_{22}	
	Mag.	Ang.	(dB)	Mag.	Ang.	(dB)	Mag.	Ang.	Mag.	Ang.
100	0.93	-11.5	16.2	6.46	168.0	-42.0	0.01	77.0	0.99	-4.0
200	0.89	-23.0	17.1	7.13	158.0	-37.0	0.01	77.0	0.97	-8.0
300	0.86	-34.0	16.4	6.58	149.0	-34.0	0.02	66.0	0.94	-12.0
400	0.83	-44.0	15.9	6.26	142.0	-32.0	0.03	60.0	0.92	-16.0
500	0.79	-54.0	15.6	6.02	135.0	-30.0	0.03	55.0	0.89	-19.0
600	0.75	-65.0	15.4	5.91	128.0	-29.0	0.04	51.0	0.87	-21.0
700	0.71	-73.0	15.0	5.62	121.0	-29.0	0.04	48.0	0.85	-24.0
800	0.68	-81.0	14.4	5.25	116.0	-28.0	0.04	45.0	0.84	-25.0
900	0.65	-91.0	14.0	4.99	111.0	-28.0	0.04	43.0	0.83	-27.0
1000	0.62	-97.0	13.5	4.72	106.0	-27.0	0.04	41.0	0.81	-28.0
1500	0.52	-129.0	11.4	3.71	84.0	-27.0	0.05	32.0	0.74	-35.0
2000	0.50	-151.0	9.3	2.93	69.0	-26.0	0.05	31.0	0.72	-43.0
2500	0.50	-169.0	7.8	2.45	55.0	-26.0	0.05	31.0	0.69	-51.0
3000	0.49	175.0	6.5	2.12	42.0	-26.0	0.06	33.0	0.68	-57.0
3500	0.54	165.0	5.4	1.87	29.0	-25.0	0.06	35.0	0.65	-68.0
4000	0.52	156.0	4.5	1.67	19.0	-24.0	0.06	37.0	0.68	-76.0
5000	0.53	140.0	2.6	1.35	-3.0	-23.0	0.08	35.0	0.71	-96.0
6000	0.48	120.0	0.9	1.11	-22.0	-21.0	0.09	34.0	0.73	-112.0

Figure 8–17.   S-parameter data.

8.4.3   Data Sheet Example 2

Fig. 8–17 shows how another manufacturer lists S-parameters. The main differences from Fig. 8–14 are that no Smith chart or polar plot is presented and that this manufacturer supplies S_{21} and S_{12} in both voltage (magnitude and angle) and in decibels. The value of these parameters in decibels is calculated by first taking the log of the parameter and multiplying it by 20. (It is multiplied by 20 instead of 10 because the S-parameters are expressed in voltage.) Expressed mathematically, the calculations are

$$S_{21} \text{ (dB)} = 20 \log_{10} |S_{21}| \qquad (8.26)$$

and

$$S_{12} \text{ (dB)} = 20 \log_{10} |S_{12}| \qquad (8.27)$$

To illustrate these relationships, consider the following example.

EXAMPLE 8.1

With the frequency of 2.0 GHz, the S-parameters are

$$S_{21} = 2.93 \text{ at } 69°$$

and

$$S_{12} = 0.05 \text{ at } 31°$$

To find the values in decibels, we will take the absolute value of the parameters and perform our operation on them.

$$S_{21} \text{ (dB)} = 20_{\log} \left| S_{21} \right|$$
$$= 20_{\log} (2.93)$$
$$= 20 \, (0.46)$$
$$= 9.34 \text{ dB}$$

and

$$S_{12} \text{ (dB)} = 20 \log \left| S_{12} \right|$$
$$= 20 \log (0.05)$$
$$= 20 \, (-1.30)$$
$$= -26 \text{ dB} \qquad \blacklozenge$$

You can see that these numbers match those in Fig. 8–17 for 2000 MHz, or 2.0 GHz. Note once again that the parameter in Fig. 8–17 also applies for a specific set of conditions. Remember that in Chapter 5, we emphasized the point of operating conditions many times. These conditions are of equal importance here. The conditions used were $V_{CE} = 10$ V and $I_C = 3$ mA. Considering the conditions is an important point to remember whenever dealing with S-parameters. It should become second nature to ask for test conditions whenever a manufacturer quotes S-parameters. Without these conditions, the S-parameters are almost totally useless.

8.5   S-Parameter Measurements

We have been looking at S-parameters, defining them, and finding how to use them throughout this chapter. An obvious question now arises—how do we measure them?

To understand the measurements to be taken, we must go back to a previous definition. Remember from Section 8.2 that the following relationships describe S-parameters:

$$b_1 = S_{11} \, a_1 + S_{12} \, a_2 \tag{8.28}$$

$$b_2 = S_{21} \, a_1 + S_{22} \, a_2 \tag{8.29}$$

These relationships are illustrated in Fig. 8–18. Further expansion of these relationships resulted in the individual parameters being defined (see page 338).

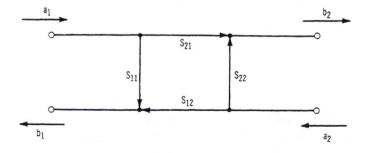

Figure 8–18.   S-parameter relationship.

$$S_{11} = \frac{b_1}{a_1} \Big| \; a_2 = 0 \qquad\qquad (8.30)$$

$$S_{21} = \frac{b_2}{a_1} \Big| \; a_2 = 0 \qquad\qquad (8.31)$$

$$S_{22} = \frac{b_2}{a_2} \Big| \; a_1 = 0 \qquad\qquad (8.32)$$

$$S_{12} = \frac{b_1}{a_2} \Big| \; a_1 = 0 \qquad\qquad (8.33)$$

To achieve the condition where $a_2 = 0$, a termination at the output of the device equal to the characteristic impedance, Z_0, of the system must exist. This value is 50 ohms in the typical microwave system. Conversely, to achieve the conditions where $a_1 = 0$, the input must be terminated in 50 ohms.

8.6   S-Parameters and Test Sets

In order to create the conditions described above, we must have equipment available that is capable of applying the appropriate signals and terminating the proper ports at the proper time. Fig. 8–19 shows the conditions that must be achieved. In Fig. 8–19a it can be seen that the signal is applied to the input of the device under test with the output port terminated in the system characteristic impedance (Z_o), with dual-directional couplers on either side of the unit being tested. If we now look at equations 8.28 and 8.29, we see that with this condition $a_2 = 0$. Thus, we can find the value of S_{11} and S_{21} from this setup since the equations now become:

$$b_1 = S_{11} \, a_1 \qquad\qquad (8.34)$$

and

$$b_2 = S_{21} \, a_1 \qquad\qquad (8.35)$$

Since the "a" terms represent the power entering a port and the "b" terms represent power leaving a port, we can relate these to Fig. 8–19a and the indicators on the dual-directional couplers marked A, B, C, and D. Since S_{11} is the ratio of b_1 to a_1, it can be seen that the measure of input power, a_1, is at point A and the measure of the reflected power, b_1, is at point B. So it stands to reason that we can write

$$S_{11} = B/A \qquad\qquad (8.36)$$

Similarly, S_{21} is the ratio of b_2 to a_1. The term b_2 can be measured as the power leaving the output, or that read at point C. We have already determined that a_1 was referred to point A, so we can write

$$S_{21} = C/A \qquad\qquad (8.37)$$

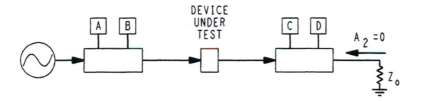

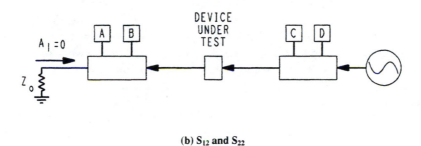

(a) S_{12} and S_{21}

(b) S_{12} and S_{22}

Figure 8–19.   S-parameter testing.

(Point D in Fig. 8–19a is not read since it measures the reflected power, a_2, which we have already determined to be zero.)

In a similar manner, we can measure the remaining parameters using the test setup in Fig. 8–19b. With this setup, the input is terminated and $a_1 = 0$. With this condition equations 8.28 and 8.29 become

$$b_1 = S_{12}\, a_2 \qquad\qquad\qquad (8.38)$$

and

$$b_2 = S_{22}\, a_2 \qquad\qquad\qquad (8.39)$$

Once again, when relating these equations to a test setup (Fig. 8–19b) we see ports A, B, C, and D and will express them in a form that defines S_{12} and S_{22}. Since S_{12} is the ratio of b_1 to a_2, it can be seen that b_1 is measured at point B and a_2 is measured at point D. So we can write

$$S_{12} = B/D \qquad\qquad\qquad (8.40)$$

In the same way, S_{22} is the ratio of b_2 to a_2, with b_2 being measured at point C and a_2 at point D. With this information we can write

$$S_{22} = C/D \qquad (8.41)$$

(In this case point A in Figure 8–19b is not used since it measures reflected power, a_1, which has already been determined to be equal to zero.)

The test setups shown in Figure 8–19 are representative of the basic components and procedures that would be used to measure S-parameters. Equipment is available that combines the dual-directional couplers, terminations, and a series of switches that accomplish all of the steps we previously went through. Such pieces of equipment are S-parameter test sets. The test sets generally operate from 45 MHz to 50 GHz, depending on which model is chosen. Typical individual unit ranges are 45 MHz to 20 GHz, 45 MHz to 26.5 GHz, 45 MHz to 50 GHz, 2 GHz to 20 GHz, and 45 MHz to 2 GHz. These units are designed to work with a vector network analyzer (to be covered in Chapter 9). A test setup showing the S-parameter test set in conjunction with a vector network analyzer is shown in Fig. 8–20. In this setup two conditions will occur. First, the signal will originate from Port 1 on the test setup, with Port 2 terminated in the system characteristic impedance (Z_o). This is the condition where $a_2 = 0$. Measurements for this condition will be S_{11} and S_{21}. The second case is where the signal originates from Port 2 and Port 1 is terminated in Z_o. This is the condition where $a_1 = 0$. Measurements are now made for S_{12} and S_{22}. All of these functions and switching operations are taken care of by the network analyzer. All the operator need do is tell the analyzer which parameters are desired.

Typical data taken with this test procedure are shown in Fig. 8–21. Tabular information is shown in Table 8–1. These data are shown in polar plots of S_{11}, S_{22}, and S_{21}. The values

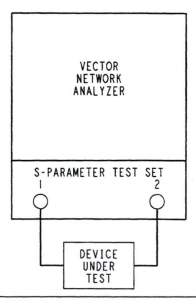

Figure 8–20.   S-parameter test setup.

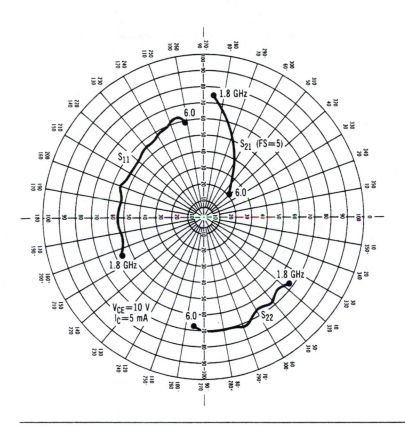

Figure 8–21.   Measured S-parameter data.

of S_{12} are shown only in the table since they are small values and would clutter up the plot as it stands. As we noted in earlier sections, S_{11} and S_{22} are usually on one chart and S_{21} and S_{12} are shown on a separate chart. This division of the data makes each parameter much easier to read and distinguish.

8.7   S-Parameter Design Example

With the S-parameters defined and measured, the next step is to put them to use for us. There is no better way to put them to work than to use them in an amplifier design example. Our example will consist of a low-noise amplifier to operate at 4 GHz. Fig. 8–22 shows curves for minimum noise figure and associated gain for an I_C = 4 mA (V_{CE} is given as 10 V). These figures will result, you will note, in a circuit gain of approximately 11 dB when the amplifier is designed for the 2.5-dB minimum noise figure. (It is understood that there are many design packages available which that prompt the designer to put numbers in blanks to result in a finished design for an amplifier or other circuit. What this

Table 8–1. Transistor S-Parameters, V_{CE} = 10 V, I_C = 5 mA

Freq. (GHz)	S_{11}	S_{21}	S_{12}	S_{22}
1.8	0.58 −156°	3.7 85°	0.048 15.2°	0.68 −38°
2.0	0.51 −163°	3.3 80°	0.051 13.5°	0.66 −42°
2.2	0.54 −173°	3.2 79°	0.052 11.8°	0.66 −45°
2.4	0.54 −176°	2.9 75°	0.053 10.5°	0.66 −48°
2.6	0.54 179°	2.7 72°	0.055 19.2°	0.67 −50°
2.8	0.54 174°	2.6 68°	0.057 8.8°	0.68 −53°
3.0	0.54 168°	2.3 66°	0.058 7.7°	0.66 −60°
3.2	0.55 165°	2.2 63°	0.060 6.7°	0.67 −62°
3.4	0.55 163°	2.1 62°	0.062 15.4°	0.68 −64°
3.6	0.54 160°	1.9 59°	0.064 14.0°	0.69 −65°
3.8	0.52 145°	1.7 56°	0.066 12.5°	0.70 −66°
4.0	0.51 142°	1.6 55°	0.068 1.0°	0.71 −67°
4.2	0.52 140°	1.5 52°	0.070 −0.50°	0.72 −70°
4.4	0.53 132°	1.4 51°	0.073 −2.1°	0.71 −72°
4.6	0.54 128°	1.4 50°	0.076 −3.6°	0.70 −75°
4.8	0.54 126°	1.3 49°	0.078 −5.3°	0.70 −78°
5.0	0.55 123°	1.3 48°	0.081 −7.0°	0.70 −81°
5.2	0.56 122°	1.3 47°	0.084 −8.8°	0.70 −84°
5.4	0.57 115°	1.3 45°	0.086 −10.7°	0.70 −87°
5.6	0.60 110°	1.2 143°	0.089 −12.7°	0.70 −90°
5.8	0.62 107°	1.1 140°	0.091 −14.8°	0.70 −93°
6.0	0.58 103°	1.1 137°	0.094 −17.0°	0.69 −95°

example is doing is showing the reader the steps that need to be taken and helping show why certain parameters are used to achieve the *proper* design. What is hoped is that designers are created, not just terminal operators.)

With bias conditions set at V_{CE} = 10 V and I_C = 4 mA, the S-parameters and noise parameters at 4 GHz are found to be as follows:

$$S_{11} = 0.552\ \underline{/169°}$$

$$S_{12} = 0.049\ \underline{/23°}$$

$$S_{21} = 1.681\ \underline{/26°}$$

$$S_{22} = 0.839\ \underline{/-67°}$$

$$F_{min} = 2.5\ \text{dB}$$

$$\Gamma_O = 0.475\ \underline{/169°}$$

where, Γ_O is defined as the optimum source reflection coefficient.

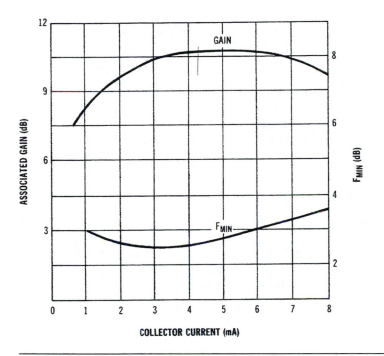

Figure 8–22. F_{min} and gain at 4 GHz.

Γ_O is basically the S_{11} of the device at its optimum noise figure. Note that there is a slight variation from the given S_{11} because S_{11} is used to design the input match of a device to get maximum gain and Γ_O is used to get a minimum noise figure. This value can be found on the data sheet for low-noise devices along with all S-parameters so that either minimum noise or maximum gain may be used as a design criterion.

8.7.1   Matching

Let us now begin to match our transistor for minimum noise figure operation. We will begin by matching the input. The main purpose of the input matching network is to provide the best source impedance for the minimum noise figure, which is not to say that the transistor will be perfectly conjugately matched. In all likelihood it will not be conjugately matched because, as you will recall, we are not looking for a maximum transfer of power as we would for maximum gain; we are looking for the match that will result in minimum noise. To realize such a match, we must proceed as follows:

1.   Convert Γ_O (optimum source reflection coefficient) to impedance.
2.   Determine an equivalent admittance for this impedance.

3. Realize a susceptance component with a short-circuited eighth-wavelength stub ($\lambda/8$).

4. Realize a conductance component with a quarter-wavelength transformer ($\lambda/4$).

The use of $\lambda/8$ shorted stubs and the $\lambda/4$ transformer is only one method of providing a match for the transistor. Other methods are available and may be used in a variety of applications. This method was chosen because it seemed appropriate for our requirements and was a convenient method to obtain the necessary reactance and impedance transformations. It also illustrates very well the use of S-parameters and the Smith chart.

Fig. 8–23 shows what we are trying to accomplish, how we will do it, and the circuit configuration. We start out with the input impedance (obtained from the optimum source reflection coefficient), find the admittance that corresponds to it (Y_{NF}), calculate the short-circuited stub from this value since parallel stubs can be added directly when admittance and susceptance are used, transform back to impedance (Z_A), and calculate the quarter-wave transformer that will bring the impedance to the center of the circle (50 ohms). The $\lambda/8$ stub and $\lambda/4$ transformer are used because the stub adds susceptance, which cancels the imaginary part of the transistor's input/output impedance, and the transformer moves the impedance to 50 ohms.

8.7.2 Designing the Input Matching Network

With our approach explained, let us proceed to design the input matching network.

1. To find the impedance Z_{NF}, we need to plot the value of the reflection coefficient given for optimum noise (0.475 @ 169°). To plot these values we will mark off a distance of 0.475 on the radial parameter of the Smith chart on the *reflection coefficient voltage scale*. Transfer this distance to the Smith chart by placing the point of a compass at the center of the chart and drawing a complete circle having a radius of 0.475 (see Fig. 8–24). By setting the edge of a ruler at the center of the circle and at 169° (read on the outside edge), we can draw a line that intersects the circle at the value of impedance we are looking for. Reading directly from the chart, we get approximately 0.36 + j0.105 ohms, which is a normalized value. To get the true value in our 50-ohm system, this value must be multiplied by 50. We, therefore, have $Z_{NF} = 18 + j5.25$ ohms.

2. With a value of Z_{NF} now available, we can now find a value for Y_{NF}. Mathematically expressed,

$$Y_{NF} = \frac{1}{Z_{NF}}$$

We could go through the calculations to determine Y_{NF} with this equation, but the task is much easier with the Smith chart. The determination is simply a matter of starting at Z_{NF} and moving completely across the chart until the circle (0.475 radius) is intersected. We then simply read the normalized value (2.55 − j0.76) and multiply it by 0.02 (since $Y = 1/Z = 1/50 = 0.02$). The value for Y_{NF} is 0.051 − j0.0152.

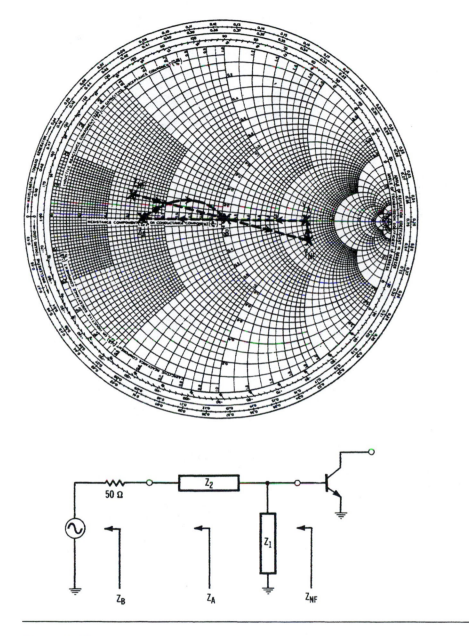

Figure 8–23. Input matching network.

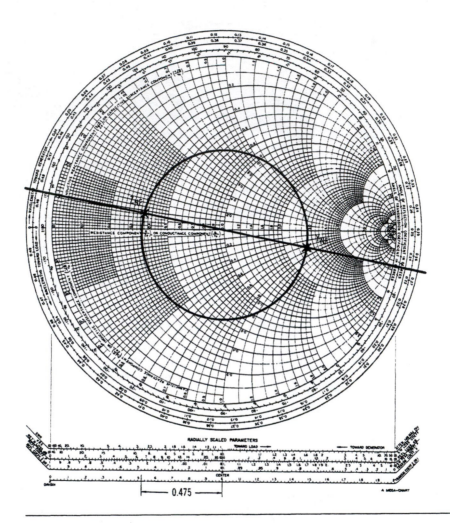

Figure 8–24.   Smith chart plotting.

3.   The next task is to design the short-circuited eighth-wavelength stub. A short-circuited eighth-wavelength stub looks like a shunt inductor with an impedance of jZ_0. We would, therefore, calculate the impedance of the stub as

$$Z_1 = \frac{1}{\text{imaginary part of } Y_{NF}}$$

$$= \frac{1}{0.0152}$$

$$= 65.8 \text{ ohms}$$

4. The input match will be completed as far as impedance by calculating the impedance of the quarter-wave transformer. This impedance is obtained as follows:

$$Z_2 = \sqrt{Z_0 \left(\frac{1}{\text{real part of } Y_{NF}} \right)}$$

$$= \sqrt{50 \left(\frac{1}{0.051} \right)}$$

$$= \sqrt{50 \, (19.6)}$$
$$= \sqrt{980}$$
$$= 31.3 \text{ ohms}$$

The lengths of the stub will be determined by the substrate used to build the circuit. We will use Teflon® fiberglass material for this application. This material has a dielectric constant (ϵ) of 2.55. (This is a relative dielectric constant. The effective dielectric constant for a 31-ohm line is 2.25.) A wavelength calculation will be as follows:

$$\lambda = \frac{C}{f \sqrt{\epsilon}}$$

where, λ is the wavelength,
       C is the speed of light (3×10^{10} cm/sec),
       f is the frequency,
       ϵ is the dielectric constant of the material.

A wavelength calculation for 4 GHz would then be

$$\lambda = \frac{3 \times 10^{10} \text{ cm/sec}}{4 \times 10^9 \text{ Hz} \sqrt{2.25}}$$

$$= \frac{3 \times 10^{10}}{1.524 \times 10^{10}}$$

$$= 1.968 \text{ inches}$$

Solving for $\lambda/4$ gives us 0.492 inch, and solving for $\lambda/8$ gives us 0.246 inch.

So the final input circuit is as is shown in Fig. 8–25. The numbers given are in inches, with the centimeter number in parentheses.

8.7.3   Designing the Output Matching Network

With the input matching circuit completed, the next step is to work with the output. To provide a design with a low value of VSWR on the output (in the order of 1.5:1), we will design the output matching section to provide a conjugate match with the source impedance referred to earlier, Z_{NF}. A conjugate match is an impedance with the same resistance

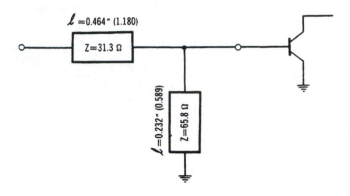

Figure 8–25.   Final input circuitry.

component as the device and a negative reactive component of the original. If, for example, we found that we had an output impedance of $10 - j32$ ohms, the conjugate match would have to be $10 + j32$ ohms.

Our design of the output network will be broken into three steps:

1. Calculate a new value of S_{22} using the source reflection coefficient, Γ_O, called S_{22}'.

2. Provide a conjugate match for the output as discussed earlier. The load reflection coefficient will be S_{22}'.

3. To increase the tuning capability of the output sections, we will place an additional section of matching along with the short-circuited stub and series transformer as used at the input.

To begin the design, we will determine S_{22}'. (*Reference: Hewlett-Packard Journal, Feb., 1967, Vol. 18, No. 6.*)

$$S_{22}' = S_{22} + \frac{S_{21} S_{12} \Gamma_O}{1 - S_{11} \Gamma_O}$$

This calculates out to be

$$S_{22}' = 0.846 \underline{/-72°}$$

From our previous definition, we will need a load reflection of $0.846 \underline{/72°}$ for a conjugate match. Fig. 8–26 shows the point now called Z_{22}' and how the match is to take place. The layout of the matching return is shown below the Smith chart.

The first step is to move away from the transistor to allow room to solder the collector lead. A 50-ohm line 0.15 inch long is used. This line works out to be 59° long. If we move 59° from Z_{22}', we will be at Z_1, which can be read as $5.03 - j22.6$ ($0.1 - j0.45$ on the normalized chart). By crossing the Smith chart, we can read the admittance at point Z_1, which is

$$Y_1 = \frac{1}{Z_1} = 0.009 + j.0422$$

As in the input matching case, we can determine the impedance of the short-circuited stub by using the imaginary portion of Y_1. The purpose of the stub is to tune out most of the susceptance component in Y_1. In our output matching section, a tuning section will be added later to further match to the output. The impedance of the stub is calculated as follows:

$$Z_{stub} = \frac{1}{\text{imaginary } Y_1}$$

$$= \frac{1}{0.0422}$$

$$= 23.7 \text{ ohms}$$

We can now find the value of Z_2 by moving from Y_1 along the pure resistance circle (0.4) 90° around the chart to the point Y_2 in Fig. 8–26. We can make this 90° move because we are using a $\lambda/4$ transformer. We can move along the pure resistance circle (0.4) because we are only matching conductance and not susceptance. This point (Y_2) is now:

$$Y_2 = 0.009 + j0.0077$$

If we go directly across the chart, we will find Z_2, which is $63.7 - j52.5$ ($1.27 - j1.05$ on the normalized chart).

To obtain the value of the following series line, we will now use the value $Y_2 = 0.009 + j0.0077$. In this case we will use the real portion of Y_2.

$$Z_{line} = \sqrt{50 \left(\frac{1}{\text{real } Y_2} \right)}$$

$$= \sqrt{50 \left(\frac{1}{0.009} \right)}$$

$$= \sqrt{50 \,(111)}$$
$$= \sqrt{5550}$$
$$= 74.5 \text{ ohms}$$

The length of the transmission line is adjusted to obtain an admittance, with its real part matched to a 50-ohm load. The use of the 74.5-ohm line calculated here results in a line length of 0.179 to satisfy these requirements.

$$Z_3 = Z_0 \frac{(Z_2 + jZ_0 \tan \beta l)}{(Z_0 + jZ_2 \tan \beta l)}$$

(*Reference: Hewlett-Packard Journal, Feb. 1967, Vol. 18, No. 6.*)

We now have an impedance at Z_3 of $39.12 + j21.1$ ohms. The admittance, Y_3, can be read directly across from Z_3 or can be calculated as $\frac{1}{Z_3}$. Either way, $Y_3 = 0.02 - j0.0107$.

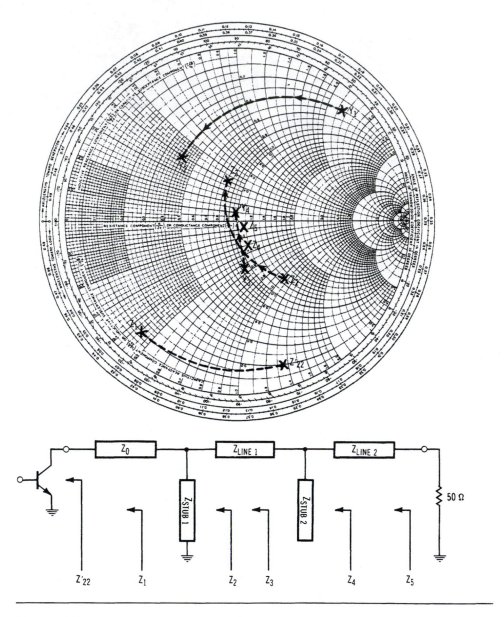

Figure 8–26. Output matching network.

The open-circuit stub shown in Fig. 8–26 must have some tuning capability. An open-circuit stub less than a quarter-wave long looks like a shunt admittance equal to jY_0 tan βl. We can accomplish our matching task if we have a stub of 0.039 mhos and a length of 0.053 (or 38°). These stub and length measurements come about by the relationship that $Y = jY_0$ tan βl. Also, we can move Y_3 along the resistance circle until we cancel the $-j0.0107$ mhos of susceptance from Y_3. This will be Y_4. Moving directly across the chart yields Z_4, which results in an open stub that is 25.6 ohms $\left(Y = \dfrac{1}{0.039} \right)$ and is 0.053 λ long. The results of this line for Z_4 and Y_4 are as follows:

$$Z_4 = 50 - j7.1$$

$$Y_4 = 0.02 + j0.0029$$

Our last objective is to cancel out $-j7.1$ ohms of reactance to produce a pure 50-ohm output. This will be done by staying with a 75-ohm line as used to obtain Z_3 and have a line 18° or 0.025 λ long obtained by moving along the resistance circle. These amounts will result in a Z_5 in Fig. 8–26 equaling exactly 50 ohms.

This output matching procedure has been long and involved, but in many areas this arrangement has great tuning flexibility:

1. Adjusting the length of the series lines (both 75-ohm lines can be adjusted).

2. Changing the widths (impedance) and length of the open circuit stub is a valuable method to allow for tuning of the amplifier.

Fig. 8–27 shows the final output configuration for the amplifier design. All dimensions are in inches, with millimeters shown in parentheses. The short-circuit stub on the output is not a dc short but a microwave short. Between the stub and ground is placed a capacitor with a low reactance. A 1000-pF capacitor, for example, will have an X_C of 0.004 ohm at 4 GHz, which is an excellent microwave short and will not short the dc bias for the transistor to ground. (These chip capacitors are discussed in Chapter 4.)

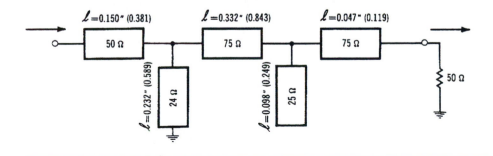

Figure 8–27. Final output circuitry.

8.8   Chapter Summary

This chapter has been an explanation of the most useful microwave parameters available—S-parameters. We have compared them with Z, Y, and h parameters; defined their operation in microwaves; measured them; and, finally, used them to design a single-stage transistor amplifier. From this discussion, how these parameters fit into microwave component and system designing should be clear.

In the introduction, we made the comment that for someone else to understand an item you are talking about, you both must talk the same language. If you talk S-parameters to someone in microwaves, you can be guaranteed that you are both talking the same language.

8.9   References

1. *Diode and Transistor Designer's Catalog.* Hewlett-Packard Co., Palo Alto, CA, 1980.

2. *S-Parameter Circuit Analysis and Design.* Application Note 95, Hewlett-Packard Co., Palo Alto, CA, 1968.

3. *S-Parameter Design.* Application Note 154, Hewlett-Packard Co., Palo Alto, CA, 1972.

4. *Stripline Component Measurements with the 8410A Network Analyzer.* Application Note 117-2, Hewlett-Packard Co., Palo Alto, CA, 1971.

5. Vendelin, G. D. *Design of Amplifiers and Oscillators by the S-Parameter Method.* New York: John Wiley, 1980.

Questions

8.1   Introduction

   1.   What are Y-parameters?
   2.   What are Z-parameters?
   3.   What are h-parameters?
   4.   Why can't Y, Z, and h parameters be used at microwave frequencies?

8.2   S-Parameter Definition

   5.   What is the "a" designation for signals?
   6.   What is the "b" designation for signals?
   7.   What condition is necessary to measure S_{11} and S_{21}?
   8.   What condition is necessary to measure S_{12} and S_{22}?
   9.   Define S_{11}, S_{12}, S_{21}, and S_{22}.
   10. List two advantages of using S-parameters at microwave frequencies.

8.3   S-Parameter Uses

   11. Why do S-parameters for chip devices look different from packaged devices?

12. What is the most important piece of information needed when using S-parameters?

8.4 S-Parameters and the Transistor Data Sheet

13. Define f_t.
14. Define f_s.
15. Define f_{max}.

8.5 S-Parameter Measurements

16. Why are dual-directional couplers used in S-parameter test sets?
17. Explain how S-parameter test sets work with vector network analyzers.

8.7 S-Parameter Design Example

18. How do the noise figure curves figure into an S-parameter design example?
19. When using a device for low noise, which parameters are used to match the device?
20. When using a device for maximum gain, which parameters are used to match the device?

Microwave Test Equipment

Objective

This chapter presents some representative types of microwave test equipment. The types to be covered are signal generators, sweep generators, power meters, spectrum analyzers, frequency counters, noise-figure meters, network analyzers, and high frequency oscilloscopes. Each of these will be introduced and discussed, and a typical data sheet will be presented.

Key Terms

Buffer Amplifier
Network Analyzer
Noise Figure
Peak Power
Power Meter
Prescaling
Reference Oscillator
Sensitivity

Sensor
Spectral Purity
Spectrum Analyzer
Sweep Generator
Synthesized Generator
Thermistor
YIG Oscillator

9.1   Introduction

The design of a particular circuit, or system, will involve some, or all, of the previous eight chapters. A solid-state amplifier, for example, would involve the chapters on S-parameters, the Smith chart, and microwave transistors and material at the very least. If the input needed to be filtered, the chapter on microwave components would also be used. All the chapters to this point have been aimed at producing a product. The task to be performed now is to test that product.

Before you can *properly* test the product you have created, you must have test equipment, and the equipment needed will vary with the type of test to be run. The important thing to do in each case is to first understand the test equipment available for testing and then to understand how to run these tests. Since both steps are important, we are devoting a complete chapter to each area—test equipment and testing. This chapter, of course, is on the test equipment.

The test equipment used in microwave testing can be classified as falling into three general classes: *signal generation, signal indication,* and *test systems.*

Under *signal generation* we will cover

- Signal generators
  - Conventional (signal oscillator)
  - Synthesized
- Sweep generators

Under *signal indication* we will cover

- Power meters
- Spectrum analyzers
- Frequency counters
- Noise-figure meters
- High frequency oscilloscopes

We will complete our discussion on microwave test equipment with a look at microwave network analyzers from

- Hewlett-Packard Co.
- Anritsu Wiltron
- Marconi Instruments

Each type of equipment covered will have a discussion of the instrument and a typical data sheet describing the terms for that particular piece of equipment.

9.2   Signal Generators

With the exception of microwave oscillators, every microwave component requires a signal source to operate or be tested. The requirements for this source will vary depending on particular applications and equipment availability, but the basic requirement for a source will not change. Still, microwave energy is needed at an input to get microwave energy at an output.

We will cover two types of signal generation: conventional type, or *signal oscillator*, and the *synthesizer*.

9.2.1   Signal Oscillator

The signal oscillator for many years was the main type of signal source for microwave signals. It was a basic method of generating a stable signal to be used as a carrier, local oscillator, or test signal. In present-day laboratories this source has been replaced in many instances with synthesized generators, which will be covered shortly. There are, however, still many applications where the basic signal oscillators are still preferred and used. We will cover this type of microwave oscillator here to give the reader an insight into the present generation of microwave signals. This basic coverage will also aid in understanding the theory behind more sophisticated generators that may be used at a later date.

Microwave signal generators are available in various bands (depending on the manufacturer) in frequencies up to 40 GHz. Usually they will supply a single frequency output (with some small bandwidth associated with it) that can be varied both in frequency and amplitude. Also, there are various internal modulations that may be obtained with these generators. External jacks are usually available for you to insert your own particular type of modulation.

In order to understand this basic type of microwave test equipment, refer to Fig. 9.1. It shows five distinct areas that may or may not be a part of a microwave frequency generator. We will now look at all five of these areas to aid in a better understanding of the generators that can be used.

Variable Reference Oscillator   The first block, Fig. 9–1A, the heart of the signal oscillator, is the variable reference oscillator, which could be a kylstron, a YIG-tuned oscillator, voltage-tuned oscillator, or sometimes a mechanically tuned cavity oscillator. Any oscillator whose frequency can be changed would be acceptable in this spot, and it, as we have said, is the heart of the generator. Like most hearts, however, it is not much good by itself. The variable oscillator as it appears in Fig. 9–1 (block A) would exhibit very loose specifications for output level, frequency stability, and output level stability. Just as a human heart will not function without something to drive and control, so too must the oscillator have additional circuitry to achieve the most efficient system that will perform the necessary test functions.

Buffer Amplifier   Block B is a buffer amplifier that provides the same function that an isolator would, only that it has the additional property of increasing the oscillator signal level in the process. The buffer is needed since any change in load will cause the level and frequency of the oscillator to deviate from the preset points. We say that the frequency of the oscillator will *pull*. The buffer isolates the oscillator from these load changes and maintains the frequency stability of the device.

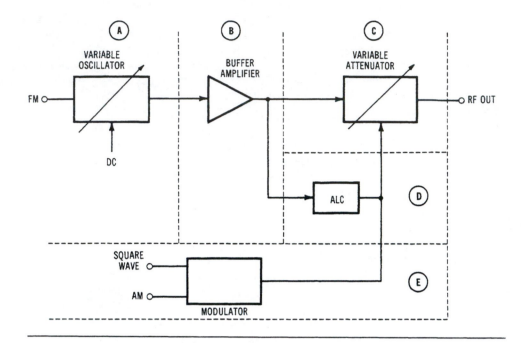

Figure 9–1. Signal oscillator block diagram.

Amplitude Control Block C is the amplitude control you turn on the front of the generator to increase or decrease the output level. Variations in this attenuator can cause the load changes that we just referred to. These load changes necessitate the need for the buffer amplifier. This attenuator may be a step attenuator, a continuously variable attenuator, or both. This attenuator may also be a programmable attenuator in modern signal generators. Since many test systems are automatic systems, it is necessary to control such functions as attenuation with a computer program.

Automatic Leveling Circuit A constant output level can be obtained by using block D, an *ALC* (automatic leveling circuit) in conjunction with the variable attenuator. The system shown uses a *feed-forward method* of leveling, which is accomplished by sampling the output level of the buffer amplifier, comparing it to a reference level, and adjusting the variable attenuator to maintain the constant output level. In this scheme the difference signal (difference between the output of the buffer amplifier and the reference signal) is *fed forward* toward the output, with all the control taking place at the output. If, for example, the output of the buffer amplifier increased, the ALC circuitry would produce a difference signal that would put attenuation in the variable attenuator to keep the output constant. If the buffer output decreased, the ALC would work in reverse—that is, take attenuation out of the system to bring the output back up, an instantaneous process that cannot be seen.

As we have stated, our system uses a feed-forward method for leveling. A *feed-back system* could also be used. In this system the signal level is sampled directly at the output.

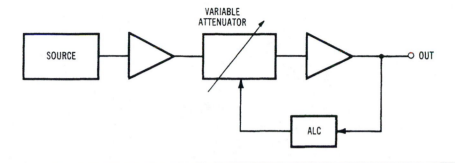

Figure 9–2.   Feed-back leveling system.

Fig. 9–2 shows the feed-back leveling system. The sampled signal is *fed back* through the ALC circuitry, where it is compared to a reference level just as in the feed-forward system. The necessary level adjustments are made back towards the input. If the output level drops, for example, the ALC circuitry would produce a resultant level that would take attenuation out of the system and, thus, increase the output to keep it leveled. This type of leveling is used to level sweep generators in test systems and is the method used in receiver AGC (automatic gain control) systems.

Modulator   The final block, E, allows the continuous wave (cw) signal to be modulated. The modulation may be square wave, amplitude modulation (AM), frequency modulation (FM), or some digital modulation schemes used in some generators available today. (The source shown in Fig. 9.1 shows an FM scheme.) Phase and pulse modulation is also available on some generators. GMSK (Gaussian Minimum Shift Keying), which is a continuous-phase frequency shift keying (FSK), can also be obtained. So it can be seen that the modulation capabilities of a straightforward signal generator can many times be varied.

With the basic blocks now more understandable, it is time to look at a typical microwave signal generator data sheet.

9.2.2   Signal Generator Data Sheet

The signal generator data sheet shown below is a composite of generators that results in a representative set of specifications.

Frequency Range	0.1 MHz to 6 GHz
Spectral Purity	SSb Phase Noise (−117dbc/Hz)
	Harmonics: −75 dBc to 1.5 GHz, −40 dBc to 3 GHz, and −50 dBc to 6 GHz
	Residual FM: <75 Hz rms

Output Level	+13 to −140 dBm
Modulation	AM, FM, square wave, external connection

These terms are defined as follows:

Frequency Range   This term, of course, is the range of frequencies over which the generator will operate. If at all possible, you should take precautions not to operate a generator at its upper or lower frequency limits. Operation in these areas may result in instabilities or level drop-off that will make good measurement very difficult. Try to use a generator that has your operating range near the center of its range.

Spectral Purity   It can be seen that this one term is divided into three separate categories: SSB (Single Sideband) Phase Noise, Harmonics, and Residual FM. The SSB phase noise is characterized in dBc/Hz, which is dB below the carrier per Hertz. This is a reading of the noise component of the signal a certain number of Hertz from the center frequency, or carrier. Harmonics are the signals generated that are twice the fundamental (second harmonic), three times the fundamental (third harmonic), and so on. These signals must be considerably lower than the desired fundamental frequency so that harmonic distortion of that signal does not occur. Finally, residual FM is the range of frequencies that the fundamental signal "wobbles" around the desired frequency since it is not a stable single frequency component. Ideally, this would be zero. Practically, every signal has a certain amount of residual FM. Some generators are better than others.

Output Level   The RF output level is the amount of power delivered to a 50-ohm load. Note that there is a range of dBm, from +13 dBm (20 mw) to −140 dBm (10 pw). This is a continuously variable level in some generators and a step and fine adjustment in others.

Modulation   This term tells the reader what modulation is available within the generator and if external connectors are available for applying different types of modulation.

9.2.3   Synthesized Signal Generator

One of the most sophisticated generators and, certainly, one with the best spectral purity is the synthesized signal generator. The word *synthesize* conjures up many things to people, mostly implying that an item is phony or artificial if it is synthesized. We think of synthetic fabrics, synthetic foods, and synthetic colors as being inferior to material fabric (cotton, wool, etc.), natural foods (vegetables, fruits, etc.), and natural colors (red, blue, green, etc.). Although they may, in fact, not be inferior to natural items, anything synthetic is created by artificial means. A synthetic item has changed in some way from the basic original substance so that it does not resemble the original substance in its final form. In a microwave synthesizer, the final output (or product) does not resemble the original oscillator signal (original substance) because the output is much higher in frequency. These frequencies are created by using a very stable low-frequency oscillator and phase locking to the harmonic produced by it. The final result is a series of noise-free, stable frequencies that would require hundreds or thousands of crystals or separate tuned circuits to produce if conventional signal generator methods were used.

Probably the most important term to understand in the explanation of synthesizing signals is *phase locking*. Phase locking is the technique of making the phase of an oscillator signal follow exactly the phase of a reference signal. This technique is accomplished by comparing the phases between the two signals and using the result to adjust the frequency to keep the oscillator tracking the reference. Phase locking is very similar to an ALC loop, which keeps a constant amplitude at the output of a circuit.

Most synthesizers used for microwave applications use a technique called *indirect synthesis* (also called *phase-locked loop*). Such a system is shown in Fig. 9–3. The circuit in Fig. 9–3 is one that is called a *single loop* synthesizer. This is because it contains only one PLL (phase locked loop). There are more complex synthesizers available, but we will look only at the single loop device at this time. If the theory of the single loop synthesizer is understood, the reader should be able to relate it to other systems.

The single loop synthesizer has a reference that is a single crystal-controlled oscillator. The range of frequencies and the resolution of the system depend on the divider network and the gain of the loop (open loop gain). The divider is a "divide by n" circuit that is usually controlled by an operator to adjust the frequency or by a control circuit that automatically sets a frequency or range of frequencies. The synthesizer can basically be considered to be a times-n multiplier circuit. This can be seen as being true since the output of the synthesizer is actually n times f_{ref}.

The statement has been made that the frequency range and resolution depend on the n-divider and the open loop gain of the system. To investigate further, we can say that the open loop gain is equal to the product of the gain of the comparator (A_d), gain of the amplifier (A_a) and the transfer function of the VCO (H_o). For the synthesizer in Fig. 9.3, the open loop gain can be said to be

$$A_v = (A_d A_a H_o)/n \tag{9.1}$$

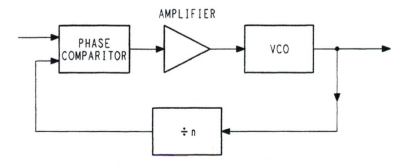

Figure 9–3.   Indirect synthesis.

EXAMPLE 9.1

A synthesizer exhibits the following parameters: $A_d = 0.7$ v/rad, $A_d = 8$, $H_o = 12$ kHz/volt, and $n = 2$. The n-factor is inadvertently changed so that the open loop gain of the synthesizer is 16.8. What is the new n-factor?

Solution   In order to determine the new n-factor, the initial open loop gain must be found. This is found by using equation 9.1.

$$A_v = (A_d A_a H_o)/n$$
$$A_v = [(0.7)\,(8)\,(12)]/2$$
$$A_v = 67.2/2 = 33.6$$

We can use the same equation to find the new n-factor.

The new gain is 16.8, and the numerator of the equation is 67.2. Thus, we can now find the new n-factor as follows:

$$A_v = 67.2/n$$
$$16.8 = 67.2/n$$
$$n = 67.2/16.8$$
$$n = 4 \hspace{3cm} \blacklozenge$$

It can be seen from equation 9.1 that as n changes the open loop gain changes at an inversely proportional ratio. To eliminate this reliance of open loop gain on the n-factor of the synthesizer, a common practice is to have the amplifier gain programmed to the divider ratio also. This makes the gain of the amplifier equal to $n(A_a)$. The open loop gain of the synthesizer is now

$$A_v = A_d A_a H_o \hspace{3cm} (9.2)$$

Equation 9.2 ensures that the changing divider ratio in the synthesizer will *not* affect the open loop gain of the system. Thus, only the divider ratio will be used to control the frequency range and resolution.

EXAMPLE 9.2

If we take the circuit of Fig. 9.3 and apply a 10 kHz signal from the reference oscillator and our divider ratio is equal to 10, what will the frequency range be?

Solution

$$F_o = n\,F_{ref}$$
$$F_o = (10)\,(1\text{ kHz})$$
$$F_o = 10\text{ kHz range}$$
$$F_o = 1\text{ kHz to }10\text{ kHz in }1\text{ kHz steps} \hspace{1.5cm} \blacklozenge$$

There are other methods of designing frequency synthesizers. Some use a scheme called *prescaling* (to be covered in Frequency Counters later in this chapter), while others use a series of harmonic generators and filters to create a specific range and resolution. The method used depends on the particular application of the synthesizer.

At microwave frequencies, one of the advantages of indirect synthesizers is their low phase noise. They have better wideband phase-noise performance than either direct synthesizers or lower-frequency synthesizers multiplied to microwave frequencies. Indirect synthesizers accomplish this phase-noise performance by taking advantage of the difference in noise characteristics of the crystal reference and the VCO for optimum noise performance.

With indirect synthesizers, the output phase noise is that of the reference multiplied up to microwave frequencies within the phase-locked-loop bandwidth. As the offset from the carrier increases, the effects of the phase-locked loops decrease and the noise performance approaches that of the VCO only. The result is an overall improvement in noise performance because close to the carrier, the multiplied reference has the lower phase noise (although at larger offsets, the VCO is actually cleaner). The synthesizer then includes the best regions of both signals for optimum noise performance.

9.2.4   Synthesized Signal Generator Data Sheet

Our next step in discussing microwave synthesizers is to present a typical data sheet and explain the terms.

Frequency Range	2–20 GHz
Frequency Resolution	100 kHz
Spectral Purity	SSB phase noise, −65 dBc
	offset 30 Hz from carrier
	Residual FM, 20 Hz from
	2–8 GHz, 40 Hz from 8–20 GHz
RF Output Power	+10 dBm leveled

Frequency Range   Frequency range is the range of operation of the synthesizer. It is the frequencies the unit is capable of producing.

Frequency Resolution   Frequency resolution is how accurately the synthesizer frequency can be read. It is not the display resolution but the instrument resolution.

Spectral Purity.   As in the case of the signal oscillator, there are three components of this specification: Harmonics (multiples of the desired output frequency), SSB Phase Noise (a measure of the noise energy in a specific bandwidth at a specified frequency offset from the carrier), and Redidual FM (the amount the carrier frequency varies around the desired output.) Fig. 9.4 shows some synthesized sources that are available.

Output Power   Output power is the RF output of the synthesizer. This measurement was given in both dBm and milliwatts for convenience.

9.3   Sweep Generators

In the previous sections we covered generators that produce a single-frequency output. These generators are fine and most adequate for many applications. There are times, however, when the response of a component or system over a band of frequencies is needed. Getting this response over a band of frequencies would be a time-consuming procedure if done one frequency at a time. For these occasions the microwave sweep generator (sweeper) is used.

The sweep generator gains its wide use not only because of its capability to cover a band of frequencies but also because of its versatility. Most sweepers consist of a main frame and a series of plug-ins that will give the exact band of frequencies needed for the application. Getting the exact band of frequencies is much better than having one generator go from 1 to 18 GHz (or higher). This wide of a bandwidth will cause other parameters to be degraded: power output, accuracy, and stability, to name a few. If we go into narrower bands with individual plug-ins, we will have units that are covering a variety of precise bands simply by unplugging one unit and plugging in another. A typical sweeper block diagram is shown in Fig. 9–5. Notice the two individual sections of the sweeper and how each has its own function. The main frame primarily generates the ramp function and linearizes it. It also provides dc power for both the main frame and the plug-in. The plug-in contains the oscillator, modulation capability, leveling (if desired), and output power-level control. The main frame is like a universal socket, and the plug-ins are the specialized units that fit the socket.

Although the main frame of the sweeper is of vital importance, the plug-in is usually the heart of the generator. The plug-in ultimately gives the swept signal of the proper frequency, stability, and power level for your particular application.

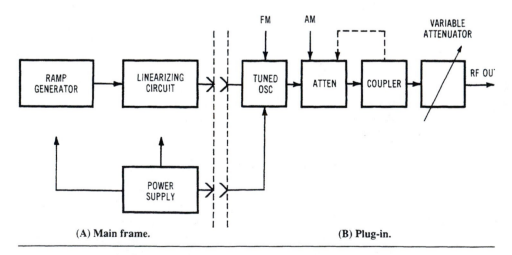

Figure 9–5.   Sweeper block diagram.

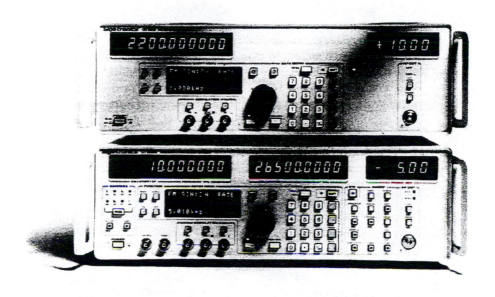

Figure 9–4A. Synthesized microwave source *(Courtesy of Giga-tronics Incorporated).*

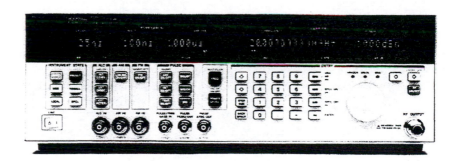

Figure 9–4B. Synthesized microwave source *(Courtesy of Hewlett-Packard Co.).*

9.3.1 Varactor and YIG-Tuned Oscillators

The previous sections covered signal generators that provided single-frequency outputs by the use of *voltage-tuned oscillators*. These oscillators may have been a kylstron, cavity, YIG (yttrium iron garnet), or other type that could be tuned to a single frequency. All that these oscillators were required to do was to provide a single frequency with a specific stability and power output. The sweeper oscillator has many more requirements put on it. These requirements are generally fulfilled by using varactor-tuned oscillators or YIG-tuned oscillators (YTO). The circuit shown in Fig. 9–5 uses a varactor-tuned oscillator and thus requires the linearizing circuit shown. The linearizing circuit is needed because the characteristics of a varactor exhibit a large degree of nonlinearity. Since a sweeper must be exceptionally linear across its entire operating band, a circuit must be used to eliminate these nonlinearities in the final output.

The YIG-tuned oscillator (YTO) does not use a voltage applied to a nonlinear device (diode) but has a current applied to YIG spheres to determine the frequency of operation. An RF interaction with the spheres causes a resonance that produces a frequency dependent only on the magnitude of the dc field. A change in current causes a change in frequency, which is a linear function. This circuit, then, obviously does not require the linearizing circuitry needed for a varactor-tuned circuit. The YIG does, however, require a driver unit.

The YIG driver circuits, as well as the associated circuitry, are shown in Fig. 9–6. Fig. 9–6 is a block diagram of a microwave sweep generator. The figure shows the

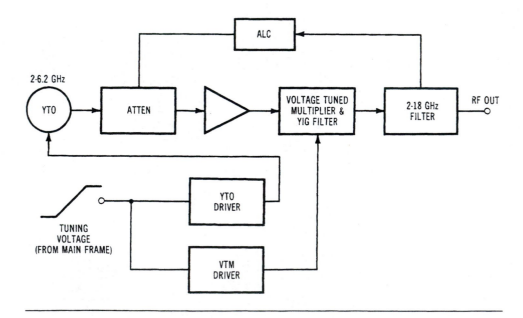

Figure 9–6. Plug-in block diagram.

fundamental oscillator as a YIG-tuned oscillator (YTO) that tunes from 2 to 6.2 GHz directly. This oscillator output is amplified to a high power level (in the range of 100 mW), which is needed to provide the proper input to the YIG-tuned multiplier (YTM). This multiplier contains a step-recovery diode that produces harmonics of the fundamental band (2 to 6.2 GHz) when hit with the high-power input signal. A tracking YIG filter is used in the same block to select the proper frequency for your application. This filter has very steep skirts that allow only the desired frequency to pass.

Output power considerations demand exceptionally close tracking between the two YIG circuits. This close tracking is made possible through development of extremely linear magnet structures for the YIG devices, which, in turn, results in excellent frequency accuracy. Even at 18 GHz, frequency can be set from the dial scale to ±20 MHz, which is better than frequency meter accuracy. Signal drift in cw is typically less than 0.005% per 10 minutes, which means narrow bandwidth measurements can be made reliably. Temperature compensation also minimizes frequency drift with ambient changes. Hysteresis effects in the tuning are imperceptible. The majority of the sweep generators available are synthesized sweepers. These sweepers use indirect synthesis, previously covered, and range over frequencies of 10 MHz to 20 GHz, 2 to 26.5 GHz, and 2 to 40 GHz. Figure 9.7 shows some synthesized sweepers commercially available. It can be seen from this figure that there is a wealth of information available on the front panels of the equipment. Displays show start frequency, stop frequency, output power level, sweep time and modulation type, and characteristics, to name only a few. The sweep generator has come a long way from the hand crank that set the frequencies and a yellow light that told you your output was not leveled. The advances that have been made have greatly improved the accuracy and repeatability of these sources.

9.3.2 Sweep Generator Data Sheet

To wrap up the sweep generator, we will now present a typical data sheet for both plug-in and synthesized sweepers and explain the terms used.

	Plug-in Sweeper	Synthesized Sweeper
Frequency Range	8–20 GHz	2–40 GHz
Frequency Accuracy	±50 MHz	±30 MHz
Frequency Stability	±800 kHz/C	±200 kHz/C
Maximum Leveled Power	20 dBm	10 dBm
Spurious Signals		
Harmonic	< −20 dBc	< −60 dBc 2–20 GHz; < −40 dBc 20 to 40 GHz
Non-Harmonic	< −50 dBc	< −60 dBc

Figure 9–7A.   Synthesized sweep generator *(Courtesy of Anritsu Wiltron).*

Figure 9–7B.   Synthesized sweep generator *(Courtesy of Hewlett-Packard Co.).*

Frequency Range   This range is the band of frequencies over which the sweeper will operate. Careful consideration should be given to this figure since a variety of plug-ins are available with many combinations of frequency ranges, as well as many synthesized models.

Frequency Accuracy   Frequency accuracy is how close to the set frequency the output actually is. It is expressed as a positive or negative figure to allow drift in either direction.

Frequency Stability   Frequency stability is how well the sweeper maintains its frequency when subjected to various conditions. Such factors as temperature, voltage and power changes, and VSWR at the output are all considered when specifying frequency stability.

Maximum Leveled Power   Maximum leveled power is the maximum amount of power out of the sweeper when it is in a *leveled* condition. The wording of this specification is very important because if the sweeper is unleveled it will be capable of producing more power than in a leveled condition. So be sure to know which condition is referred to when specifying output power.

Spurious Signals   These signals are any harmonics or nonharmonics present at the output of the sweep generator. The tracking filter will eliminate the biggest majority of these signals because of the steep skirts.

The data sheets shown are an interesting comparison of plug-in and synthesized single-unit sweep generators. Notice, the comparison of accuracy, stability, and leveled power outputs.

9.4   Power Meter

The microwave power meter is a device that is capable of measuring either CW (continuous wave) or pulsed peak power levels (some instruments will measure either in one unit). These power levels usually range from -70 dBm to $+44$ dBm (100 pw or 1×10^{-10} watts to 25.1 watts) for CW measurements, and the peak power levels useful with the appropriate power meter will range from -40 dBm to $+20$ dBm (0.1 µw to 100 mw). Frequency ranges for these meters are usually from the low MHz range (10 to 20 MHz) or kHz (100 kHz for example) to 40 GHz, with certain units being capable of going to 110 GHz. When you stop to think of this, it is a pretty remarkable instrument. The power meter by itself, however, is like having a rowboat without oars. It looks nice and has tremendous capability, but is almost useless. The power meter needs a *power sensor* to be used along with it. Maybe the *power meter* would be more appropriately called a *power system* rather than just a power meter. We will, however, refer to the instrument as a power meter with the understanding that we are referring to *both* the *meter* and the *sensor*.

Most power meters operate on the principle of a balanced bridge, which can be seen in Fig. 9–8. You can see the prominence of the bridge circuit and how it is used both for measurement and for compensation. The upper bridge circuit in Fig. 9–8 is used to measure the power presented to the power sensor. This power reading unbalances the bridge and causes a movement of the meter (or a digital readout if a digital display is used) proportional to the power at the sensor. The lower bridge is for temperature compensation of circuits. Notice that this voltage is combined with the voltage difference as a result of the RF bridge unbalance to give a true compensated power reading.

As we said previously, the power sensor is a very integral part of the power meter "system." You can see from Fig. 9–8 how much of a part it plays in the overall operation of the power meter. (It is the section marked *thermistor*.) The term *thermistor* is a broad term for a power sensing unit. This is a contraction of *thermal resistor*. The device makes use of the change in resistivity of a semiconductor as the temperature changes. Some power sensors rely on only a thermistor for sensing the temperature that coincides with a particular power level. This is a device that is very linear in operation. Other sensors are the diode sensor (very fast speeds possible with these devices), thermocouple sensor (consisting of two dissimilar metals exposed to the power being measured), and peak power sensor (designed specifically for pulse power measurement). Some sensors are also a combination of types, which results in the linearity of a thermistor and the speed of a diode sensor.

Fig. 9–9 shows some power meters with and without power sensors that are available.

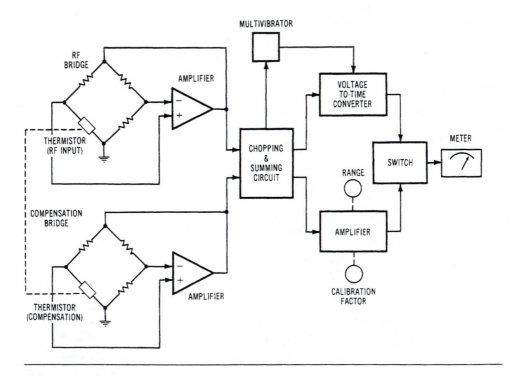

Figure 9–8.   Power meter block diagram.

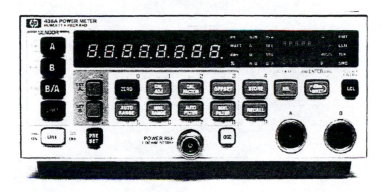

Figure 9–9A.   Power meter *(Courtesy of Hewlett-Packard Co.)*.

Figure 9–9B.   Power meter *(Courtesy of Hewlett-Packard Co.)*.

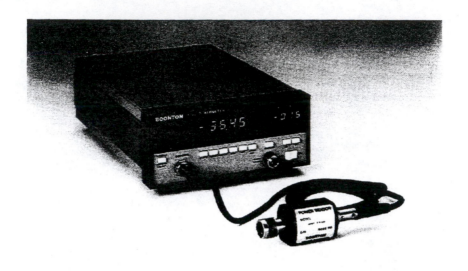

Figure 9–9C. Power meter (Courtesy of Boonton Electronics Corp.).

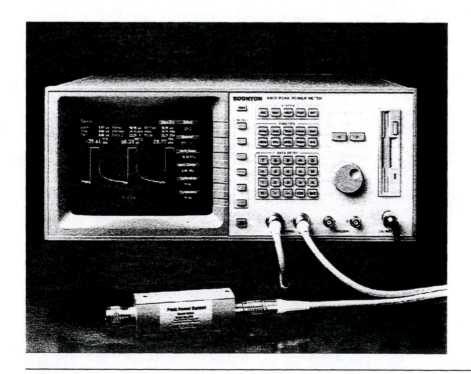

Figure 9–9D. Power meter (Courtesy of Boonton Electronics Corp.).

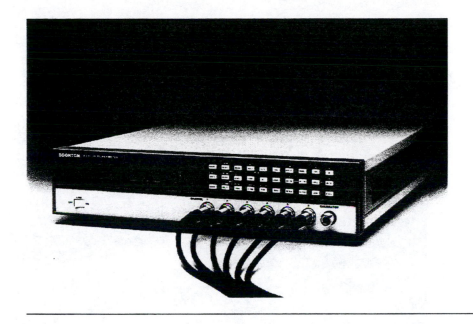

Figure 9–9E. Power meter *(Courtesy of Boonton Electronics Corp.).*

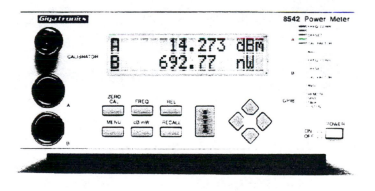

Figure 9–9F. Power meter *(Courtesy of Giga-tronics Incorporated).*

Figure 9–9G.   Power meter *(Courtesy of Giga-tronics Corp.)*.

9.4.1   Power Meter Data Sheet

To further explain power meters and sensors, we will present the following data sheets, first for the meter and then for the sensor. For the power meter (including a standard sensor):

Frequency range	100 KHz to 40 GHz
Power range	−70 dBm to +44 dBm

Accuracy	±0.5%
Noise	30-mW peak
Drift	±100 pW (over 1 hour)

Frequency Range   Frequency range is the band of frequencies over which power can be accurately measured. It is the range that both the meters and the sensor were built to operate over.

Power Range   This range is the maximum and minimum amount of power that can be measured with this meter. The range is broken up into a variety of steps so that each step contains a small power range that increases measurement accuracy.

Accuracy   This figure tells how close to the actual measured power the meter or digital display is reading. If, for example, we were measuring 10 mW (+10 dBm), the 0.5% accuracy would tell us that 9.95 to 10.05 mW could be read and still be within the accuracy of the meter.

Noise   This parameter is very important, especially at low power levels. If the noise of the instrument is high, the low power range will be limited because the meter will not be able to distinguish between the power to be read and the noise. This noise should be as low as possible so that accurate readings may be obtained at all power levels.

Drift   Drift is the amount the instrument will vary away from the normal with time. It may be due to temperature shifts, voltage changes, or other internal parameters.

9.4.2   Power Sensor Data Sheet

The power sensor also has specifications of its own. The terms and typical values are

Frequency range	50 MHz to 18 GHz
Power range	−33 dBm to +20 dBm
Max SWR	1.6:1

Frequency Range   This range must be the same band as for the power meter so that the meter and sensor will be compatible. It is the frequency range where all of the following specifications hold true.

Power Range   This range is the minimum and maximum value of power that the sensor can handle and still meet the specifications.

Maximum SWR   Maximum SWR tells how good a match the sensor is for a circuit being measured. You will sometimes see a variety of standing-wave ratios on a data sheet for various frequency ranges. The one SWR given here shows SWR as an overall figure for a frequency range of 10 MHz to 18 GHz.

9.5   Spectrum Analyzer

Perhaps the most versatile and most widely used piece of test equipment used in microwaves is the spectrum analyzer. This instrument is capable of measuring

- Power (relative and absolute)
- Frequency
- Noise
- Spectral purity
- Sensitivity
- Modulation
- Distortion

and many parameters too numerous to mention. The spectrum analyzer can make all these measurements because it operates in the *frequency domain* rather than the *time domain* of an oscilloscope. To distinguish the difference between these two, consider that when a signal is displayed on an oscilloscope, its amplitude and period are seen with respect to time. If the same signal were displayed on a spectrum analyzer, once again its amplitude would be seen, but its frequency components also would be seen—that is, its fundamental frequency and any harmonics it may contain. So you can see the difference in the domains and how the frequency domain gives much more information.

A block diagram of a typical spectrum analyzer is shown in Fig. 9–10. This arrangement can be recognized as a swept-front-end superheterodyne receiver, which makes a lot of sense since the spectrum analyzer is actually a swept-tuned receiver. Signals to be analyzed are applied at the analyzer input. There is usually a step attenuator prior to the mixer that protects the mixer and allows signals of different levels to be checked. The voltage-tuned local oscillator that supplies the LO to the mixer is swept at the same rate as the horizontal deflection system of the display. Both of these circuits are controlled by the sawtooth generator. The mixed signal is amplified (IF amp) and detected before being applied to the vertical deflection system of the display. The result is a display of signal amplitude as a function of *frequency*, that is, a display of the *frequency domain*.

A spectrum analyzer available commercially is shown in Fig. 9–11.

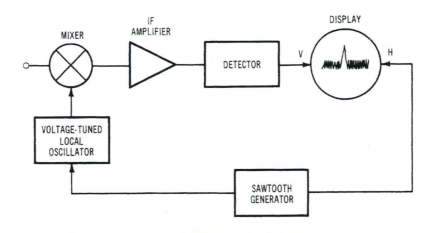

Figure 9–10. Spectrum analyzer block diagram.

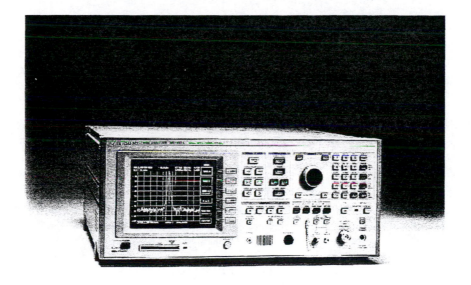

Figure 9–11. Spectrum analyzer *(Courtesy of Anritsu Wiltron).*

9.5.1 Spectrum Analyzer Data Sheet

A portion of a typical data sheet for a spectrum analyzer is shown here:

Frequency range	100 Hz to 32 GHz
Frequency spans	
Full band	Band selected swept in one band
Per division	10 Hz to 3 MHz/division in 1–3 sequence
Resolution bandwidth	10 Hz to 3 MHz in 1–3, sequence
Measurement range	Noise level <-103 dBm to $+30$ dBm damage level
Input SWR	$<1.5:1$ to 1.8 GHz; $<2.0:1$, 1.7 to 32 GHz

There are many parameters that characterize a spectrum analyzer that are not presented. Most analyzer specifications are grouped into frequency, amplitude, and input characteristics. The parameters listed are a sort of summary of these characteristics. Keeping in mind that these parameters do not completely characterize an analyzer, we will provide a brief description of each parameter presented.

Frequency Range   Frequency range is the range of frequencies over which the spectrum analyzer will accept frequencies and analyze them. This range can be extended upwards to 300 GHz by use of external mixers.

Frequency Spans   This parameter tells how much of the total frequency capability of the analyzer is displayed on the CRT. In some cases the entire range is displayed. Other times a single band, selected portions of a band, or a single fixed-tuned setting is displayed.

Resolution Bandwidth   This parameter tells the bandwidth of internal filters used in the analyzer. These filters allow the operator of the equipment to select a filter that will best show the signal being analyzed. Most analyzers have selectable bandwidths in 1 and 3 sequences.

Measurement Range   This parameter shows the lowest level the analyzer, will accept and still be able to distinguish between the signal and the internal noise level of the instrument. It also shows the maximum level that may be applied to the analyzer without damaging the input circuitry.

Input SWR   This parameter is the input match of the analyzer, presented in two figures in our case. This method is common for specifying SWR over a broad range of frequencies.

9.6   Frequency Counters

Frequency counters are a vital part of many microwave test setups. They ensure that the test setup is set at the correct frequencies for the testing needed. Correct frequencies are essential for conducting the *proper* tests on your component or system. Some commercially available counters are shown in Fig. 9–12.

Direct-reading counters are limited to about 500 MHz, which can hardly be called microwaves. There must be, therefore, other methods by which higher frequencies can be measured. Three methods are commonly used:

- Prescaling—frequency ranges to 1.3 GHz
- Heterodyne down-conversion—frequencies to 20 GHz
- Transfer oscillator down-conversion—frequencies to 40 GHz and higher

The prescaling techniques use a process of dividing the input frequency to produce a lower frequency that can be read by direct counting as previously mentioned. The key

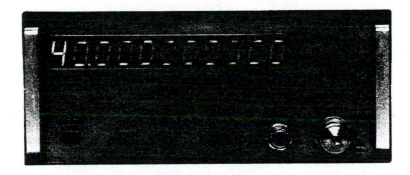

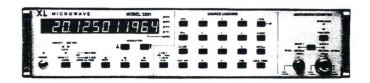

Figure 9–12.   Frequency counters *(Courtesy of XL Microwave).*

element in prescaling is the integer N. This number usually ranges from 2 to 16 and is used to either increase the gate time of the counter by this integer or to multiply the content of the counter by N.

Heterodyne down-conversion involves a mixer that beats the incoming frequency against a high-stability local oscillator. Once again, this lower frequency is counted by a direct-reading counter once this mixing has occurred. This version of counting has a basic layout similar to the spectrum analyzer. In one case the frequency is read on a digital display (counter), and in the other case the signal is displayed on a CRT (analyzer).

The heart of the transfer oscillator technique is the phase-locked loop. Remember the discussions of the loop in our discussions on synthesizers. A low-frequency oscillator is phase locked to the incoming signal in this application. Once again a direct-reading counter can be used to read the frequency. This technique has many similarities to the previously mentioned synthesizers. It even has been referred to as a "synthesizer-in-reverse" technique of frequency counting.

The counters referred to are all for cw signals. When pulse signals must be measured, a special type of instrument is needed. This type of instrument uses a combination of prescaling and heterodyne down-conversion to accomplish these readings.

9.6.1   Frequency Counter Data Sheet

A typical data sheet for a microwave frequency counter is shown here:

Frequency range	120 MHz to 20 GHz
Sensitivity	−25 dBm
Maximum input	+10 dBm continuous

Frequency Range   This range is the limit of frequencies the counter will read accurately. If this range is violated, either on the high or low end, the counter will display random numbers that may be harmonics, subharmonics, or have nothing at all to do with the signal at the counter input. Also note any frequency ranges printed over different connectors. Some counters have more than one input for different frequency bands. So be sure to put the signal into the proper input connector.

Sensitivity   This parameter is the minimum signal level the counter needs at its input for proper operation. If the counter does not have this proper level, it will read erratically similar to the situation described previously when we were under or over the counter's operating frequency range.

Maximum Input   Maximum input is one of the most important parameters found on a data sheet. It is the highest signal level that can be applied to the counter without damage to the instrument. *You should never exceed this level.* Some counters will also display the level of the input signal.

9.7   Noise-Figure Meter

In some of the previous discussions on test equipment, we have mentioned such areas as sensitivity, minimum signal level, and measurement range. Each of these parameters is a description of low-level signals. These low-level signals can be completely lost, or masked, if the *noise* of the particular instrument is too high. A high noise level means low sensitivity, high minimum-signal level, and a limited measurement range for the instrument.

To characterize each instrument or system, we must know what the noise level is. Fig. 9–13A shows a block diagram of a common noise-figure meter. Probably the most critical block in Fig. 9–13 is that of the *gating source*. This block virtually controls the meter and allows noise-figure measurements to be taken. This statement can be more fully realized by looking at the actual operation of the meter.

9.7.1   Noise-Figure Meter Operation

To begin with, we must first present the basic relationship that defines noise figure. This relationship is

$$F_{dB} = 15.2 - 10 \log (N_2/N_1 - 1) \qquad (9.3)$$

where, F_{dB} is the noise figure in decibels,
N_2 is the noise power with the noise source on (hot),
N_1 is the noise power with the noise source off (cold).

N_1 and N_2 are graphically illustrated in Fig. 9–13B. N_1 will remain constant since the instrument and termination noise will not change. N_2 is the variable since the excess noise with its associated gain is what the meter is attempting to read.

Obviously, N_1 and N_2 and their associated timing will determine the operation of the noise-figure meter. This operation is as follows:

The gating source pulses the noise source; N_1 and N_2 pulses arrive at the IF amplifier. Noise sources have a finite noise build-up time, so the IF amplifier is gated to pass only the final amplitudes of N_1 and N_2 to the square-law detector. The detected N_2 pulse is switched to an automatic gain control (AGC) integrator where a voltage for gain control of the IF amplifier is derived. The time constant of this circuit is made long enough to control the IF amplifier gain even when the N_1 pulse is passing through it. Since the AGC action keeps the detected N_2 pulse at a constant level, a measurement of the detected N_1 pulse is, in effect, a measurement of the pulse ratio. The N_1 pulse is measured by switching it to the meter integrator and meter.

Convenient internal calibration of the meter is accomplished by artificially creating readings of "+∞" and "−∞." By pulsing the noise source during both the N_2 and N_1 time periods, we obtain a condition of $N_2 = N_1$. In the formula $F_{dB} = 15.2 - 10 \log (N_2/N_1 - 1)$, this condition results in a noise figure of +∞. The artificial condition of F = −∞ would correspond to an "N_1" value of "0." This condition can be created by gating "off" the IF amplifier during the "N_1" time period. If the metering circuit is designed to be a linear indicator of the

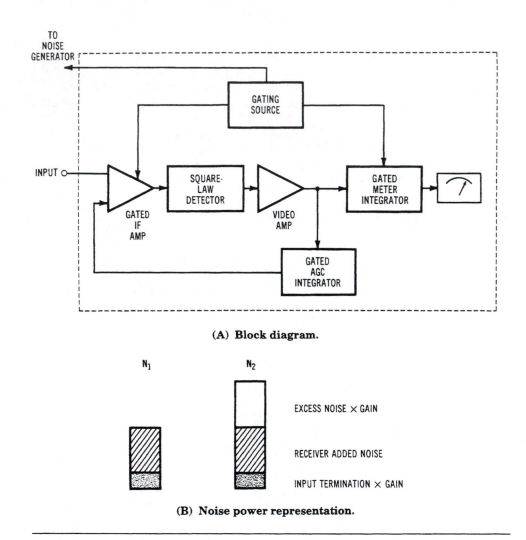

(A) Block diagram.

(B) Noise power representation.

Figure 9–13.   Noise-figure meter.

power of "N" (square-law detector) and the meter minimum position is calibrated as $-\infty$ and the full-scale deflection as $+\infty$, all other points on the meter face can be calculated by the formula $F_{dB} = 15.2 - 10 \log (N_2/N_1 - 1)$. For example, an "$N_1/N_2$" ratio of 1/2 would bring about a mid-scale reading. From the formula this mid-scale reading is calculated to be 15.2. In a similar fashion the balance of the scale is calibrated.

Notice from this operational discussion how the gating source controls the meter and its operation. By various gating sequences the values of N_1 and N_2 can be used to provide accurate and repeatable noise figure readings. Fig. 9–14 shows an available noise-figure measurement system. One of the problems you run into when measuring noise characteristics is

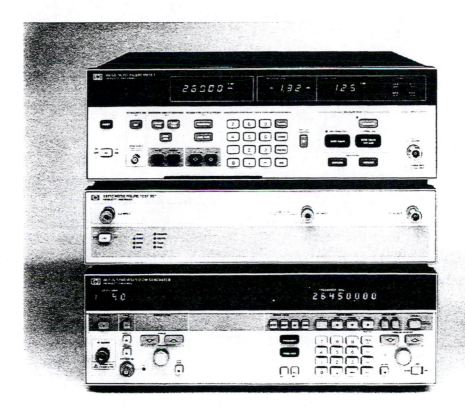

Figure 9–14.   Noise figure meter and test set *(Courtesy of Hewlett-Packard Co.).*

the correlation of noise figure with gain. Many times a circuit is set to a required gain, and when the noise figure is read it is out of spec. Adjustment of the circuit for the required noise figure results in the gain now being out of spec. The noise figure measurement system in Fig. 9.14 eliminates all of this. It can be seen that the noise figure meter (top part of the equipment) displays the frequency, insertion gain, and noise figure simultaneously. This is much better than having to have two setups, one for gain and one for noise figure.

As accurate as the noise-figure meters may be, they still rely on noise sources for their ultimate operation. There are a variety of noise sources available. There are hot/cold sources, gas-discharge sources, and solid-state sources.

The hot/cold noise sources are highly accurate sources capable of producing noise measurements with better than ±0.1-dB accuracy. The noise source standard employs two resistive terminations: one is immersed in liquid nitrogen (77.3K or −195.7°C); the other is in a proportionally controlled oven that is set to the temperature of boiling water (373K or 100°C). Terminating impedances are carefully controlled so that they track each other, in both magnitude and phase, over the full frequency range of the instrument. Thus, mismatch uncertainties are virtually eliminated. These sources are used as standards.

Solid-state noise sources are available to cover the range from 10 MHz to 26.5 GHz. These devices are both small and high in reliability. The noise-generating diode used is isolated from the output by 12 to 20 dB; consequently, there are almost no mismatch errors as the source is switched on and off, and no destructive transients are present in the output. Noise sources are shown in Fig. 9–15.

9.7.2 Noise-Figure Data Sheet

A typical data sheet for a noise figure will wrap up our discussions of the instrument.

Frequency range	10 MHz to 26.5 GHz
	(dependent on noise source)
Noise-figure range	0 to 30 dB
Input frequency	10 to 1600 MHz
Input power level	−10 dBm max

Frequency Range Frequency range is the range over which the meter will read the noise figure. Note that it depends on the noise source since the source is used as a generator in the same manner as if it were one of the previously discussed generators. Be careful when checking frequencies that the specifications deal with the microwave range and not the input frequency range. These are two drastically different numbers.

Noise-Figure Range The noise-figure range gives the smallest and largest noise figures the meter is capable of reading. This example range is a very typical range, although some meters will go to 33 dB in ranges rather than continuously.

Input Frequency Input frequency indicates the actual input frequency to the meter. The specification shows a range of frequencies from 10 to 1600 MHz. If the signal going into the meter falls within this range, the meter will respond. Some noise-figure meters may have fixed frequency inputs (30, 40, 60 MHz, for example). These are narrow-band inputs that require that the mixing process going from microwaves to the proper intermediate frequency be very precise.

Input Power The level shown is a maximum level, and to exceed it would mean instrument damage. There is also a *minimum* level for the input, and the particular operator's manual should be consulted. An additional 40 to 50 dB of IF gain usually must be supplied to ensure staying above this minimum level. If you are not in the operating range (level) of the meter, it will not zero or read infinity, and thus will not operate properly.

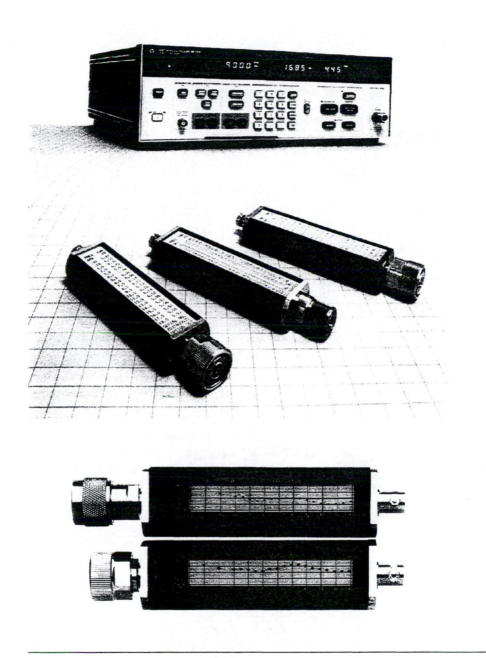

Figure 9–15. Noise sources *(Courtesy of Hewlett-Packard Co.).*

9.8   Network Analyzers

When the term *network analyzer* is used, it usually brings to mind an elaborate test setup with complicated calibration and output data that requires extended time to become useful. This is not what the microwave analyzers we will be covering here are all about. These analyzers are a combination of many signal sources, measurement systems, and indicating techniques all rolled into one physical container to provide accurate and efficient methods of characterizing microwave components and systems.

There will be two types of network analyzers discussed: *scalar* and *vector*. The scalar analyzer is concerned with the amplitude characteristics of a component or system as a function of frequency while the vector analyzer is concerned with *both* amplitude and phase.

9.8.1   Scalar Network Analyzer

As mentioned earlier, the scalar network analyzer measures only the amplitude of a signal. This does not mean that this type of analyzer is inferior in any way; it just states the capability of the system. There are, in fact, many applications where only amplitude readings are required. In this section we will not cover any specific scalar analyzers but will approach the topic in a general way by giving typical parameters and features.

Scalar network analyzers have become much more than a display unit with a few buttons to be pushed. In many cases they are complete microwave test systems containing an internal source and various measurement devices. Fig. 9.16 shows some of these scalar network analyzers.

The heart of the scalar analyzer is a component we covered in Chapter 4 (Microwave Components) and re-introduced in Chapter 8 (S-Parameters). That component is the directional coupler or, more precisely, the dual-directional coupler. You will recall that the directional coupler would pass energy easily in one direction while isolating that energy in the reverse direction. This is an ideal arrangement when comparing forward and reverse power readings, which the scalar analyzer is designed to do. By using the appropriate inputs and terminations, all four S-parameters can be obtained using dual-directional couplers and comparing the data obtained from each port.

The analyzers shown in Fig. 9.16 are available with various frequency ranges. The Anritsu Wiltron scalar analyzer is available in ranges of 0.001 to 1.5 GHz, 0.001 to 2.2 GHz, 0.001 to 3 GHz, 0.01 to 8.6 GHz, 0.01 to 20 GHz, 2 to 8.6 GHz, 8 to 12.4 GHz, 12.4 to 20 GHz, 10 to 16 GHz, 17 to 26.5 GHz, and 2 to 20 GHz. The Marconi Test Set is available in ranges of 10 MHz to 46 GHz, 10 MHz to 20 GHz, 10 MHz to 26.5 GHz, and 10 MHz to 46 GHz. And, finally, the Hewlett-Packard analyzers are available in ranges of 300 kHz to 3 GHz, 10 MHz to 20 GHz, and 10 MHz to 40 GHz. The analyzers as they are shown in Fig. 9.16 will not really accomplish much in the way of measuring parameters. What makes them useful and operational are the detectors that accompany the basic test instrument. Without the proper detector attached to the test unit, the operator will not obtain

Figure 9–16A. Scalar analyzer *(Courtesy of Marconi Instruments).*

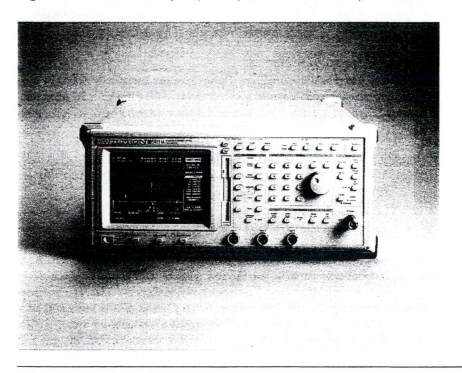

Figure 9–16B. Scalar analyzer *(Courtesy of Anritsu Wiltron).*

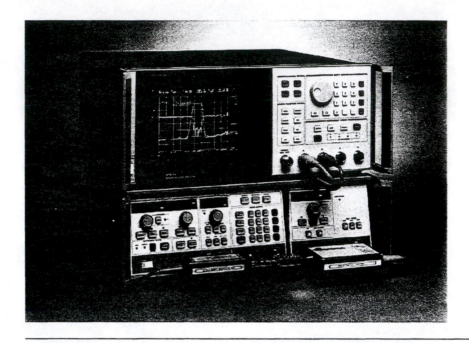

Figure 9–16C.   Scalar analyzer *(Courtesy of Hewlett-Packard Co.).*

accurate and repeatable readings. These detectors are usually Schottky diode detectors that have a measurement range of approximately −60 dBm to +20 dBm (an 80 dB dynamic range). Typical specifications for a useful microwave network analyzer detector would be

Frequency Range	0.01 to 20 GHz
Impedance	50 Ω
VSWR	1.5:1 max to 20 GHz
Input Connector	SMA (m)

The *frequency range* is the range over which all of the other parameters are valid; the *impedance* value shows that the circuit being tested is being terminated in a good characteristic impedance (50 Ω); the VSWR shows just how well the detector is being matched to the rest of the circuitry; and the *input connector* (Subminiature Type A) tells the operator which connectors are to be used so that adapters will not be necessary.

The Marconi Microwave Test Set, shown in Figure 9.16A, is a good unit to summarize how versatile scalar network analyzers have become. It is classified as having many instrument functions in one unit. It advertises a synthesized sweep generator, four input scalar analyzers, color display, accurate power measurements, integral frequency counter, real time fault location, and programmable V/I source for component evaluation. There are not many more things you could request from a test unit.

9.8.2 Vector Network Analyzer

The vector network analyzer is a test system that measures both amplitude *and* phase or, as the name indicates, a vector quantity. This combination gives a tester a complete picture of what a component or system is doing.

The vector analyzer came into existence in 1967 when Hewlett-Packard introduced the HP8410 system. This instrument remained unchallenged for a period of 20 years. From 1987 until the present, however, there have been other vector network analyzers that find applications in many different areas of microwaves. We will look at some of the more recent models and, once again, refrain from describing specific models but cover a variety of analyzers in a general approach.

Fig. 9.17 is a vector network analyzer designed to operate from 40 MHz to 40 GHz. Notice the lower portion of the analyzer. This is a built-in S-parameter test set that results in allowing an operator to obtain S_{11}, S_{12}, S_{21}, or S_{22}, whichever is needed. The display shown has all four parameters displayed. The upper left corner is S_{11}; the lower left corner is S_{21}, both amplitude and phase; the upper right corner is S_{12}, amplitude and phase; and the lower right corner is S_{22}. This entire display can be presented, or the operator can choose only one. Typical parameters for this analyzer are

Frequency Range	40 MHz to 40 GHz
Dynamic Range	>80 dB
Leveled Power at Port 1	−15 dBm (0.0316 mw)
Test Port Match	1.02:1 (38 dB return loss)
Max Input Power	+20 dBm (100 mw) max

The last two parameters shown above are important in providing accurate and repeatable measurements. First, the test port match should be as close to a perfect match (VSWR = 1:1) as possible so that you can be assured that you are really displaying the response of the device being tested, not that of the analyzer. Second, the maximum input power specification should be observed so that you can be sure you are not overdriving the front end of the analyzer or, even worse, not destroying the front end.

For the vector analyzers available there are certain standard features that should be present. Some of these are as follows:

- Fast sweeping synthesized source
- Automatic reversal of test ports when measuring S-parameters
- Clear, concise displays
- Internal hard and floppy disk drive
- GPIB, RS-232, HP-IB interfaces

These features will greatly enhance the measurements obtained from your vector network analyzer.

Figure 9–17.   Vector network analyzer *(Courtesy of Anritsu Wiltron).*

9.9   High-Frequency Oscilloscopes

Until a few years ago you would never see a special session on oscilloscopes in a microwave text. The only place you would see an oscilloscope in a microwave test setup was when you attached a detector to a microwave circuit and displayed the detected output on the scope. In the past the highest frequency a scope could handle was 100 MHz. This, thankfully, is no longer the case. This is not to say that any microwave frequency can be displayed directly but that the frequency response is getting higher. Oscilloscopes with 500 MHz, 600 MHz, 1 GHz, and 6 GHz bandwidths are available for use in microwave testing. These scopes are not the conventional oscilloscopes of years ago; they use highly refined digital technology to accomplish these bandwidths.

A representative set of specifications for a 6 GHz digitizing oscilloscope is shown below.

Signal Bandwidth	6 GHz
Rise Time	58.3 nsec
Vertical Resolution	14 bits (16,384 levels over 10.24 vertical divisions)
Sensitivity Range	2 mv/div to 200 mv/div
Random Noise	1.2 mv (max), typ. 600 µv

Many of these oscilloscopes have a variety of plug-ins that can allow the operator to switch to get a high sample rate (2 Gs/sec), high bandwidth (4 to 6 GHz), or high channel count (as many as eight channels). This adds greatly to the value of these devices. Many of the units are menu driven and will provide hard copy outputs if desired.

Thus, it can be seen that the oscilloscope has truly invaded the previously unapproachable world of microwaves.

9.10   Chapter Summary

In this chapter we have covered the three equipment groups outlined in the beginning. We discussed the generation group (signal generators and sweepers), the indication group (power meter, spectrum analyzer, frequency counter, and noise-figure meter), and the system group (the scalar and vector network analyzers), and finished up by discussing briefly the entrance of the digital oscilloscope into the world of microwaves. Each group of test equipment has had some *typical* specifications presented. It should be stressed that these are typical. Your particular requirement will determine just which parameters are important for your test setup and which are not.

9.11   References

1.   *Electronic Instruments and Systems*. Hewlett-Packard Co., Palo Alto, CA, 1982.

2.   Laverghetta, T. S. *Handbook of Microwave Testing*. Dedham, MA: Artech House, 1981.

3.   *PM1038/NS20*. Product Bulletin, Pacific Measurements, Inc., Sunnyvale, CA.

4.   *Sweep Oscillators*. Tech Data 5952–9321(D), Hewlett-Packard Co., Palo Alto, CA, 1981.

5.   *Microwave Frequency Synthesizers*. Watkins-Johnson Co., Palo Alto, CA, 1979.

6.   "7000 Series Synthesized Microwave Signal Source," Giga-Tronics, Inc., GT-082R, Pleasant Hills, CA, 1993.

7.   "7100 Series Synthesized Microwave Signal Generator," GT-083A, Giga-Tronics, Pleasant Hills, CA, 1993.

8.   "68100B Series Synthesized Sweep Generator; Data Sheet 68100B-2, Wiltron Co., Morgan Hills, CA, December, 1993.

Questions

9.2   Signal Generators

   1.   Define *signal generator*.
   2.   What is the purpose of a buffer amplifier in a signal generator?
   3.   Is automatic leveling necessary for a typical signal generator?
   4.   Define *CW*.

5. Name three parameters within the spectral purity specification.
6. What is a synthesized generator?
7. Explain direct synthesis.

9.3   Sweep Generators

8. Explain what a YIG oscillator does.
9. What is one advantage of a plug-in sweeper?
10. What is one disadvantage of a plug-in sweeper?
11. How does the frequency stability of a synthesized sweep generator compare to that of a plug-in sweeper?

9.4   Power Meter

12. Define *thermistor*.
13. Name three types of power meter sensors.
14. What is the primary purpose of a sensor?

9.5   Spectrum Analyzer

15. Name three parameters that can be measured with a spectrum analyzer.
16. Is the spectrum analyzer a time domain or frequency domain device? Explain.
17. Why is it much more acceptable to have an attenuator at the spectrum analyzer input?

9.6   Frequency Counters

18. Name three methods used when designing frequency counters.
19. What is *sensitivity* in a frequency counter?
20. Why is the maximum input parameter important in a frequency counter?

9.7   Noise-Figure Meter

21. What is the correlation of noise figure and gain?
22. What type of noise sources can be used with noise-figure meters?

9.8   Network Analyzers

23. Define *scalar network analyzer*.
24. Define *vector network analyzer*.
25. Explain how the directional coupler is used in network analyzers.
26. Why does the vector network analyzer produce more information than a scalar analyzer?

9.9   High Frequency Oscilloscopes

27. How was the oscilloscope used in microwaves when its bandwidth was limited to 100 MHz?
28. Name some high frequency oscilloscope bandwidths now available.
29. Why is *rise time* important in these high frequency scopes?
30. Is it an advantage or disadvantage to have plug-ins for the high frequency scopes?

Problems

9.2 Signal Generators

   1. Determine the output frequency range of a synthesizer that has a single crystal oscillator at a frequency of 13.3 kHz and a dividing ratio of 32.
   2. Determine the open loop gain (in dB) of a synthesizer that has $A_d = 0.35$ V/rad, $A_a = 8$, and $H_o0=45$ kHz/V.
   3. We have a synthesizer with a reference oscillator of 22.8 kHz and a divider ratio of 8. What is the frequency range?
   4. A frequency synthesizer has a comparator gain of 0.42 V/rad, an amplifier gain of 22, and a VCO transfer function of 18 kHz/volt. What is the open loop gain as a ratio and in dB?

9.7 Noise-Figure Meter

   5. We have a system that has a noise power of 5 mw when the source is off, and we want to have a maximum of 10 dB noise figure. What is the most the noise can be when the noise source is on?
   6. If the noise figure is 13.5 dB and the noise read with the source on is 25 mw, what is the required power level when the source is off?

10 Microwave Measurements

Objective

This chapter is designed to present basic microwave measurement theory and test setups so that the reader can understand how the more complex automatic test systems operate. The chapter will cover representative setups for transmission, impedance, power, noise, frequency and active testing measurements.

Key Terms

Audio Substitution
Gain Compression
High Power
IF Substitution
Impedance
Intermodulation Products
Low Power
Medium Power
Noise Factor

Noise Figure
Noise Temperature
Reflectometer
RF Substitution
Slotted Line
Thermistor
Transmission
VSWR Bridge

10.1   Introduction

In the previous chapters we have discussed various components and aids used in microwaves—transistors, diodes, materials, S-parameters, and the Smith chart, to mention a few. This chapter will tie these together by showing how to test that special device you have worked so hard to design. Until this design, whether it be a simple component or a complete system, has been tested, it is almost totally useless. Therefore, to perform the correct and proper tests that will characterize the design and prove it actually does what it is intended to do is very important.

This chapter will cover six areas of testing. For each area we will present representative test setups and explain what each part of the setup does. These setups are not intended to be memorized and treated as the only way to perform a specific test. Rather, they are examples that you can build your own special test setup around. Each of the test setups we will show is a setup consisting of individual pieces of test equipment: generators, couplers, filters, meters, or other test equipment. In many actual cases, all of these functions may be performed within a single unit where many functions are measured at the same time and automatic testing results. This is an excellent way to test microwave components or systems. The individual setups are presented here to let the reader see and understand what is going on inside these automatic units so that he or she can become a knowledgeable test person rather than simply button a pusher.

The areas to be covered in this chapter are

- Transmission measurements
- Impedance measurements
- Power measurements
- Noise measurements
- Frequency measurements
- Active testing

10.2   Transmission Measurements

The general implication when the phrase *transmission measurement* is mentioned is one of a measurement of loss through a device or system. This implication is generally true of most transmission tests but not all. There are times when a transmission measurement involves an overall gain. For this reason, we should derive a definition that will cover both cases—loss and gain.

The formal definition of the word *transmit* is "to send or transfer from one place to another." This is a good starting point for our definition since we all agree that the ultimate goal of a microwave system is to transfer energy from one place to another. However, there must be more to our definition than just saying how much energy was transferred from point A to point B. The key to the definition is what takes place between points A and B since the action between the points is what determines how much energy appears at point B. Consider the following example, which illustrates how important the idea is.

Suppose we let a hollow pipe be our device or system and a clay ball be the microwave energy. What would happen if the clay ball were rolled through the pipe? How would it appear at the other end? How long would it take to get there? Obviously, the answers to these questions would depend on how the pipe was constructed—in other words, what the inside of the pipe was like. If the pipe has a very large diameter, the ball will roll through very easily. If the pipe has a very small diameter, it will not go through at all. If the pipe has sharp edges on the inside, the ball will appear scarred and nicked when it arrives at the other end. If the inside contains pieces of wet clay, the ball will probably be larger at the output end than at the input since it will pick up small pieces of clay as it travels through the pipe. The example could go on, but the idea would be the same. The energy (or ball) is transmitted from one point to another. The transmission characteristics of this or any other system will depend on what takes place between the beginning of the transmission (the input) and the end (the output). This example should help to show the importance of what takes place within a device and how what takes place can either increase or decrease the microwave energy (or clay ball).

Up to this point we have stated a good definition vaguely many times. Now let us put it down in a good, concrete form. From all of the information, we can define a transmission measurement as follows:

> A transmission measurement is a measurement of the internal action of a device or system on a microwave signal. This internal action may act to increase, decrease, or have virtually no effect on the signal.

This *internal action* is measured by comparing the resulting output signal with the known input signal. For instance, in a cross-country race 100 runners may start (go into the input of the course). If only 25 runners finish (are at the *output* of the course), the *internal action* of the course (the difficulty of it) has resulted in a "loss" of 75 runners. The same type of comparison is used when measuring the transmission characteristic of an attenuator, amplifier, or mixer. The transmission measure is a measure of the output with respect to the input, and it describes the internal action of the device on a microwave signal.

With a definition for transmission measurements established, the next step is to proceed with some actual measurements. For the sake of simplicity, we will concentrate only on transmission measurements that involve losses or attenuation. This may seem like a contradiction of our opening statement, but the subject of transmission gain is to be covered in Section 10.7 under Active Measurements. That section is a much more appropriate place for it, and there it can receive proper coverage.

Although several procedures for making attenuation (or loss) measurements exist, the three used most generally are

1.   RF substitution method,
2.   IF substitution method,
3.   Audio substitution method.

For each of these methods we will recommend equipment to be used, go through a setup, explain each piece of equipment in it, cover the measurement procedure, point out areas where errors or ambiguities may exist, and give alternate methods where applicable. The entire procedure will be presented in such a way that you can analyze any test setup you may need for a transmission test rather than having to rely on your memory as we previously mentioned.

The three methods listed will first be covered individually for cw measurements. They then will be grouped together and presented in the final paragraphs for swept-frequency measurements. In this way we can eliminate some confusion by keeping the cw and swept measurements under separate sections. (Remember also that these setups may not be visible to the tester but are incorporated into an automatic test setup. Regardless of this fact, they should be studied and learned.)

10.2.1   RF Substitution Method

RF substitution means that all the calibration and adjustment within the test setup is done in the RF (microwave) line as shown in Fig. 10–1. Notice that the *calibration attenuator* is directly in line with the cw source, which is a microwave signal. There is no frequency conversion before the attenuator. This puts additional requirements on the attenuator if it is required to run transmission tests over a wide frequency range. Over a wide range, the attenuator must hold accurate calibration over the entire frequency range to be covered. This requirement is eased somewhat, however, by the use of the microwave network analyzers covered in the previous chapter. With some of these analyzers, the variations in attenuator calibration are put in a memory and stored. Then, they are subtracted from the final output of the device being tested to give only the response and performance of the device. There are limits, however, as to how far the calibration can vary and still be taken care of by the memory within the test unit. So obtaining accurate attenuators is advisable when using the RF substitution method to avoid problems with calibration and measurement accuracy further down the line.

Basic Setup   To understand the basic setup, we will examine each block in Fig. 10-1 and explain its purpose. We must emphasize that this diagram is only a representative setup for an RF substitution transmission test. It should not be taken as gospel and adopted as the only setup to ever use for transmission testing. It is designed to show the basic elements from which to build the right setup for your application.

The cw source could be the signal oscillator or synthesizer covered in the previous chapter. It may even be a sweep generator operated in a cw mode. The main criterion is that the source cover the frequencies needed to make your particular test. One point should be made about the cw source; that is, it is a good idea to have a source with an operating band such that your desired operating frequency is somewhere near the center of the band. This will improve the accuracy and stability of the signal and will result in more accurate readings. At either of the ends of the bands, there is a tendency for the frequency to drift and the amplitude to change with time, temperature, and other factors. So, if at all possible, use a source with the frequency near the center of its band.

The filter in Fig. 10–1 is listed as "optional" because it can be used as a matter of preference or as your application may require. The filter may be a bandpass filter or a high

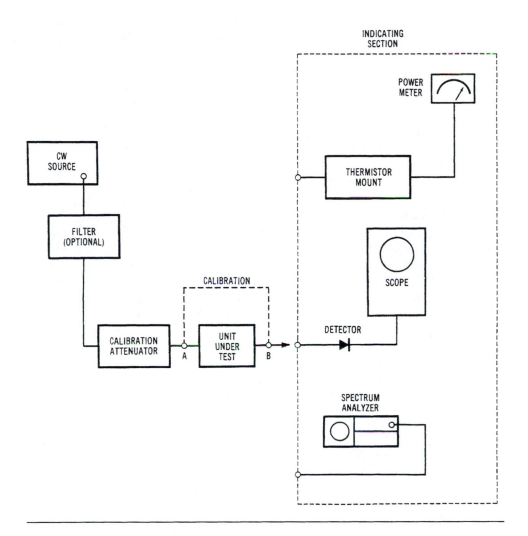

Figure 10–1.   RF substitution test setup.

pass. Its primary job is to remove any low-frequency signals, spurious responses, or harmonics that may cause problems later on in the test setup.

The calibration attenuator is one of the most important parts of the setup in Fig. 10–1. Normally a variable attenuator (either step or continuous) is used, but a fixed value could be used. Three parameters of importance are in this attenuator:

- Attenuation accuracy
- Minimum insertion loss
- VSWR

The attenuation accuracy is important since it is the governing component in the overall RF accuracy of the setup. Remember that this component determines that this setup is called an RF substitution test. The insertion loss of the attenuator should be low so that it will not be a significant loss in the overall setup. Rather, it should be a component that simply provides a calibration reference and a final attenuation (or gain) figure as an output.

VSWR is always a prime factor whenever a component is inserted into a setup. The VSWR should be as low as possible so that it does not disturb any of the surrounding circuitry. A low VSWR will enable the readings obtained and the overall setup to exhibit a much higher degree of accuracy.

Three choices are shown in Fig. 10–1 for indicating the transmission characteristics of the device being tested because there is no one best way of displaying the final output. You may have a personal preference, or your choice may be based solely on what equipment is available at the time you are running the tests.

The first indicating scheme is a power meter with its associated thermistor mount. This measurement, of course, is a direct RF reading and may be made with a variety of meters and mounts. One area to consider seriously is the input power level to the thermistor mount. It may be necessary to place an attenuator before the mount if there is a danger of exceeding the maximum input level to the mount. Be sure that the attenuator used is adequate and calibrated over the frequency ranges being used. This attenuator should be included in any calibration runs, also.

The second method was previously shown in Chapter 9. This method uses a detector and oscilloscope combination. Once again, an attenuator may be placed in front of the detector to protect it from being overdriven. All the precautions and procedures discussed previously will also apply to the attenuator placed before the detector. One statement should be added before moving on from this method: an oscilloscope display will not be a linear display and, as such, could be difficult to read if high-attenuation values are being attempted. In other words, the range of operation is limited with this type of display method. Also as discussed in Chapter 9, there are some oscilloscopes that do not require a detector in front of them. Recall that oscilloscopes with bandwidths up to 6 GHz are available. If this equipment is available in your lab, you may want to use it for a direct reading on a scope.

The final method shown uses the spectrum analyzer, and it once again is a direct RF reading method. This method will give the most information of all three methods shown.

Basic Procedures   With the test setup presented and explained, we can go through a basic procedure. It would go as follows:

- Connect the test setup as shown in Fig. 10–1 and set the cw frequency needed on the cw source.

- Connect points A and B together.

- Set a convenient level on the indicating device chosen. Remember to observe input power levels to all three methods. The reference level is set by the calibration attenuator.

- Record the reference level.
- Insert the device to be tested between points A and B, and note the new reading on the indicating device. The difference between this reading and the reference level is the transmission characteristic of the device being tested.

The RF substitution method is probably the most widely used method for making bench transmission test measurements. One reason is the availability of equipment to make such a setup. One drawback to this setup is, however, that the calibration attenuator must maintain a calibration over a wide frequency range when used in an RF substitution mode. If this range is broad, the calibration may not be accurate enough.

10.2.2   IF Substitution Method

In Chapter 9 we covered network analyzers (both scalar and vector). It was not mentioned at that time, but all of these analyzers use an IF substitution method for obtaining information on specific components or systems. This method is one of the most accurate means of measuring high values of attenuation. By *high attenuation* we mean 50 dB, 100 dB, or greater. The setup in Fig. 10–2 shows the secret of the high degree of accuracy—the process called *heterodyning*. This technique is widely used in high-sensitivity, wide-dynamic-range receivers and is the heart of many measuring devices. The technique is a mixing process in which an RF signal (the signal under test) is mixed with a local oscillator signal to form a lower, or intermediate frequency (IF). In Fig. 10–2 the RF signal is $f_1 - f_2$.

Also, an output from the mixer of $f_1 + f_2$ is not used because it is higher than either of the two RF signals, and our objective is to make our measurement at a lower frequency to enable us to use much more accurately calibrated components.

Basic Setup   The setup in Fig. 10–2 is widely used for attenuation measurements. To understand how and why the particular arrangement of equipment performs the desired task, we will go through each piece individually and explain why it is where it is. Knowing the reasoning behind a certain arrangement is a good way to remember a setup and be able to duplicate it over and over again. Do not memorize it; reason it out.

The signal source used for this setup is a signal oscillator, for we are making only cw measurements at this time. In Chapter 9 we noted that the signal oscillator is the most basic of signal sources, designed to produce a single-frequency output. This type of source is much more stable than a sweep generator. As mentioned before, the sweeper is not designed specifically for cw operation. Notice that relatively good stability must be maintained in the setup since the IF output of the mixer is a single frequency. This frequency usually can drift only a small amount before it affects the indication devices and readings. Of course, for ultrahigh stability, a microwave synthesizer can be used. This use, however, usually is not necessary.

The RF attenuator is inserted to ensure that the source for our setup maintains a good match. The output impedance of most generators is listed as *nominally* 50 ohms. The word *nominally* can cause a lot of trouble because some people substitute the word *exactly* for it, and they do not understand why a simple setup does not produce the results expected.

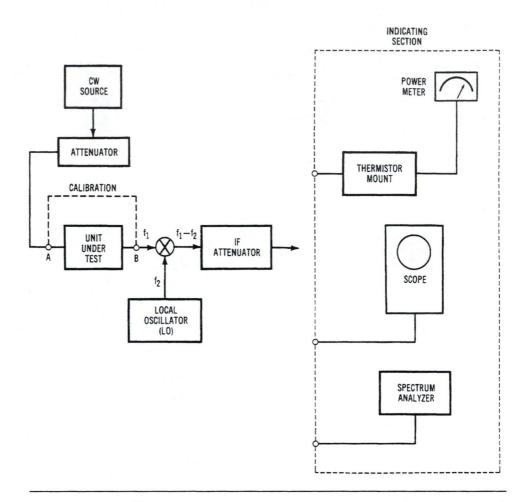

Figure 10–2.   IF substitution test setup.

Remember that the word *nominally* means that the impedance can be in the general vicinity of 50 ohms. When someone specifies an *exact* impedance for a generator output, the impedance will be listed with a specific tolerance (50 ohms $\pm$ 0.5 ohm). What the device being tested sees at the source output is important. This attenuator can be replaced by a circulator or isolator and produce the same results. Usually, however, a circulator in the range you are working is not available, and an attenuator always seems to be available.

The mixer and local oscillator will be examined together since they are required to work so closely. The mixer is a balanced type with good isolation between ports (RF–IF, RF–LO, LO–IF). Its LO drive level will usually be fairly high to yield the dynamic range necessary.

The LO usually is at +7 dBm, +10 dBm, or even +17 dBm. This requirement puts an additional burden on the local oscillator. It not only must have stability, for the same reason as mentioned in the signal source explanation, but it is required to have up to 100-mW capability in power output. These two items are many times the reason why some people do not use the IF substitution method.

The next block is what makes the whole setup able to make the wide dynamic range measurements—the *IF attenuator*. This attenuator may be any of the variable (step or continuous) attenuators discussed in Chapter 4. Whichever attenuator is used exhibits a very high degree of accuracy since it need be calibrated at only a single frequency (30 MHz, 60 MHz, etc.). This single calibration is a much easier task than trying to calibrate the same attenuator from 4 to 8 GHz, for example, to be used in an RF substitution test setup.

The final section in Fig. 10–2 is the indicating section. Once again, there are three possible methods. There may have to be an IF amplifier between the attenuator and the indicating section since the signal level from the attenuator is at a low level. (Notice that one difference between this setup and the RF substitution method when using a scope is that this method does not require a detector since it is an IF signal that can be displayed directly on the scope.)

Basic Procedures   With the test setup identified and explained, now let us run through the procedure for making an attenuation measurement. The procedure contains two phases: the calibration phase and the measurement phase.

Calibration—calibration is the first step in the procedure.

1. Set up the equipment as shown in Fig. 10–2 with points A and B tied together (device to be tested out of the circuit).

2. Turn on all power switches and set the standard IF attenuator to maximum attenuation.

3. Set the signal source to the desired RF frequency. Check with a frequency counter if desired.

4. Set the local oscillator to a frequency that is the IF frequency (30 MHz, for example) below the incoming RF signal set in Step 3. (This frequency may be checked and adjusted by placing a frequency counter at the output of the mixer and tuning the local oscillator until the IF is read on the counter.)

5. Decrease the attenuation in the standard attenuator until a reference level is on the indicator. Record the level and the reading on the attenuator—these are the reference points.

Measurement—with the system now calibrated, we can proceed with our measurement.

1. Put the RF signal source in a *standby* position if possible. If there is no *standby* position on the source, disconnect the source from the test setup by removing the RF attenuator from the source. Do not remove the cable on the test side of the attenuator since you may not get it reconnected properly and create a mismatch at the input.

2. Connect the device to be tested at points A and B.

3.   Turn on (or reconnect) the RF signal source.

4.   If the indicating device is calibrated and readable, read the attenuation directly off the scales provided. Record in a lab book for a permanent record.

5.   If the indicating device is *not calibrated,* decrease the amount of attenuation in the standard IF attenuator until the indicating device returns to its calibration reference position. Note where the attenuator is now set, subtract this amount from the reference attenuation setting, and record in a lab book as the attenuation of the device under test.

Thus, we have now covered one of the most accurate means of conducting transmission measurements—the *IF substitution method.*

10.2.3   Audio Substitution Method

The *audio substitution method* is sometimes referred to as *dc substitution.* It is basically the same type of setup as the RF substitution method, except that the calibrated attenuator used is calibrated at low frequencies or dc. At these frequencies, the accuracy of the measurement increases markedly. Remember that we said the attenuators could be calibrated so much more accurately when operating in an IF substitution mode (30 MHz, for example). Well, if we are in the kilohertz range or all the way down to dc, you can imagine how much more accurately these attenuators can be calibrated. Fig. 10–3 shows a typical audio substitution test setup. This setup, as mentioned earlier, is very similar to the one used in the RF substitution method. The major change is in an instrument known as an *attenuation calibrator.* As can be seen,

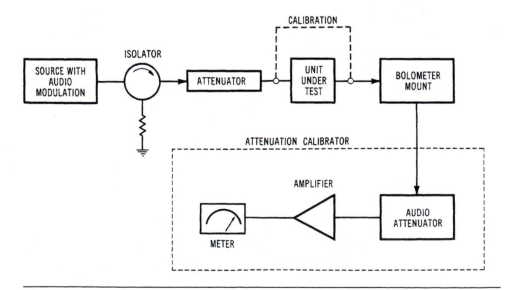

Figure 10–3.   Audio substitution test setup.

the calibrator contains a combination of measurement equipment: an indicating meter, amplifier stage, and the most important component of all—the audio calibration attenuator. The actual construction and operation will be covered in detail later in this section.

Basic Setup   Referring to Fig. 10.3, the signal source is a cw microwave generator modulated with an audio signal (usually 1.0 kHz). This could be a signal oscillator, a sweeper set on cw, or even a microwave synthesizer if accuracy and stability are that critical to your measurement. The isolator in the setup provides the generator with a constant impedance match, as discussed earlier in Chapter 4.

The attenuator following the isolator is an RF attenuator inserted to aid in matching the generator to the bolometer mount during calibration and to the device to be tested during the testing procedure. This attenuator also compensates for any connector variations that can arise when a straight-through line is used during calibration and then the device to be tested is connected for test. It also can be used to extend the measurement range of the setup, as will be discussed later.

The bolometer mount, which may contain a barretter or thermistor, is used to detect the RF power and convert it to a useful form. The frequency range of the mount should be checked to ensure that the proper one for the application is used. Also, check the power rating of the mount. Most mounts have a maximum power rating printed in a very prominent place. So take a few seconds to check it, even if the present application is low power. In this way, you will become familiar with the rating and will more readily remember it for future applications when working with higher powers.

The audio attenuator, output amplifier, and indicating meter are all combined within a commercial device termed an *attenuation calibrator*. Several commercial attenuators can be used as calibration standards. For the settings of the attenuator to hold, the input impedance to the following amplifier must be adjusted to the characteristic of the attenuator, and the impedance looking back from the standard attenuator must be matched. In case of mismatches, appropriate connections have to be made (within the instrument itself). The calibrator also provides bias current for the bolometer if the barretter is used.

In the single-frequency audio substitution method of measurement, a calibrated audio-attenuation standard is used to maintain a constant level at the output indicator with the unknown in and out of the line. The output level is obtained at the indicator with the unknown out of the system; the unknown, then, is placed in the system; and the necessary attenuation is removed from the audio standard attenuator to obtain the original reference level. Since the input power is changed when the unknown is placed in the line, the accuracy of measurement depends upon the response of the bolometer to the audiomodulated RF signal. The square-law response characteristics of bolometers are such that the decibel change on a linear output voltmeter is divided by 2 to obtain the RF power level change (in decibels) when using a square-law detector. In the present measurement the value of the unknown (RF attenuator) is obtained by taking one-half the value of the attenuation change of the standard. This adjustment is necessary because the input to the standard attenuator is ac voltage, which is proportional to the RF power level.

Basic Procedures The general procedure for performing attenuation measurements with the calibrator setup is as follows:

1. The source is adjusted to the desired frequency, the modulation frequency is adjusted to the proper value, and the necessary source and load tuning is accomplished.

2. The RF power level at the barretter is set with the unit under test out of the system. This value is obtained by setting all attenuators on the calibrator to zero, lowering the gain controls for minimum gain, and then adjusting the RF power with the level set attenuator to obtain full-scale reading on the output meter.

3. The precision attenuators are then set to a value greater than the value anticipated for the unknown. The gain on the attenuation calibrator is then increased to obtain a convenient reference level.

4. The unit under test is now placed in the system at A and B, causing a decrease in the output reference level. The precision attenuators are adjusted to return the output level to the reference point, and the value of attenuation is recorded.

This type of measurement procedure is used over a 20- to 30-dB attenuation range. The limiting factors to its range are the square-law response of the bolometer (barretter) and the internal noise level of the calibrator. This range can be extended by using an attenuator substituted in the RF line as a fixed value, with the readings obtained by the previous procedure added to this figure. For example, with a fixed 20-dB attenuation in the line during initial calibration and the calibrator equipment indicating a value of 11.2 dB, the total attenuation of the device is 31.2 dB.

Matching is a primary concern that should be watched very carefully. Errors in the audio substitution method are due to nonlinearity in the bolometer element, calibration errors in the audio attenuator, and equipment instability. The first two can be held to very good accuracies for measurements of 30 dB or less. Accuracies of about 0.1 dB are obtained with the setup shown. If source instability is accounted for by monitoring input power by one of several methods, accuracy can be improved to about 0.05 dB.

For highly accurate cw readings, the audio substitution method will provide the needed results.

10.2.4 Swept Transmission Measurements

To this point all our measurements have been single-frequency, cw measurements. Many times, however, a band of frequencies must be checked, and the swept technique must be used. The three methods covered previously for cw cases are all adaptable for swept measurements. Each of these will be briefly covered to illustrate their adaptability for swept applications.

Fig. 10–4 is a swept setup using RF substitution. Notice many similarities between this setup and that for the cw measurement (Fig. 10–1). There is a generator, calibration attenuator, and a variety of indicating devices. Although these similarities exist, there are also many differences that enable you to sweep the component being tested and to see the results.

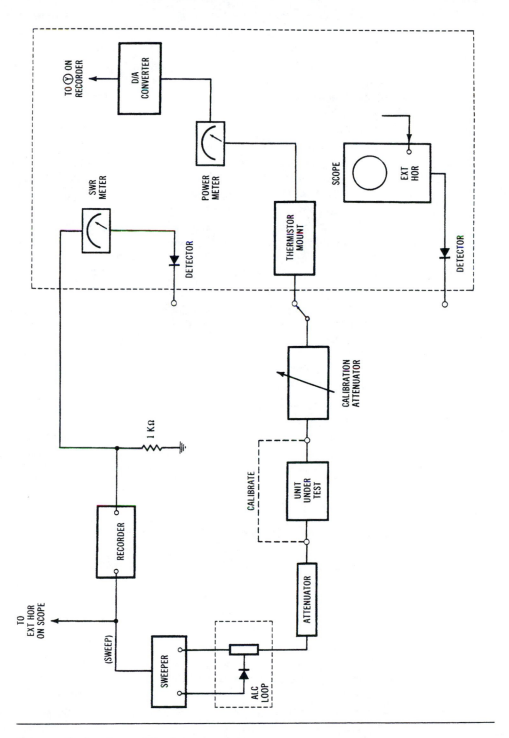

Figure 10–4. Swept RF substitution test setup.

The first major difference, of course, is the generator. It is not a single-frequency device but one that sweeps a band of frequencies. This, of course, is its prime requirement. The band it covers depends on the particular application, and its power output wil! depend on what type of device is being tested and/or how low an attenuation is to be measured. The sweeper must also have a 1-kHz modulation capability for SWR meter operation.

A second difference associated with the sweeper is the ALC (automatic level circuit) loop at the output. This circuit was covered in previous discussions and is used to provide a constant level output over the entire frequency band being swept. This ALC loop is generally incorporated within the sweeper itself. It is shown externally to point out the need for a leveled output for swept measurements. Also, this output (leveled or unleveled) may be stored in an internal memory and compared to the measured data to obtain a true reading of the characteristics of the particular component or system being measured over this wide frequency range.

Another difference is in the indicating section. In the cw setup we have a power meter, scope, and spectrum analyzer for output indication. We still maintain the power meter (plus a d/a converter to drive the recorder) and scope (with an external input from the sweep generator), but we have added the SWR meter. The 1-kHz resistor associated with the meter at the recorder Y-input matches the meter output at the meter to the input of the recorder.

One additional difference to the setup mentioned throughout is the recorder. This recorder is used to plot the test response when the SWR meter and power meter are used. The scope provides a visual display and may be preserved by taking a picture. The recorder could also be driven with the vertical input to the scope and the sweep voltage from the sweeper. It should be pointed out that the recorders shown in these figures are not always the devices used. Many times the recorder is replaced by a plotter. The plotter is used with the appropriate interface bus and is usually the preferred method. So, the reader should not become confused when the diagrams say "recorder," for this is usually a plotter.

A swept IF substitution setup is shown in Fig. 10–5. The main concern with this type of setup is to sweep both the RF and local oscillator to produce a constant IF output.

The two sweep generators are offset by the desired IF frequency, and the sweep reference of one sweeper drives the other to ensure synchronization in time. The output of the synchronous detector provides the reference by which the phase-locked box develops an error signal to keep both the sweeper frequencies as initially set and the spacing constant. This error signal is the same type as we discussed regarding ALC. The only difference is that with the ALC circuit, any change in amplitude creates an error signal; in our phase-locked system, a change in the IF frequency beyond a set tolerance causes an error. This scheme achieves a constant IF frequency that allows the use of narrow bandwidths in the IF amplifier; the result is a very wide dynamic range in the system.

With the two sweepers locked to a set IF frequency over your required frequency band, the procedure would be the same as if we were using only a single-frequency band rather than a mixer. The major difference in performance is the greatly increased dynamic range of the IF setup that we mentioned before.

Fig. 10–6 shows the last method in swept form—*audio substitution*. The major difference between this setup and that in Fig. 10-3 (cw setup), other than the use of a sweeper

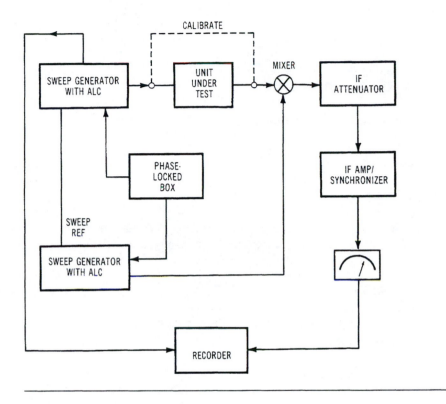

Figure 10–5. Swept IF substitution test setup.

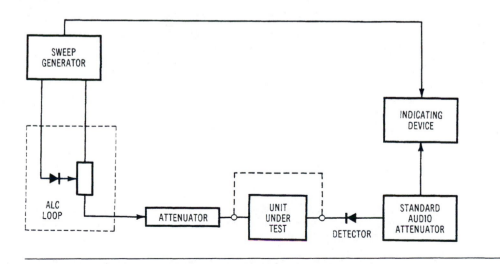

Figure 10–6. Swept audio substitution test setup.

409

and a cw generator, is that the cw setup has an isolator at the generator output to improve the match for the generator. The swept setup does not use an isolator. The ALC loop is used at the output of the sweeper and provides a good match in itself. All other components are performing the same functions as in the cw setup.

10.3   Impedance Measurements

Before we can create a meaningful method of measuring impedance, we must first understand what it is. The word *impedance* or *impede* means different things to different people. To the Bureau of Motor Vehicles it means something that blocks your vision, to a running back on a football field it is an opposing player who gets in the way, and to a doctor it means a clot or an obstruction.

In each of the cases of impedance mentioned, we have described a *resistance* of some kind. A resistance keeps a motorist from seeing the road clearly; a resistance prevents a running back from advancing down the field; and a resistance prevents the normal flow of blood or fluids through the body.

A microwave impedance is also a resistance. In most systems the impedance is 50 ohms. This impedance, however, is not resistance in the sense that we normally think of it. The resistance we normally think of is a resistance in a dc circuit when we measure a voltage and a circuit, divide E by I, and come up with a number. It is not that easy with microwave impedance. The 50 ohms mentioned previously is, in actuality, $50 + j0$ ohms—a real part (50) and an imaginary part ($j0$). Since there are two terms ($50 + j0$), each impedance has not only a magnitude (50 ohms), but a phase angle ($0°$ in this case) as well. Clearly there is more to a microwave impedance than just reading a voltage and a current.

The use of the term *"phase angle"* should give a clue as to how we can define a microwave impedance. A phase angle usually means that a situation creates two or more signals at different angles. The term we are looking for to clarify our definition of impedance is *reflection*. Reflection at the input (or output) port causes signals to come back to the source at different phase angles.

Remember our discussion of the principle of reflections applicable to microwave impedance in Chapter 3. Referring back to Fig. 3–5, you may remember that a flashlight was used to represent the microwave source, a venetian blind as the input or output circuit of a device, and a wall as the device. Referring back to this discussion of reflection principles would be helpful at this time.

After reviewing this discussion of reflections, we should be able to define just what the word *impedance* means. We can define it as follows:

> Impedance is the opposition to an RF signal that causes part of that signal to be reflected back. The magnitude and phase angle of these reflections are determined by the characteristics of the input (or output) of the device being tested.

With this definition we can see that by measuring and characterizing the reflection at the input or output of a circuit or device, we can obtain an excellent reading of the impedance. This section is designed to do exactly that. It will first consider the slotted line, reflectometer, and VSWR bridge in the cw application, and then cover each of these areas in swept frequency applications. Mention will also be made of some of the measurement systems covered in Chapter 9.

Before we get into actual impedance measurements, there is a term that has come up before that is of prime importance in understanding the measurement techniques that follow. This term is *standing-wave ratio* (SWR). Standing-wave ratio is also discussed in Chapter 3. Remember Fig. 3–4 showed the vibrations of a string tied to a wall bouncing back to the person moving the string. To refer back to this discussion in 3.1.2 titled "VSWR, Reflection Coefficient, and Return Loss" would be helpful at this time. After you have reviewed this section, we should be ready to get into the actual setups used for microwave impedance testing and understand what is actually being measured.

10.3.1   Slotted Line

The most accurate device used for manual impedance measurements is the slotted line. Only the complex computer controlling a network analyzer will yield higher measurement accuracies. A slotted line is usually described as a section of uniform, lossless (very low-loss) transmission line with a longitudinally oriented slot that provides access to a pick-up probe that detects voltage variations on its line as it slides along it. Very simply put, it is a piece of low-loss line (coaxial or waveguide) slotted in such a way to accommodate a probe that measures the voltage along the line. They are excellent devices for use in the area of 12 to 18 GHz since their size and construction make them ideal for shorter wavelength applications. They also have a low residual SWR. That is, the SWR of the slotted line itself is very low (1.01:1, typically) and does not interfere with the measurement of a component under test.

Fig. 10–7 shows a single-frequency slotted-line test setup, from which the standing-wave ratio can be read directly. The easiest way to understand the single-frequency setup is to go through a measurement procedure.

Calibration—calibration is the first step in the procedure.

1. Connect the test setup as shown in Fig. 10–7. Turn on power to the signal generator and SWR meter. Allow for any instrument warm-up time and set the generator frequency.

2. Place a short on the output port of the slotted line (point A). Adjust the probe depth just deep enough to give a good indication on the meter.

3. Run the carriage of the slotted line down the line (in either direction) until a maximum reading on the meter is found. The gain of the meter may have to be adjusted to keep the needle from being pegged. When the first maximum is found, record the position of the carriage. (By a maximum reading we mean a maximum meter deflection to the right; it actually indicates a voltage minimum.)

4. Move the carriage (in either direction) until a minimum is reached; then keep moving in the same direction until a second maximum is found. Record the position of the carriage.

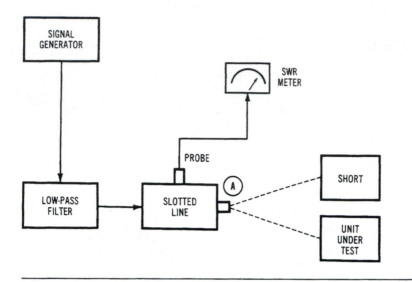

Figure 10–7.   Single-frequency slotted-line test setup.

5.   Subtract the two readings. The resulting number is one-half the wavelength of the frequency at which you are operating; two times the reading is one wavelength (λ).

Measurement—with the setup now calibrated, we can proceed with our measurement.

1.   Remove the short, and connect the device to be tested.
2.   By moving the carriage and adjusting the gain of the SWR meter, adjust the meter for its maximum reading at "1" on the VSWR scale.
3.   Move the carriage (either toward the load or toward the generator) until the meter reaches a minimum value. This figure is the VSWR of the device being tested at this particular frequency.

This VSWR figure can be used to find a value of reflection coefficient or return loss; it can also be used to obtain a direct value of impedance by employing the most useful tool available in microwave—a *Smith chart*, covered extensively in Chapter 7. How the slotted line/Smith chart combination gives an impedance reading is very significant. To demonstrate the fact, the following example is presented:

Let us suppose we have the following specifications: (with a short connected) peak 1 = 14.30 cm, peak 2 = 12.30 cm, VSWR = 1.8:1, distance to the point = 1.7 cm (toward the generator). The solution is as follows:

1. $\lambda/2$ at our frequency = 14.3 − 12.3 = 2.0 cm. Therefore, λ = 4 cm.

2. The VSWR measured was 1.8:1. We now place a compass point at the center of the Smith chart in Fig. 10–8, set the pencil on 1.8, and draw a circle. The impedance of our device lies somewhere on the circle.

3. To obtain the 1.8 value, we found a meter reading corresponding to a minimum voltage, or 1.0:1 VSWR, and moved toward the generator until we found our 1.8 reading, where the meter stopped moving to the left. This distance was measured as 1.7 cm. Therefore, we must move 1.7/4.0, or 0.425 λ, toward the generator to find the actual impedance.

4. The next step is to determine where to start so that we can move 0.425 λ toward the generator. We said that we started from a point where the meter read a VSWR = 1.0:1, a voltage minimum. Where there is a voltage minimum there must be a low resistance—at the left side of the chart where R = 0.

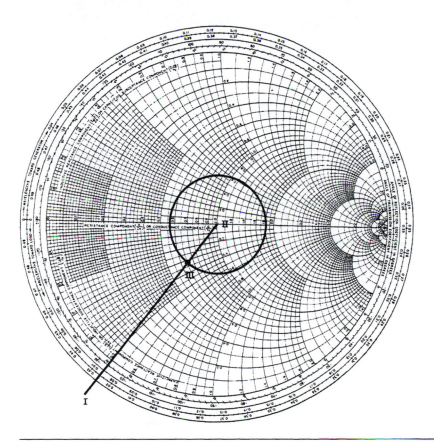

Figure 10–8.   Impedance example.

5. Starting at R = 0, move along the outside edge of the chart in a clockwise direction using the top set of numbers—"Wavelengths Toward Generator." When you reach a point where the numbers say 0.425, place a mark (I).

6. Place a straightedge through I and the center of the chart (II); connect these points together with a line. Where the line intersects the 1.8:1 circle is the actual impedance of the device at the particular frequency (III). In our case it is 0.65 − j0.32 ohms; this value is normalized so that the actual impedance is 32.5 − j16 ohms.

You can now see how handy the slotted-line setup is for single-frequency measurements. The accuracy of such a system is very high, but it does have some obvious disadvantages.

Time-Consuming for Multiple Point Measurements   Since the measurement described must be repeated at every frequency of interest, this technique is seldom used for other than single-frequency measurements (although swept measurements are becoming more and more feasible).

Loss of Measurement Continuity   When the single-frequency technique is used for broad-band measurements, only a discrete number of points can be measured. For example, if measurements are being made in X-band, the following frequencies might typically be chosen: 8.0, 8.5, 9.0, 9.5, 10.0, 10.5, 11.0, 11.5, 12.0, and 12.4 GHz. It is not uncommon for an impedance mismatch of great magnitude to occur over a narrow frequency range of 200 MHz or less. If this mismatch were centered about 11.7 GHz, for instance, it would go completely undetected.

Residual SWR   Reflections due to the measurement port connector and discontinuity at the slot end are typically 1.02 SWR at 2 GHz (40-dB equivalent directivity), 1.03 at 10 GHz (36.5 dB), and 1.04 at 18 GHz (34 dB) for slotted lines with a precision 7-mm connector. However, one precision line offers only 1.015 SWR (42.5 dB) to 18 GHz.

Error Due to Source Residual FM   The minima of a standing wave are obscured in the presence of FM, thereby producing an inaccurate ratio of maxima and minima.

Probe Tuning Errors   One of the major sources of error in standing-wave measurements is excessive probe penetration. The presence of the probe affects the VSWR because it is essentially shunting the line. This fact is why we emphasized that the probe be set only deep enough to obtain an indication on the meter.

Line Variations   Dimensional variations along the line introduce a measurement error.

Errors Due to Detector Characteristics   Since SWR indicators are calibrated in terms of square-law response, errors are introduced if the detector departs from its square-law response.

The slotted line, although it has its disadvantages, is still one of the most accurate means of manually obtaining impedance data. It is widely used for cw measurements.

10.3.2   Reflectometer

In Section 3.1.2 we explained reflection coefficient (ρ) and return loss and how they apply to impedance measurements. Nowhere do these terms apply more than in reflectometer applications, readily apparent simply by considering the name. Just as a speedometer measures speed and a thermometer measures temperature, a reflectometer measures reflections.

Fig. 10–9 is a test setup used for cw reflectometer tests. The objective of this setup is to provide an indication of both incident (forward) and reflected (reverse) powers to the operator. From these values it is possible to calculate the reflection coefficient (ρ) and VSWR at a single frequency using the relationship shown:

$$\text{Return loss (dB)} = 20 \log_{10} (\rho) \tag{10.1}$$

$$\text{VSWR} = \frac{1 + \rho}{1 - \rho} \tag{10.2}$$

(Recall that this is the method employed for testing S-parameters in Chapter 8 and was again mentioned as being used in certain test equipment in Chapter 9.)

The generator shown is one that will operate in the appropriate range of frequencies for your application. It may be a signal oscillator, a sweeper set on the cw position, or even a

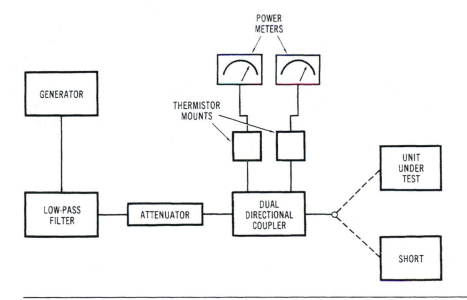

Figure 10–9.   CW reflectometer test setup.

microwave synthesizer, if such a device is necessary. A low-pass filter is sometimes used in conjunction with the generator when the generator may not be of high enough quality to have adequate output filtering. This filter reduces the harmonics and spurious signals present at the generator output so that only the desired signal is sent to the directional coupler for test purposes. Without the filter, there will be the possibility of unwanted harmonics (or spurious signal) causing false readings on the power meter. These signals may get to the power meter at a sufficiently high level to cause problems even though the directional couplers are designed to operate over a specific band of frequencies (and hopefully you are using the proper one in your setup). Once they are passed by the coupler, though attenuated more than the desired signal, the power meter cannot distinguish between them and the frequency you wish to test your component at since the meter is a wide open device with regard to frequency. Therefore, using the low-pass filter in your setup if there is one available is a good idea. The attenuator in the setup is inserted to ensure a good match between the filter and its coupler. There is a definite chance of a mismatch in the higher frequency region of the filter, and the attenuator aids in providing the coupler with a constant source impedance.

The next component in the setup in Fig. 10–9 is the heart of the system—the directional coupler. The setup shown has a dual-directional coupler in it. However, two single directional couplers placed back to back can also be used. We call the coupler the heart of the system because the accuracy of your measurements depends on a property of the coupler called *directivity*. The directional coupler and its associated parameters were discussed in detail in Chapter 4.

A very important part of the reflectometer setup is one of the diagram's smallest blocks—the one titled "short." This little device is used as a reference for all of our reflection measurements. What makes a microwave short suitable for use as a calibration standard? It certainly does not sound as though it would work very well; most people think of a short in an electrical circuit as something very dangerous and undesirable. However, to clarify matters, remember that it is called a *microwave short;* that is, at microwave frequencies it is a short circuit. (Granted, at dc you also read a short with an ohmmeter; this short circuit affects you, though, only if you have some active device in your setup that does not have a dc block at its input or output.) The microwave short causes all RF energy from the incident wave to be reflected (100% reflection, $\rho = 1$). The value of a device with a known reflection coefficient is obvious, but an open circuit also has a reflection coefficient of 1.0—wouldn't it be easier to leave the coupler port open and eliminate the possibility of any mismatches between the coupler and the short? The answer is yes, it would be simpler, but it would not be as accurate. With an open circuit a fringing capacitance is set up between the center conductor and the outside shield, making it difficult to tell just how good an open circuit you have. To obtain highly accurate readings, the capacitance must be calibrated out. But how much capacitance is there in a type N, SMA, or TNC connector at 1 GHz, or 18 GHz? And even if the proportions of the capacitance were known, how would it be calibrated out of the setup in Fig. 10–9? Thus, we use the short instead. There is no capacitance to consider, the reflection coefficient is 1.0, and you can be sure that it has a good ground (because it is a plane and not a point, a prerequisite in microwaves). By using computer-aided measurements the open circuit capacitance can be calibrated out, but for general test setups, the short is much easier to work with, and it produces more accurate and reliable results.

With the setup of Fig. 10–9 explained, let us proceed with a test procedure.

Calibration—calibration is the first step.

1. Connect the test setup as shown in Fig. 10–9, with the short at the coupler output.
2. Apply ac power to the generator and power meters and allow sufficient warm-up time, as required.
3. Turn up the input power level to check to see how closely the input level meter and the reflected level meter track one another. This checking will give you any calibration differences in the setup. Record any differences noted so that they can be used in your calculations later on.
4. Remove the short from the coupler.

Measurement—with calibration finished, the measurements may be taken.

1. Place the device to be tested, in place of the short, on the output of the coupler.
2. Record the values of the power, incident and reflected, as read on the decibel scales of the power meter. The difference is the return loss of the device. For example, if you have a $+10$-dBm incident level and a -10 dBm reflected, the return loss is 20 dB. A reflection coefficient of 0.1 is achieved from the 20-dB figure by applying the equation: $dB_{return\ loss} = -20 \log_{10} \rho$. With a value of $\rho = 0.1$ the VSWR is 1.22:1, using the equation $VSWR = (1 + \rho)/(1 - \rho)$.

The procedure outlined is used for single-frequency measurements. A band of frequencies can be measured by processing each frequency individually, then plotting the results.

10.3.3   VSWR Bridge

A bridge circuit is one that performs a comparison reading on a parameter to be tested. Many times a Wheatstone bridge, such as the one shown in Fig. 10–10, will have a thermistor in one of the legs and then be balanced at a certain ambient temperature. Any variation in the temperature will cause the bridge to become unbalanced, producing a voltage to exist across it that makes the circuitry compensate for the variation. This idea of comparison of readings is very helpful to us for measuring impedances, as we have previously seen.

Figure 10–11 is a schematic representation of a commercially available VSWR bridge. For the bridge to have good accuracy, it is necessary that its components be matched throughout the designated operating frequency range. A particularly important factor is the impedance in the lower arm opposite the test port, the one marked "$R_c = 50\ \Omega$ (reference)." Since this arm is the one to which the measured load is compared, it must have excellent characteristics over the entire frequency range of operation. This termination may be physically integrated into the bridge both to achieve maximum bridge accuracy and so that its performance can be part of the bridge specifications.

Physical integration of the reference termination also guards against damage, poor contact, unknown performance, and a loss of the termination; all of these are far more likely if the arm is constructed as an accessory to the bridge.

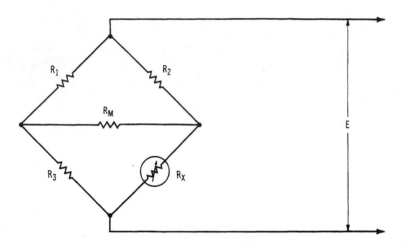

Figure 10–10. Wheatstone bridge.

The bridge in Fig. 10–11 is just what it appears to be—a resistive Wheatstone bridge that is used to measure reflections on an unknown load. The bridge operates to 12 GHz and exhibits less than a 1.02:1 residual VSWR over most of the range. Such a low VSWR is attained in slotted-line sections only at great expense; here it is achieved in a unit that is inexpensive and small enough to fit in the palm of your hand. These bridges also exhibit the characteristic of being broad-band devices that are suited to sweeping and reflectometer work, which slotted sections are not.

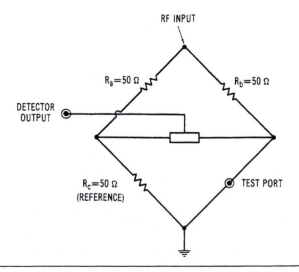

Figure 10–11. VSWR bridge schematic.

To understand the VSWR bridge and how it works, let us investigate a typical test setup. Generally, the VSWR bridge is used in swept setups. These setups will be covered later on in this chapter. We will, however, cover a basic cw setup in this section to give you an idea of its applications.

Fig. 10–12 is a block diagram of a cw setup using a VSWR bridge. Two points to remember about the setup are

- The calibration attenuator is an RF attenuator and must be calibrated over the RF range being used.

- The indicating device may be a meter or an oscilloscope. The output of the VSWR bridge is a dc voltage and, as such, requires an indicating device that will respond to that dc voltage.

The test procedure for a cw measurement is as follows:

1. Connect the test setup as shown in Fig. 10–12. Do not attach the device to be tested.

2. Turn on the ac power to the generator and indicating device. Allow for any time delays or warm-up periods.

3. Apply RF power and set in a level on the calibration attenuator that will give a convenient reading on the indicating device. This is a 0-dB return loss since there is an open circuit on the test port. Record where this 0-dB point is on the indicating device and what value is set on the calibration attenuator.

4. Connect the device to be tested and note the new value on its indicating device. This new reading plus the calibration attenuator reading is the value of return loss of the device under test.

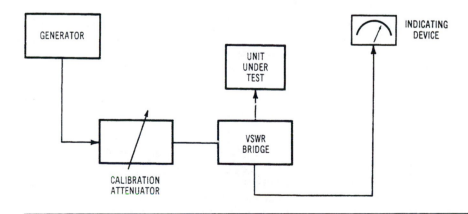

Figure 10–12.   CW setup using a VSWR bridge.

An example of a test would be as follows:

- The attenuator initially has 10 dB set in during calibration.
- When the device is connected, the indicating meter is deflected 8 dB from the calibrated mark for 0-dB return loss.
- The result is a device with a return loss of 18 dB ($\rho = 0.13$).
- The VSWR at the particular frequency tested is, therefore, 1.30:1 since

$$\text{VSWR} = \frac{1 + 0.13}{1 - 0.13} = \frac{1.13}{0.87} = 1.30$$

Compared to couplers, the bridges have wider frequency ranges, lower coupling loss, superior reflection performance, and lower cost. Another advantage is that they are available with a wide variety of connectors; their high performance includes the effects of a given connector. In other words, the bridge performance is not degraded by adapters needed to fit your particular system.

10.3.4 Swept Setup

All of the test setups described so far have been for cw measurements. Many times, however, a single-frequency measurement is just not adequate. In times like these, some means of sweeping your test setup is needed. Each of the systems described earlier for cw can be swept with a high degree of accuracy and repeatability. These systems will now be covered individually for swept applications.

Swept Slotted Line   Remember from our discussions of cw slotted-line measurements that the slotted line was one of the most accurate means of manually obtaining impedance data. This statement has been true for many years and still is true. Until the emergence of the storage scope on the market there was never any thought given to sweeping a slotted line. The measurement complexity was too high. Actually, an XY recorder is also used, but once again, was not previously considered.

A swept slotted-line setup is shown in Fig. 10–13. Note that either an XY recorder or storage scope may be used with this setup. The sweeper must have a 1-kHz modulation capability because of the operating requirements of the ratio meter. This modulated swept signal is divided equally to provide a *test* and *reference* channel for the setup. The attenuators in each channel are for matching purposes. One matches the power divider to the reference detector in the reference channel, and the other matches the power divider to the slotted line in the test channel. The ratio meter is the heart of the swept system. This instrument does practically every necessary calculation needed to display the swept slotted line response. It takes the *reference* and *test* signals, compares them in a log ratio detector, and produces a dc voltage proportional to the resulting ratio. The unit contains an audio attenuator in the *test* channel that is an adjustable precision attenuator,

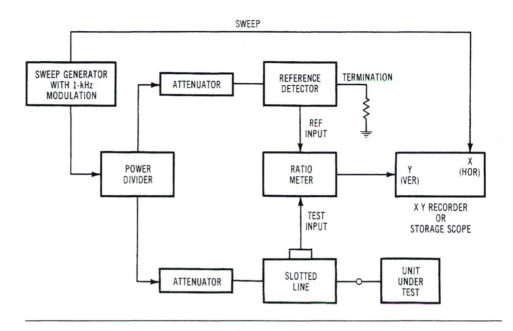

Figure 10–13.   Swept slotted-line test setup.

which results in an overall attenuation range of 50 dB with resolutions on the order of 0.005 dB.

With a test setup explained and ready to work, let us see how we obtain a swept measurement from the method. For initial considerations we will use the XY recorder as an indicating device and explain the resulting measurement procedure. Measurements employing this technique are made in basically three steps:

1. The XY recorder is adjusted for the desired sensitivity over the selected frequency range (a sensitivity of 1 dB/inch, for example).
2. Repeated swept recordings are made to produce an envelope of the VSWR of the full frequency range.
3. VSWR is read in decibels at any frequency in the swept range.

The audio attenuator of the ratio meter is used to calibrate the vertical displacement on the XY recorder to the desired resolution in decibels. Then, with the slotted-line probe in a set position, the frequency band of interest is swept repeatedly, each time with the probe reset to a slightly different position. This action eventually results in clearly visible maxima and minima envelopes. The VSWR, in decibels, at any given frequency is equal to the vertical distance between the maxima and minima envelopes, as illustrated in Fig. 10–14.

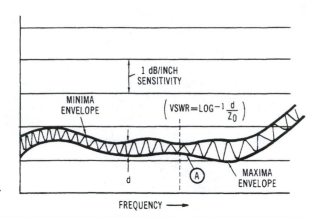

Figure 10–14. Swept slotted-
line display.

To illustrate how you convert the display to a VSWR, consider an example. Suppose we have chosen the point in the figure marked A. The distance, D, between the maxima and minima envelopes appears to be on the order of 0.5 dB. Our formula states that

$$\text{VSWR} = \log^{-1}d/20$$
$$= \log^{-1}0.5/20$$
$$= \log^{-1}(0.025)$$

The VSWR at A is equal to the number whose log is 0.025; consulting a calculator or a book of tables shows that number is 1.06. Thus, the VSWR at point A is 1.06:1.

The actual measuring time can be greatly reduced by using a memory scope that can store each of the many sweeps necessary for a good measurement. This process saves the time otherwise taken as the XY recorder repeatedly moves mechanically from one side to the other. By the use of a signal processor, the VSWR can be displayed directly without any calculations necessary. Getting the VSWR without calculations is a reality with the automatic test setups used today.

Swept Reflectometer   A typical swept reflectometer setup is shown in Fig. 10–15. In this setup the heart is the dual-directional coupler. This component was discussed in detail in Chapter 4. Its primary parameter must be high *directivity*. This will ensure good isolation between the incident and reflected signal. With good directivity you can be sure that when a reflected signal is read, it will be only the reflected signal.

To understand how the setup works, we will go through a measurement procedure. Since calibration is as important as the actual measurement, we will cover both procedures.

Calibration—calibration is the first process.

1.   Connect the test setup as shown in Fig. 10–15.

2.   Turn on the ac power to the sweep generator, XY recorder, and SWR meter.
     Keep the sweep generator in a *standby* or *RF off* position initially. (Once again,

the XY recorder is shown, but a plotter with the appropriate interface connections is also a viable indicating device for such a setup.)

3. Place the short on the output port of the directional coupler.

4. Set the variable attenuator to its maximum attenuation setting.

5. Set the SWR meter to a 0-dB setting.

6. Set the *start* and *stop* frequencies on the sweeper that coincide with the lowest and highest frequencies you will be testing at. Then turn the sweeper to the cw position.

7. Turn the *RF power on* to the sweep generator. Adjust the RF level until the *ALC* loop is locked. This leveling may be indicated by a light, meter, or some other means that indicates when the RF output is leveled.

8. Adjust the variable attenuator for an indication on the SWR meter. Set a convenient level that will allow enough range on the meter to make your measurement. (If a VSWR of 1.1:1 is expected, for example, you would need at least 26 dB of range on the meter. You probably would adjust for 30 dB.) Note the settings on the meter and the attenuator.

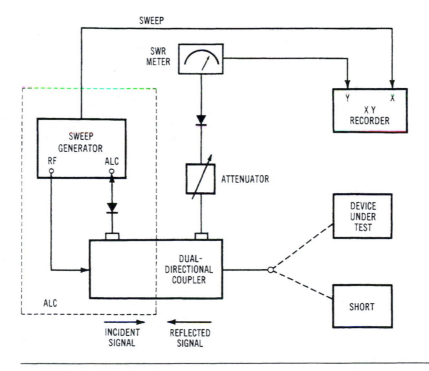

Figure 10–15.   Swept reflectometer test setup.

9. Set the end limits on the recorder by putting the sweep generator on a *manual* position and running back and forth through the range of frequencies.

10. Set the recorder at the top of the paper, set the sweep generator to a *trigger* mode, set the sweep speed for a medium speed run across the paper, and run the recorder across the paper by dropping the pen and pressing the *trigger* button. This reference will represent a 0-dB return loss, or all energy reflected back (as is the case for a short circuit). If you do not have an automatic pen lift on your recorder, be sure to watch your plot and manually lift the pen to prevent a re-trace from being marked on the plot.

11. Set in various values of attenuation and run them on the recorder as outlined above. Typical values are 1, 2, 3, 5, 10, 20, and 30 dB. This covers a wide range of return losses and, thus, VSWRs. You now have a set of calibration lines.

12. Set the SWR meter and attenuator back to their original setting as outlined in Step 8.

13. Put the sweep generator back into a *standby* or *RF off* position and disconnect the short from the coupler.

Measurement—you now have a calibrated test setup and are ready to take the measurements.

1. Place the device to be tested at the directional coupler output, as shown in Fig. 10–15.

2. Turn the *RF power on* and note the reading on the SWR meter and the XY recorder.

2. Put the sweeper in the *trigger* mode, and press the *trigger* button.

4. Put the sweep generator in *standby* or *RF off*, remove the data from the recorder, place a clean sheet on the recorder, remove the tested device, and you are now ready to perform a measurement on another component.

The swept reflectometer is widely used because many of its components are readily available, everyday components. Be sure to check coupler directivity and attenuator accuracy before putting just any coupler or attenuator in the setup. These are the two most critical components used and the ones that can be the largest source of errors.

Swept VSWR Bridge   The swept VSWR bridge test setup is not that much different than that of the cw setup. Fig. 10–16 verifies this statement. The main difference is the use of a sweep generator with an ALC loop to keep the source leveled.

A calibration and measurement procedure will illustrate how the test setup works.

Calibration—calibration is the first part of the procedure.

1. Connect the test setup as shown in Fig. 10–16.

2. Adjust the calibration attenuator for its maximum attenuation and apply ac power to the sweep generator and oscilloscope. The sweep generator should be in a

standby or *RF off* position. Set the *start* and *stop* frequencies on the sweeper to correspond to your low and high frequency limits, and set the sweeper to the *sweep position.*

3. When the sweeper and scope have power (after any time delay), turn the sweeper to *RF on.* Be sure that the RF output is leveled. This point usually can be checked by observing the "RF unleveled" light on the front of most sweepers. When the light is on, the output is unleveled.

4. The *test* port of the bridge should now be left *open.* This corresponds to an infinite VSWR (a 0-dB return loss). Set the sweeper *sweep speed* to obtain a solid sweep on the scope with no flickering.

5. Using the gain of the oscilloscope and the calibration attenuator, set the 0-dB return loss (VSWR = ∞) line to the bottom of the scope face.

6. Turn the sweep generator to the *RF off* or *standby* position, and prepare for the measurement.

Measurement   Connect the device to be tested to the test port of the bridge; the display moves to some level on the scope corresponding to the return loss, and thus VSWR, of the device. Overlays for the oscilloscope relating return loss to VSWR are available. They are transparent so that the value of the reflection can be read on the face of the scope directly in either VSWR or decibels of return loss.

To illustrate how the scales would work, consider an example. Suppose we calibrated the setup with 10 dB of attenuation set in at the calibration attenuator. When our device is

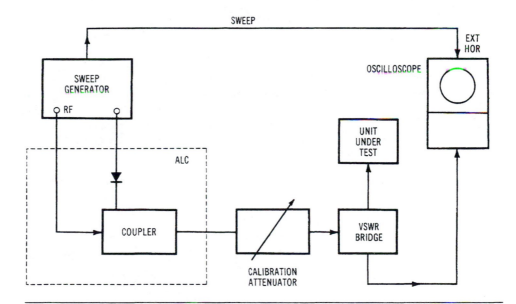

Figure 10–16.   Swept VSWR bridge test setup.

Table 10–1. Relationship between VSWR and return loss.

VSWR	Return Loss (dB)	VSWR	Return Loss (dB)
1.01	46	1.30	18
1.02	40	1.50	14
1.04	34	1.80	11
1.066	30	1.93	10
1.10	27	3.00	6
1.15	23	5.00	3.5
1.22	20		

attached, the trace on the scope sweeps from 5 db at the low frequencies down to the 3-dB level at the high end. Thus, the VSWR is approximately 1.42 at the low end (15-dB return loss) and 1.58 at the high (13-dB return loss). These values can be verified by Table 10–1. Table 10–1 shows the relationship between VSWR and return loss in decibels.

Two factors are important when considering either a swept or cw VSWR bridge setup:

1.   Bridge directivity—should be as high as possible.
2.   Test port VSWR—must be low to eliminate erroneous reflections.

These parameters will either make or break your testing.

10.3.5 Swept Systems

In Chapter 9 we covered what we called *microwave network analyzers*. These analyzers were considered to be classified as *systems*. Remember that we justified this title by saying that the analyzer was one instrument capable of making a variety of measurements that previously took many instruments and auxiliary components (couplers, detectors, etc.) to accomplish. One of the measurements they can perform is reflection.

All these systems covered will make reflection measurements, both cw and swept.

10.4   Power Measurements

When speaking of measuring microwave power, we are doing more than measuring the voltage across a component, measuring the current through it, and multiplying the two values together ($P = E \times I$). What we are actually measuring is the amount of heat generated within a microwave circuit when the energy flows through it. This heat is detected by thermocouples or bolometers. The thermocouple is formed when two wires of different metals have one of their junctions at a higher temperature than the other. The difference in temperature produces a voltage proportional to that difference.

Bolometers are classed as barretters or thermistors. The *barretter* consists of an appropriately mounted short length of fine wire, usually platinum, with sufficient resistance to enable it to be impedance matched as a termination for a transmission line. This resistance increases proportionally with an increase in temperature (positive temperature coefficient).

The *thermistor* consists of a tiny bead of semiconductor material that bridges the gap between the fine wires that are closely spaced. All the resistance of the thermistor is concentrated in the bead material. The resistance of the thermistor decreases proportionally with an increase in temperature (negative temperature coefficient).

Changes in resistance resulting from the dissipation of microwave power (heat) in the bolometer are commonly measured by using the devices as one arm of a dc or audio frequency Wheatstone bridge circuit. These circuits are found in power meters covered in Chapter 9. (Many were called sensors.)

For purposes of definition we will cover three power levels when discussing power measurements—low, medium, and high. These power levels are defined as follows:

Low power—1 mW and below ($<$ 0 dBm)

Medium power—1 mW to 10 W (0 dBm to +40 dBm)

High power—10 W and up ($>$ +40 dBm)

10.4.1   Low-Power Measurements

Components used for low-power applications should be chosen very carefully. Attenuators, in particular, should be given very careful consideration since too large an attenuator will reduce the generator level to a level that will not allow your device being tested to be characterized.

A typical low-power test setup is shown in Fig. 10–17. This setup resembles very closely some of the transmission setups shown in Section 10.2. There is a low-level generator that has a low-pass filter (this may also be a bandpass filter) at its output. This filter eliminates any harmonics that may be present at the generator output. The isolator takes the place of the attenuator that was shown in the transmission setups. It is used because of its low insertion loss ($<$ 0.5 dB) and matching qualities. This, in fact, is the main job of the isolator, to match the generator/filter to the device being tested. The attenuator at the output may or may not be needed. If a low-level signal is being used, the attenuator may reduce the level so far that it will not show on the power meter or may be down in the "grass" on the spectrum analyzer. So check your particular application before using this attenuator.

With the basic setup presented, let us now go through a test to show how all this equipment is used.

Calibration—calibration is the first part of the procedure.

1. Connect the setup as shown in Fig. 10–17, with the device to be tested removed and a calibration cable inserted.

2. Turn on the ac power to the generator and power meter or spectrum analyzer. Be sure that the generator output is turned to its minimum output.

3. If a relative power reading is desired, turn the RF generator output until a convenient level is indicated on the power meter or analyzer. Record this level as a reference.

4.   If an absolute reading is desired, turn the RF generator output until the power meter or analyzer reads 0 dBm. (If an active device such as an amplifier is to be tested, check to see what the maximum input level is to the device. The 0-dBm signal may overdrive it. A −5-dBm or a −10-dBm signal may have to be used as a reference.)

Measurement—now with calibration completed, the measurement can be taken.

1.   Remove the calibration cable and insert the device to be tested. If an active device is to be tested, be sure it has the proper dc voltages applied before inserting it into the test setup.

2.   If a relative reading is desired, record the reading on the power meter or analyzer and subtract it from the previous reference reading.

3.   If an absolute reading is desired, record the reading on the power meter or analyzer directly as your results.

Before leaving low-level power, we should mention that the matching between components should be carefully studied. When measuring power levels such as −20 dBm

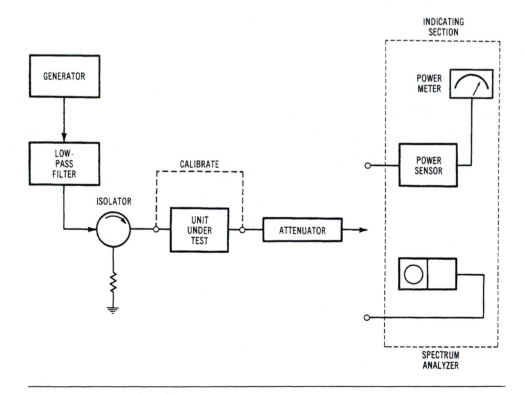

Figure 10–17.   Low-power test setup.

or −30 dBm, you do not have a lot of margin for error since you do not have a lot of power to work with. So extra care should be taken when making low-level (<0 dBm) power test setups.

10.4.2   Medium-Power Measurements

Probably 90% of the power measurements you will ever take in your career will be in the medium-power range. Chances are that if you have recently used a power meter or spectrum analyzer to check a power level, it fell somewhere within the medium-power region. This region involves many areas of consideration. You must take into account the precautions for low-level testing previously covered as well as those for high-level testing that will be covered in the next section.

A typical medium-power setup is shown in Fig. 10–18. The first part of the setup is basically the same as the low-level setup. Once the device is inserted, however, other considerations are involved that are different from the low-level case. Such things as attenuator power ratings, termination power ratings, and thermistor mount (power sensor) maximum input levels become primary considerations.

Two methods of handling the increased power level are shown in Fig. 10–18. The first uses an output directional coupler. The one-half watt signal (+27 dBm) is fed to the coupler. The insertion loss of the coupler is low, so the termination on the coupler will have to handle close to the full half-watt. The termination, therefore, is a 1-W device. The coupler attenuates the signal 20 dB at the coupled port so a +7-dBm signal is present at the attenuator. A 10-dB attenuator lowers the level to −3 dBm, which is a very safe level for the thermistor mount.

The second method is direct reading. A 1-W, 30-dB attenuator is placed in series with the amplifier output. This placement results in the same −3-dBm level to the thermistor mount as with the coupler.

Measurements with this setup are taken in the same manner as with low-power setups. The device under test is removed, the system is calibrated, the device is reinserted, and data is taken on the device. The only difference between this setup and that of the low-power case is the level out of the generator and the unit under test.

10.4.3   High-Power Measurements

We have now come to the area where the largest number of precautions are necessary—high-power testing. This area requires that *all* component ratings be checked prior to being inserted into a test setup. The maximum input ratings of test equipment must also be checked. Careful consideration of these ratings will save time and a lot of expensive equipment. It is not a pleasant experience to turn a setup on and watch over $900 worth of power sensor breathe its last. It is a sinking feeling, and one that can be avoided very easily: *check all power ratings before inserting the equipment or component into the test setup.*

The test setup usually used for high power is a version of the coupler output shown in Fig. 10–18 for medium-power applications. The main concern is the rating of the termination put on the coupler. It must be adequate to handle the total power coming from

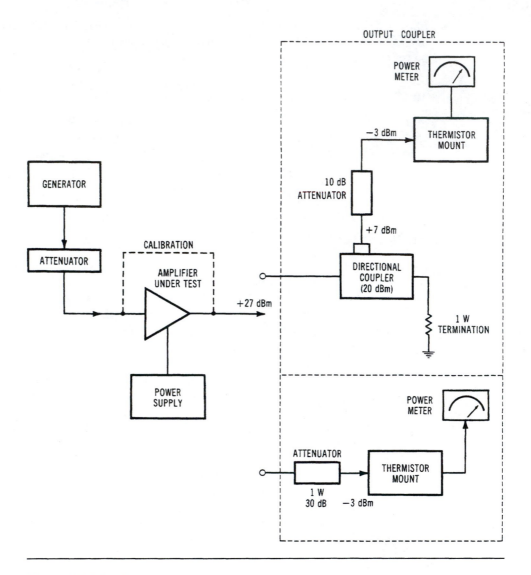

Figure 10–18.   Medium-power test setup.

the device being tested. There usually is no problem with the rating of the coupler, but it would be a good idea to check it anyway.

In addition to the precautions previously listed, the following precautions should be taken when working with high-power test setups:

- Do not touch the center conductor of a coaxial cable or connector operating at a high power level. These areas are very concentrated, and they can injure you severely.

- Tighten down all connectors and adapters. This procedure will reduce losses through the system.

- Do not touch finned attenuators or terminations. They are probably carrying a considerable amount of power and are hot. They will burn your hand in a hurry.

- Do not touch exposed microstrip circuitry that is carrying high-power energy—once again, it will burn severely.

- Operate in a shielded area if possible. High-power circuits have a tendency to radiate and may interfere with other circuits in the area.

- Always check the power handling capability of your components *before* applying RF power.

- Check the losses of all components (couplers, attenuators, etc.) when making up a test setup. The importance of accounting for all losses can be realized when considering that a 0.5-dB loss at a 100-W level is a loss of 11 W.

Obviously, great care must be taken when conducting high-power microwave tests. This statement cannot be overemphasized. Lack of care can cause great equipment damage and can also cause personal injury.

10.4.4   Peak Power Measurements

For many years there was a variety of test setups used for measuring the peak power of pulses. Even when a specific peak power meter was used, there were additional components (couplers) to be used, and still some calculations were needed. It was seen in Chapter 9 that there now is equipment available that will directly read peak power, and there is also equipment which will read either CW or peak power. So it can be seen that peak power measurement systems have come a long way. To measure this peak power, you simply need to place the proper meter in one of the previous setups.

10.5   Noise Measurements

Noise can be expressed in many ways. It may be expressed as *noise factor,* F, which is defined as the ratio of the input signal-to-noise ratio to the output signal-to-noise ratio, that is,

$$F = \frac{S_i/N_i}{S_o/N_o} \tag{10.3}$$

where, F is the noise factor,

S_i/N_i is the input signal-to-noise ratio,

S_o/N_o is the output signal-to-noise ratio.

Noise may also be expressed as *noise figure,* which is the logarithmic equivalent of *noise factor* and is expressed as

$$NF = 10 \log_{10} F$$

$$= 10 \log_{10} \frac{S_i/N_i}{S_o/N_o} \qquad (10.4)$$

This term is used widely throughout the microwave industry. Amplifier or mixer catalogs generally show the noise characteristic of those devices expressed as a certain number of decibels *noise figure*.

We said that noise figure is measured in decibels. It may be expressed also as noise temperature in kelvins (K). The relationship between noise figure and noise temperature is

$$NF = 10 \log_{10}\left(1 + \frac{T_N}{290}\right) \qquad (10.5)$$

where, NF is the noise figure in decibels,

T_N is the noise temperature in kelvins.

Noise temperature may also be expressed in terms of noise factor as

$$T_N = 300\,(F - 1) \qquad (10.6)$$

where F = noise factor

T_N = noise temperature in Kelvin

With the method of expressing noise explained, we will now proceed with test setups to measure this noise. There are both *manual* and *automatic measurement procedures*. For our purposes we will look at only the automatic procedures. These procedures are faster and more accurate than those classified as manual. Also, the automatic measurements use the noise-figure meters covered in Chapter 9.

The first setup for a cw noise-figure reading is shown in Fig. 10–19. The noise-figure meter provides both an output and receives an input. It supplies control for the solid-state noise source to gate it on and off to allow measurements to be taken. The input is from the unit under test. It is the final numbers that we are working towards when we make the test setup.

The mixer-preamp and LO blocks in the setup are termed "if necessary" because most noise-figure meters do not have inputs in the microwave frequency range. They may be a

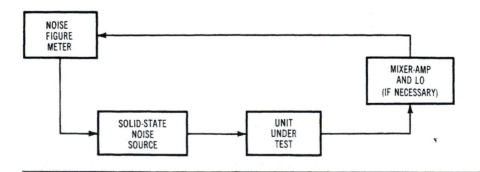

Figure 10–19. CW noise-figure test setup.

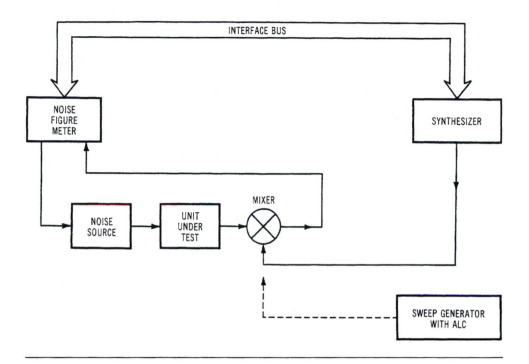

Figure 10–20.   Swept noise-figure test setup.

single frequency (30, 60, 90 MHz, for example) or accept any frequency from 10 MHz to 1800 MHz. This means that some mixing is necessary to have the setup act properly. If, however, a receiver with a 1-GHz if output is being checked, the mixer would not be needed because the IF output could be fed directly into the noise-figure meter.

Care must be taken to be sure you know the operation of the noise-figure meter and the solid-state noise source. Specific bias may be needed for specific frequencies, so be sure to check the instruction manuals for both the meter and the noise source being used.

Fig. 10–20 shows a setup that is used for swept noise-figure measurements. The noise figure, once again, provides gating pulses to the noise source and receives its input from the unit under test through a mixer (once again, if necessary). The frequency of the local oscillator, when the synthesizer is used, is entered on the keyboard of the noise-figure meter. The meter sends commands down the bus to the synthesizer and tunes the LO across the desired band. In the case where the sweeper is used, the limits are set on the generator and swept at an appropriate speed. Notice that the sweep generator is leveled with an ALC loop. This leveling ensures that the mixer LO port is receiving a constant-level input to maintain it in a saturated mode.

Test setups such as the ones just presented are used up to 18 GHz with no special equipment other than the mixer used to down convert to the noise-figure meter input range. Consult the individual manual for the meter being used to see if some additional IF gain may be needed. Some meters do require a higher level input than that available at the mixer output.

10.6   Frequency Measurements

There are many times when a test setup is made, measurements are taken, and the data does not come out the way it is expected. Sometimes there is a problem with the test setup. Usually, what has happened is that the entire setup is simply *off frequency*. To understand just what this phrase means, the term *frequency* needs to be understood.

We have used the term *frequency* many times to this point. Until now, the only emphasis placed on it was to say that the generator used should be in the *proper frequency range*. That advice really does not tell how important *frequency* is to test setups or to microwaves. You probably understand the importance of frequency the most when, for example, a filter must be checked from 1 to 2 GHz and the only available generator operates from 4 to 8 GHz. However, frequency goes much deeper than simply having the proper generator. It is the one parameter that governs the operation of every microwave component. Every component operates over a specific range. It may be 1.0 to 1.1 GHz or dc to 12.4 GHz, but it has a specific *frequency* range of operation. Outside that range it does not operate as specified.

The two classes of frequency measuring devices are *active* and *passive*. Active devices are also referred to as *electronic* since they operate on a frequency-beating technique as a form of measurement.

Similarly, passive devices are called *electromechanical* because they measure frequency by measuring a length—the wavelength. Each of these classes will be presented and investigated.

10.6.1   Active (Electronic) Devices

Purely electronic frequency measurements are those in which a standard is compared to an unknown by a null-beating technique. This technique is used to calibrate wavemeters because of its high accuracy. Typically, the signal from a frequency standard is taken and amplified, then connected into a harmonic generator to provide a so-called "comb" of standard frequencies. Ahead of the harmonic generator, this signal contains the frequency from the standard, for example, 10 MHz.

When amplified and connected into the harmonic generator, its waveshape is altered, resulting in harmonic distortion. This distortion creates a spectrum out of the harmonic generator consisting of harmonics of the original frequency. Good generators provide harmonics up to the 100th or higher harmonic number. Using a mixer, blending the harmonics with a swept-frequency signal, and connecting the result through a detector to an indicator such as an oscilloscope shows beat-frequency notes. Since a mixer passes the sum and the difference of the frequencies, a low-pass filter is usually connected to the end of the mixer to allow only the difference of the frequencies to go through. If the signal is connected to the vertical amplifier of an oscilloscope having a few megahertz frequency response, indication in the vertical direction occurs only when the beat frequency is in the passband of that amplifier. Consequently, when a difference frequency from the sweeper and the particular harmonic generator begins to approach a few megahertz, indication is given. As the difference frequency decreases, the amplitude of the signal becomes larger and larger until zero beat occurs. There is no amplitude at zero beat, of course—as the sweeper frequency passes through zero beat, a mirror image occurs and the signal decreases from maximum amplitude to zero. These beat notes appear at every harmonic number on the spectrum being swept. If the fundamental frequency being multiplied is 100 MHz, then a beat note is obtained at every 100 MHz on the spectrum.

Another method of measuring the frequency of a device is through the use of a device, previously covered in Chapter 9, called a frequency counter: a unit that counts the frequency, then displays it. Although it may not be as accurate in some cases as the frequency conversion technique discussed earlier, it is much more convenient and can be left in the circuit while all the measurements are being taken. Fig. 10–21 shows two ways of connecting a

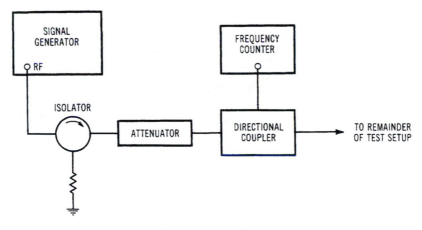

(A) Input frequency test.

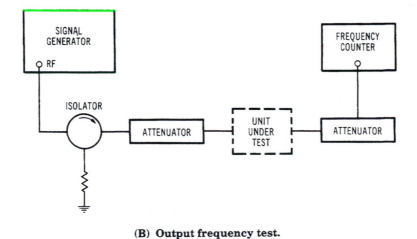

(B) Output frequency test.

Figure 10–21.　Frequency testing using a frequency counter.

test setup when using a frequency counter. In Fig. 10–21A a directional coupler is used—first to sample the RF energy going down the line, then to feed it to the counter. The coupler can be 10 or 20 dB to ensure the proper level going to the counter. There is, of course, a maximum level for the counter input, but there also is a minimum level required for counter operation. Some counters have a meter that signifies the proper level; others have an indication that says "lock" to tell that the level is of a sufficient value to make your measurements valid. Fig. 10–21B uses an attenuator to guarantee the proper level at the counter input.

The counter is the most widely used form of electronic frequency measuring in everyday systems. If super accuracy were needed, the frequency-beating technique would be used, but the accuracies needed for most test setups are met very nicely by using the counter. Some counters give a direct reading on the display; with others you must go through the process of heterodyning, mentioned earlier. In the second case, a little figuring must be done to find your frequency. Counters are also available that read pulse signals directly.

10.6.2 Passive (Electromechanical) Devices

When the accuracy of active devices is not required, passive devices offer direct readout at a considerable savings in cost. The most common type of passive frequency measuring device is the *wavemeter* (frequency meter). Passive transmission-type frequency meters are two-port devices that absorb part of the input power in a tunable cavity. When the cavity is tuned to resonance, a dip occurs in the transmitted power level that can be observed on a meter or oscilloscope display of the detected RF voltage. Frequency is then read from a calibrated dial driven by the cavity-tuning mechanism.

The wavemeter, also termed a cavity resonator or resonant cavity, is a dielectric region completely surrounded by conducting walls. It is capable of storing energy and is very similar to low-frequency LC resonant circuits. An essential property of the resonant cavity is that every cavity with a highly conducting boundary can be excited in an infinite sequence of resonant modes. Each mode is characterized by a particular standing-wave distribution of the surface current. The frequencies at which resonance occurs depend upon the shape and the size of the enclosed cavities. This is why the wavemeter in the lab that is used at X-band is a different size than the one that reads S-band frequencies.

The accuracy of cavity frequency meters depends upon the cavity Q, dial calibration, backlash, and the effects of temperature and humidity variations. Waveguide and coaxial passive frequency meters achieve accuracies of a few parts in 10^4.

We referred to the frequency meter (also called wavemeter) as a passive transmission type, which is one of three classifications given to such devices. The three classifications are transmission, reactive, and absorption. Fig. 10–22 shows diagrams of each of these, with power out versus frequency for two.

The transmission and absorption type wavemeters should be used in a matched system and may require isolation when the operation of the particular system is affected by the variations of cavity impedance as the wavemeter is tuned.

A less-often used mechanical frequency measuring technique is done with a slotted line. The setup is a generator, slotted line, and SWR meter. The probe is set to a depth only deep enough to get an indication on the meter, and a short is connected to the output of the

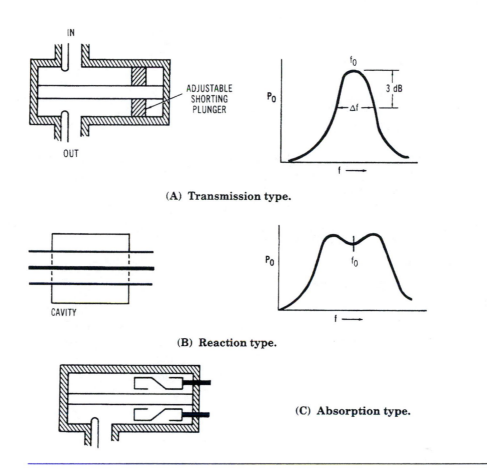

(A) Transmission type.

(B) Reaction type.

(C) Absorption type.

Figure 10–22.   Wavemeter diagrams.

slotted line. By moving the carriage and adjusting the gain of the SWR meter, we set up for a maximum deflection to the right on the meter. Record the position of this maximum (call this point x_1). Move the carriage either toward the load (short) or toward the generator until going through a dip and coming to another maximum. Record the location of this second maximum (x_2). We have now gone through one-half the wavelength of the frequency being generated. To find this frequency, subtract the two readings ($x_1 - x_2$) to get $\lambda/2$. As an example, suppose $x_1 = 12.2$ cm, $x_2 = 10.2$ cm, and $V = 3 \times 10^{10}$ cm/sec (a standard value). Then solve this equation:

$$f = \frac{V}{2|x_1 - x_2|}$$

$$= \frac{3 \times 10^{10}}{2(12.2 - 10.2)}$$

$$= \frac{3 \times 10^{10}}{4}$$

$$= 7.5 \text{ GHz}$$

A slotted line thus gives a frequency reading. It is not a direct reading because some figuring must be done, but it gives a reading with an accuracy on the order of 1%.

The measurement of a specific frequency within a microwave system can be accomplished in many ways. A direct reading counter may be used, or maybe only a wavemeter or slotted line is available for your use. Whatever way is chosen, there will be different accuracies involved. Your particular application and requirement will dictate which method will be the best. Whichever method is used, they all have one thing in common: they all are measuring the number of times a microwave signal repeats itself within a specified period of time; they are measuring *frequency*.

10.7   Active Measurements

All the measurements covered in this chapter thus far, with the exception of some power measurements, have been on passive devices. That is, these devices did not require any dc power for their operation. This section will concentrate on those measurements that characterize the components that do require dc for operation—*active devices*. We will concentrate on amplifier parameters and in particular *gain, gain compression,* and *intermodulation*.

10.7.1   Gain

The term *gain* is a relative term. That is, it is not a measure of some absolute level, or frequency, or impedance. It is a comparison of the output level of an amplifier to the input level supplied to that amplifier. Expressed mathematically it is

$$\text{Gain} = P_1/P_2 \tag{10.7}$$

where, P_1 is the output power,
$\qquad P_2$ is the input power.

This number is a ratio. For gain it is a number greater than one. To obtain the gain in decibels, as it usually is expressed, apply the following relationship:

$$\text{Gain (dB)} = 10 \log_{10} (P_1/P_2) \tag{10.8}$$

Fig. 10–23 shows a basic gain setup for cw applications. Notice the great similarity to some of the transmission test setups back in Section 10.2 because gain is a transmission test. Remember that when we began discussing the transmission test setup, we said we would be concerned only with losses (or attenuation) at that time. Gain measurements would be covered later in the chapter.

All the components in Fig. 10–23 serve the same functions as they did in the attenuation measurements. The generator, of course, supplies the microwave signal at the desired frequency; the two attenuators match the generator to the test device and the indicating device

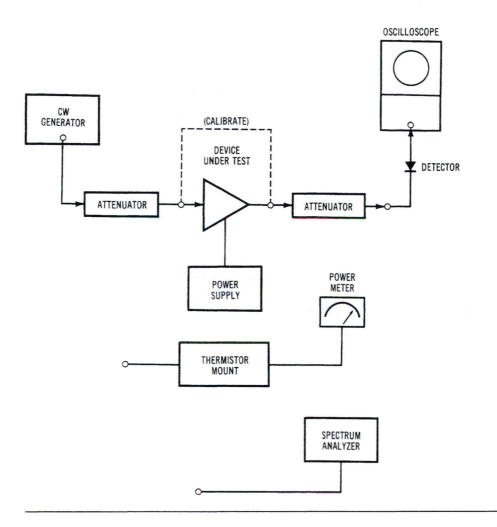

Figure 10–23.   Basic gain test setup.

to the test device, respectively; and the scope, power meter, or spectrum analyzer gives an indication of the gain.

To illustrate how a basic gain measurement is made, we will go through a procedure using the setup just described.

1.   Connect the test setup as shown in Fig. 10–23 in the *calibrate* position, with ac power applied to the cw generator and oscilloscope, power meter, or spectrum analyzer. Be sure that the generator is in a *standby* or *RF off* position.

2.   Place the generator in an *RF on* position following any built-in delays in the generator and/or oscilloscope.

3. Set up a convenient level on the oscilloscope, power meter, or spectrum analyzer, and record this as your reference level by recording the voltage from the scope, by tracing the response on the face of the scope with a grease pencil, or by recording or noting the level on the power meter or spectrum analyzer.

4. Place the generator in the *standby* or *RF off* position and insert the device to be tested.

5. Apply the necessary dc voltages and check to see that the current drain is at the proper level.

6. Place the generator in an *RF on* position once again and note the "change" in level on the scope. This change is the gain of the device under test. This change is a voltage change and may be converted to decibels by using this formula: decibels = 20 log (voltage change). Gain can be read directly on the power meter or spectrum analyzer.

Note: To obtain a direct reading on the scope, place a variable attenuator at the output of the device in place of the fixed attenuator. With a variable attenuator in the circuit, set a level on the scope during calibration, note the level by marking the face of the scope, insert the device to be tested, and adjust the variable attenuator until the display returns to the original calibration level. The value read off the attenuator is the value of the gain for the device under test. This test is very similar to the previous transmission testing.

By making adjustments to the setup in Fig. 10–23, we can perform swept-gain measurements. These changes are incorporated in Fig. 10–24. The obvious change is the sweep generator replacing the cw generator. Also, we have used just a scope for an indication device and added a recorder. The procedure would be the same as previously described, but this time there are many frequencies displayed instead of only one.

10.7.2 Gain Compression

The power input versus power output characteristics of most amplifiers are linear functions over most of the operating range of the amplifiers. That is, for a change in input level, there is a corresponding change in the output level. As an example, if we increased the level of the input to an amplifier by 10 dB, the output level would indicate an additional 10 dB if we were still in the linear region of the amplifier characteristics.

There is a limit, however, to how high an input an amplifier can take and still produce a corresponding change in the output. When the output change no longer follows the input, the amplifier is said to be in *compression*. That is, as the input to the amplifier is increased, the output level is being compressed and remains at a set level since the amplifier circuit is capable only of putting out a certain level, and that level has been reached. No matter how much the input to the amplifier is increased, the output will not increase (since gain is P_{out}/P_{in} and P_{out} is remaining the same as P_{in} increases). Eventually, the transistor will be stressed to such a level that it will be destroyed. So careful attention should be given to where an amplifier goes into gain compression.

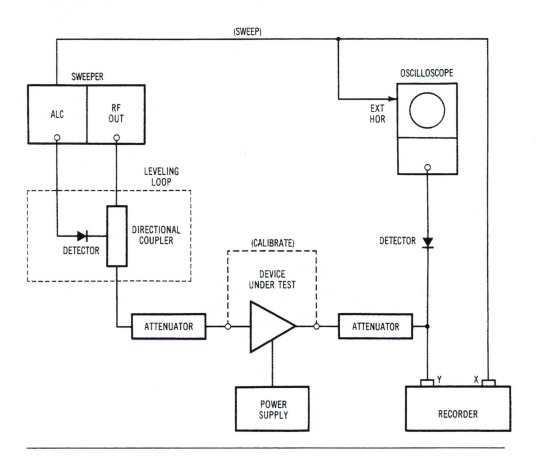

Figure 10–24. Swept gain test setup.

A very basic setup for compression would be a generator, the amplifier being tested, and a power meter. The generator output would be calibrated as to power level, and various levels would be applied to the amplifier. The output power would be recorded from the power meter, and P_{in} versus P_{out} would be plotted. As we have said, this setup is very basic and very time consuming.

10.7.3   Intermodulation

Intermodulation products (as they are called) result from the mixing of two or more input signals of different frequencies. The mixing is a consequence of the nonlinearity of the amplifier gain as a function of input power. This explanation should give a large clue about how to measure intermodulation. Apply two signals of different frequencies to the input of an amplifier, and monitor the output on a spectrum analyzer. Fig. 10–25 shows this exact setup. The two generators are adjusted a few MHz apart so that they have the same level as

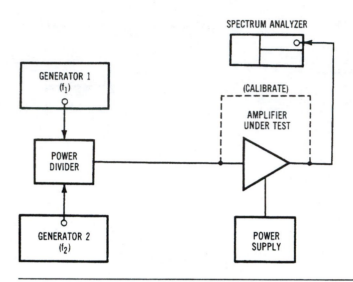

Figure 10–25.   Intermodulation test setup.

they are applied to the amplifier. The response on the spectrum analyzer is then read to determine the *third order intermodulation (IM)* responses. A more detailed procedure would be as follows:

1. Connect the test setup as shown in Fig. 10–25, with the device to be tested removed and a cable from the power divider to the spectrum analyzer.

2. Apply ac power to the generators and the spectrum analyzer, and be sure the generators are in the *standby* or *RF off* position.

3. Following any necessary time delays, place the generators in the *RF on* position. Set the frequency of generator 1 to a predetermined frequency and offset the frequency of generator 2 by an appropriate amount.

4. Set the levels of the two signals just made available so that they appear of equal amplitude on the spectrum analyzer. (These levels set the generator up so that the two signals go to the amplifier at the same level.)

5. Place the generators in the *standby* or *RF off* position and insert the amplifier to be tested.

6. Apply the necessary dc voltage(s) to the amplifier and check to see that it is drawing the right value of current.

7. Place the generators in the *RF on* position and read the *IM* product levels on the analyzer. They will be the signals that are on either side of the main two signals applied to the input. A good amplifier will have these *IM*s down at least 40 to 45 dB.

The measurement procedure presented is for what are called *third order intermodulation products*. These are the most common IMs to be present in an amplifier. These products must be characterized because they tell what type of distortion will be produced when two equal, level, closely spaced signals are applied simultaneously to the amplifier.

10.8   Chapter Summary

This chapter presents a very important part of microwaves, the part that brings all of the theory to life—testing. Without *proper* test methods, the most ideal design is totally useless. It must ultimately be tested to prove its validity in the real world.

We have presented six areas of microwave testing. Each has representative test setups presented. We emphasized in the beginning of the chapter and will re-emphasize that you should not memorize these setups. Use them only as a guide for your particular applications so that you will not be stuck with only one means of testing devices. You will have versatility and will be assured of conducting the *proper* and correct tests on all of your microwave equipment. It was also mentioned in the beginning of this chapter that these test setups may very well be a part of a more sophisticated automatic test system. It is important, however, to be aware of the basic setups in order to completely understand the more complex ones.

10.9   References

1.   Laverghetta, T.S. *Handbook of Microwave Testing*. Dedham, MA: Artech House, 1981.

2.   Laverghetta, T.S. *Microwave Measurements and Techniques*. Dedham, MA: Artech House, 1976.

3.   *Noise Figure Meter Model 8970A*, Preliminary Data, Hewlett-Packard Co., Palo Alto, CA, 1981.

4.   *Solid-State Microwave Amplifier Handbook*. Amplica, Inc., Newbury Park, CA.

Questions

10.2   Transmission Measurements

1.   Define *transmission*.
2.   What is RF substitution?
3.   What is IF substitution?
4.   What is audio substitution?
5.   Explain the operation/purpose of the attenuator calibrator in the audio substitution setup.
6.   What is a critical part of the swept IF substitution test setup?

10.3   Impedance Measurements

7.   Define *impedance*.
8.   What is a slotted line?

9. Explain the steps taken to make a slotted line measurement.
10. What is a reflectometer?
11. Describe a microwave short.
12. Why is a short used in microwave measurements rather than an open?
13. Explain the operation of a VSWR bridge.
14. How can a slotted line be used in a swept setup?

10.4 Power Measurements

15. Define the three power levels.
16. Why is an isolator used in a low power setup, rather than a filter or an attenuator?
17. Sketch a test setup for measuring a +30 dBm signal.
18. Is the test setup for a peak power measurement any different from that for cw? Explain.

10.5 Noise Measurements

19. What is noise factor?
20. What is noise figure?
21. What is noise temperature?
22. What is another name for an active frequency measurement device?
23. What is another name for a passive frequency measurement device?
24. Name two types of wavemeters.

10.7 Active Measurements

25. Define *gain*.
26. Is gain compression good or not? Explain.
27. Define *intermodulation products*.
28. Why are two generators needed when measuring intermodulation products?

Problems

10.3 Impedance Measurements

1. A return loss of −28 dB has been measured. What is the reflection coefficient of the circuit?
2. In problem 1, what is the VSWR of the circuit?
3. It has been determined that the maximum VSWR that can be tolerated is 1.62:1. We have found that our circuit has a reflection coefficient of 0.35. Will this meet the requirements? Explain.

10.5 Noise Measurements

4. Our circuit has an input signal that is 0.55 volts, with an input noise being 0.004 volts. If the gain of the circuit is 10 and the output noise is measured as 0.009 volts, what is the noise factor?
5. In problem 4, what is the noise figure?

6. In problem 4, what is the noise temperature?
7. We have found that a certain circuit needs an amplifier with a noise temperature no more than 552K. The noise factor we have to work with is 2.3. Will this be acceptable? Explain.

10.7 Active Measurements

8. We have an output power of 5 watts and an input power of 62 mw. What is the gain of the circuit?
9. In problem 5, what is the gain in dB?
10. A certain circuit has an input of 1.3 watts, and it is known that the circuit has a gain of 14.65 dB. We have an upper limit of output power imposed on us of 38 watts. Will this circuit work? Explain.

Appendix A

Microwave Bands

OFFICIAL JCS BAND DESIGNATION

A	Band	0	MHz	to	250	MHz
B	Band	250	MHz	to	500	MHz
C	Band	500	MHz	to	1000	MHz
D	Band	1.0	GHz	to	2.0	GHz
E	Band	2.0	GHZ	to	3.0	GHz
F	Band	3.0	GHZ	to	4.0	GHz
G	Band	4.0	GHZ	to	6.0	GHz
H	Band	6.0	GHZ	to	8.0	GHz
I	Band	8.0	GHZ	to	10.0	GHz
J	Band	10.0	GHZ	to	20.0	GHz
K	Band	20.0	GHZ	to	40.0	GHz
L	Band	40.0	GHZ	to	60.0	GHz
M	Band	60.0	GHZ	to	100.0	GHz

OLD JCS BAND DESIGNATION

VHF	Band	100	MHz	to	300	MHz
UHF	Band	300	MHz	to	1.0	GHz
L	Band	1.0	GHZ	to	2.0	GHz
S	Band	2.0	GHZ	to	4.0	GHz
C	Band	4.0	GHZ	to	8.0	GHz
X	Band	8.0	GHZ	to	12.4	GHz
Ku	Band	12.4	GHZ	to	18.0	GHz
K	Band	18.0	GHZ	to	26.0	GHz
Ka	Band	26.0	GHZ	to	40.0	GHz
Millimeter		40.0	GHZ	to	100.0	GHz

U.S. MICROWAVE BAND

P	Band	225	MHz	to	390	MHz
L	Band	390	MHz	to	1.55	GHz
S	Band	1.55	GHZ	to	5.20	GHz
C	Band	3.90	GHZ	to	6.20	GHz
X	Band	5.20	GHZ	to	10.9	GHz
K	Band	10.9	GHZ	to	36.0	GHz
Ku	Band	15.25	GHZ	to	17.25	GHz
K_1	Band	15.35	GHZ	to	24.5	GHz
Q	Band	36.0	GHZ	to	46.0	GHz
V	Band	46.0	GHZ	to	56.0	GHz
W	Band	56.0	GHZ	to	100.0	GHz

Appendix B

Microwave Formulas

WAVELENGTH (λ)

$$\lambda \text{ (centimeters)} = \frac{3 \times 10^{10}}{f}$$

$$\lambda \text{ (meters)} = \frac{3 \times 10^{8}}{f}$$

where, f is the frequency (hertz).

DECIBELS (POWER AND VOLTAGE)

$$dB \text{ (power)} = 10 \log_{10} \frac{P_1}{P_2}$$

$$dB \text{ (voltage)} = 20 \log_{10} \frac{E_1}{E_2}$$

where, P_1 and P_2 are the system powers,
E_1 and E_2 are the system voltages.

CHARACTERISTIC IMPEDANCE (Z_0) OF RF CABLE

$$Z_0 = \frac{138}{\sqrt{\epsilon}} \log_{10} \frac{D}{d}$$

where, ϵ is the dielectric constant,
D is the inside diameter of the outer conductor,
d is the outside diameter of the inner conductor.

PERCENTAGE VELOCITY OF PROPAGATION (V)

$$v = \frac{1}{\sqrt{\epsilon}} \times 100$$

where, ϵ is the dielectric constant.

EFFECTIVE DIELECTRIC CONSTANT (ϵ_{eff})

$$\epsilon_{eff} = 1 + q(\epsilon_r - 1)$$

where, ϵ_{eff} is the effective dielectric constant,
ϵ_r is the relative dielectric constant,
q is the filling factor (ratio of the dielectric area to the total area that would be used if the medium was stripline).

NOISE FIGURE (NF$_{dB}$)

$$NF_{dB} = 10 \log_{10} \frac{S_i/N_i}{S_o/N_o}$$

where, NF_{dB} is the noise figure in decibels,
S_i/N_i is the input signal-to-noise ratio,
S_o/N_o is the output signal-to-noise ratio.

REFLECTION COEFFICIENT (ρ)

$$\rho = \frac{SWR - 1}{SWR + 1} = \frac{Z_L - Z_0}{Z_L + Z_0}$$

where, SWR is the standing-wave ratio,
Z_L is the load impedance,
Z_0 is the characteristic impedance.

RETURN LOSS IN DECIBELS

$$dB = -20 \log_{10} |\rho|$$

where, ρ is the reflection coefficient.

VSWR

$$VSWR = \frac{1 + \rho}{1 - \rho}$$

where, ρ is the reflection coefficient.

Or

$$VSWR = \frac{E_{max}}{E_{min}} = \frac{E_i + E_r}{E_i - E_r} = \frac{I_{max}}{I_{min}}$$

where, E_{max} is the maximum voltage on the line,
     E_{min} is the minimum voltage on the line,
     E_i is the incident voltage,
     E_r is the reflected voltage,
     I_{max} is the maximum current on the line,
     I_{min} is the minimum current on the line.

Appendix C

Microwave Terms

Attenuation—the decrease in amplitude of a signal during its transmission from one point to another is attenuation. It may be expressed as a ratio or in decibels.

Characteristic Impedance—this term is the driving-point impedance (or resistance) of a line if it were of infinite length. Microwave systems are usually 50 ohms.

Coaxial Cable—coaxial cable is also called coaxial line, coaxial transmission line, and concentric line. It is a transmission line in which one conductor completely surrounds the other; the two are coaxial and separated by a continuous solid dielectric or by dielectric spacers.

Cutoff Frequency—the frequency that marks the edge of the passband of a filter and the beginning of the transition to the stopband is the cutoff frequency. Also, it is the upper frequency limit, usually of a loaded transmission circuit, beyond which the attenuation rises very rapidly. In waveguide, it is the lowest frequency at which the waveguide will propagate energy in some particular mode without attenuation.

DC Return—the dc return is a path back to ground for a microwave semiconductor. It usually consists of a low value of resistance or a coil that can be used as an RF choke also.

Dielectric—a dielectric is a material that obstructs microwave energy as it passes through. This obstruction causes a reduction in speed of the energy.

Directivity—the property that causes an antenna to radiate or receive more energy in some directions than others is directivity. Also, it is the property that causes energy to be coupled in one direction and not in another.

Dissipation Factor—a measure of the ability of a microwave material to release microwave energy is the dissipation factor. It is, in a sense, the loss through the material.

Gigahertz (GHz)—gigahertz is a term for 10^9 cycles per second. It replaces the more cumbersome and obsolete term kilomegacycle.

Harmonic—a wave having a frequency that is an integral multiple of the fundamental frequency is harmonic. For example, a wave with twice the frequency of the fundamental is called the *second harmonic*.

Iris—iris is also called a diaphragm. In a waveguide, a conducting plate (or plates) that is very thin (compared to the waveguide) and occupies part of the cross section of a waveguide.

Microstrip—a microwave transmission component in which a single conductor is supported above a ground plane.

Noise—any unwanted electrical disturbance or spurious signal that modifies the transmitting, indicating, or recording of desired data is termed a noise. Also, random electrical variations generated internally in electrical components are considered noise.

Noise Figure—the noise figure is the ratio of the signal-to-noise ratio at the input of a device to the signal-to-noise ratio at the output. If the log is taken of this ratio and multiplied by ten, the quantity is changed to decibels.

Octave—the octave is a band of frequencies the limits of which have a ratio of 2 to 1, for example, 1 to 2 GHz, 2 to 4 GHz.

Passband—the band of frequencies that will pass through a filter with essentially no attenuation is called passband.

Reflection Coefficient—at any specified plane in a uniform transmission line, the vector ratio between the electrical fields associated with the reflected and incident waves is the reflection coefficient.

Return Loss—the difference between the power incident upon, and the power reflected from, the load is the return loss.

Ripple—the wave-like variations in the amplitude response of a filter are the ripple.

Skin Effect—this term may be referred to as an RF resistance. The tendency of RF current is to flow on the outside edge of a conductor rather than in the center.

Stripline—the stripline is a transmission component in which a single conductor is surrounded by ground plane.

VSWR (Voltage Standing-Wave Ratio)—the ratio of the amplitude of the electrical field or voltage at a voltage maximum to that at an adjacent minimum is the VSWR.

Wavelength—the distance traveled by one cycle of microwave energy is a wavelength. Its exact length depends on the medium through which it is traveling.

Appendix D

Decimal-to-Metric
(Metric-to-Decimal)
Conversion

To Convert	To Obtain	Multiply by
feet	centimeters	3.048×10^1
feet	meters	3.048×10^{-1}
inches	centimeters	2.54
inches	meters	2.54×10^{-2}
yards	meters	9.144×10^{-1}
miles (statute)	kilometers	1.609
pounds	kilograms	4.536×10^{-1}
pounds	grams	4.536×10^2
grams	ounces	3.527×10^{-2}
ounces	grams	28.35
centimeters	inches	3.937×10^{-2}
centimeters	feet	3.281×10^{-1}
centimeters	yards	1.094×10^{-2}
centimeters	millimeters	10
meters	inches	39.37
meters	feet	3.281
meters	yards	1.094
meters	centimeters	100
kilometers	miles (statute)	6.214×10^{-1}
kilograms	pounds	2.2046

Appendix E

Cable Data

Dielectrics

PE . . . . Solid polyethylene
PTFE . . . . Solid polytetrafluoroethylene
PIB . . . . Polyisobulylene, Type B per MIL-C-17
Rubber . . . . Per MIL-C-17D
Sil. . . . . Silicon rubber
PS. . . . . Polystyrene

Conductors and Braid Materials

AL . . . . Aluminum
SCCAI . . . . Silver-coated copper-covered aluminum
BC . . . . Bare copper
SC . . . . Silver-covered copper
CCS. . . . . Copper-covered steel
TC . . . . Tinned copper
SCCS. . . . . Silver-covered copper-covered steel
SCCadBr . . . . Silver-covered cadmium bronze
GS . . . . Galvanized steel
TCCS. . . . . Tinned-copper-covered steel
SSC. . . . . Silver-covered strip
HR . . . . High-resistance wire
SA . . . . Silver-covered alloy

Jacket Material

PVC-I   Black polyvinylchloride contaminating Type I, per
                         MIL-C-17D

PVC-II	Grey polyvinylchloride, noncontaminating Type II, per MIL-C-17D
PVC-IIA	Black polyvinylchloride, noncontaminating Type IIA, per MIL-C-17D
PE-III	Clear polyethylene
PE-IIIA	High molecular weight black polyethylene, Type IIIA, per MIL-C-17D
FG Braid V	Fiberglass, impregnated, Type V, per MIL-C-17D
FEP-IX	Fluorinated ethylene propylene, Type IX, per MIL-C-17D
PUR	Polyurethane, black specific compounds
SIL/DAC-VI	Dacron braid over silicone rubber Type VI, per MIL-C-17D
Rubber	Per MIL-C-17D

Cable Specifications*

RG/U Type	Inner Conductor	Dielectric Material	DOD (Inch)	Number & Type of Shielding Braids	Jacket Material	O.D. (Inch)	Weight (lbs/ft)	Nominal Imped. (ohms)	Nominal Capacitance (pf/ft)	Max. Oper. Temp. Range (°C)	Max. Oper. Voltage (Volts RMS)	Comments
1-3	WAVEGUIDE											
4	0.0320" BC	PE	0.116	2: BC	PVC-I	0.226	0.025	50.0	30.8	−40 +80	1,900	Use RG58C
5	0.0508" BC	PE	0.185	2: BC	PVC-I	0.332	0.088	52.5	28.5	−40 +80	3,000	Use up to 100 MHz
5A	0.0508" SC	PE	0.181	2: SC	PVC-II	0.328	0.088	50.0	30.8	−40 +80	3,000	Use RG212
5B	0.0508" SC	PE	0.181	2: SC	PVC-IIA	0.328	0.087	50.0	30.8	−40 +80	3,000	Use RG212
6	0.0285" CCS	PE	0.185	2: Inner SC Outer BC	PVC-II	0.332	0.081	76.0	20.0	−40 +80	2,700	Use RG6A
6A	0.0285" CCS	PE	0.185	2: Inner SC Outer BC	PVC-IIA	0.332	0.082	75.0	20.6	−40 +80	2,700	Good attenuatio stability
7	0.0359" BC	Air-space PE	0.250	1: BC	PVC-I	0.370	0.080	95.0	12.5	−40 +80	1,000	Use RG63B
8	0.0855" 7/0.0285" BC	PE	0.285	1: BC	PVC-I	0.405	0.106	52.0	29.5	−40 +80	4,000	Use RG213
8A	0.0855" 7/0.0285" BC	PE	0.285	1: BC	PVC-IIA	0.405	0.106	52.0	29.5	−40 +80	5,000	Use RG213
9	0.0855" 7/0.0285" SC	PE	0.280	2: Inner SC Outer BC	PVC-II	0.420	0.140	51.0	30.0	−40 +80	4,000	Use RG214
9A	0.0855" 7/0.0285" SC	PE	0.280	2: SC	PVC-II	0.420	0.140	51.0	30.0	−40 +80	4,000	Use RG214
9B	0.0855" 7/0.02850 SC	PE	0.280	2: SC	PVC-IIA	0.420	0.150	50.0	30.8	−40 +80	5,000	Use RG214
10	0.0855" 7/0.0285" BC	PE	0.285	1: BC	PVC-II w. Armor	0.475 max.	0.146	52.0	29.5	−40 +80	4,000	Use RG215
10A	0.0855" 7/0.0285" BC	PE	0.285	1: BC	PVC-IIA w. Armor	0.475 max.	0.146	52.0	29.5	−40 +80	5,000	Use RG215
11	0.0477" 7/0.0159 TC	PE	0.285	1: BC	PVC-I	0.405	0.096	75.0	20.6	−40 +80	4,000	Use up to 100 MHz
11A	0.0477" 7/0.0159" TC	PE	0.285	1: BC	PVC-IIA	0.405	0.096	75.0	20.6	−40 +80	5,000	Use up to 1000 MHz
12	0.0477" 7/0.0159" TC	PE	0.285	1: BC	PVC-II w. Armor	0.475 max.	0.141	75.0	20.6	−40 +80	4,000	Use RG12A
12A	0.0477" 7/0.0159" TC	PE	0.285	1: BC	PVC-IIA w. Armor	0.475 max.	0.141	75.0	20.6	−40 +80	5,000	Use up to 1000 MHz
13	0.0477" 7/0.0159" TC	PE	0.280	2: BC	PVC-I	0.420	0.126	74.0	20.8	−40 +80	4,000	Use RG216
13A	0.0477" 7/0.0159" TC	PE	0.280	2: BC	PVC-IIA	0.420	0.126	74.0	20.8	−40 +80	5,000	Use RG216
14	0.102" BC	PE	0.370	2: BC	PVC-II	0.545	0.216	52.0	29.5	−40 +80	5,500	Use RG217
14A	0.1020" BC	PE	0.370	2: BC	PVC-IIA	0.545	0.216	52.0	29.5	−40 +80	7,000	Use RG217
15	0.0571" CCS	PE	0.370	2: BC	PVC-I	0.545	0.197	76.0	20.0	−40 +80	5,000	Use up to 1000 MHz
16	0.1250" BC tube	PE	0.460	1: BC	PVC-I	0.630	0.254	52.0	29.5	−40 +80	6,000	Use up to 1000 MHz

Cables recommended for use per supplement 1A Mil-C-17D.

---

*From "RF Transmission Line Catalog and Handbook," Times Wire & Cable Co.

Cable Specifications (cont.)

RG/U Type	Inner Conductor	Dielectric Material	DOD (Inch)	Number & Type of Shielding Braids	Jacket Material	O.D. (Inch)	Weight (lbs/ft)	Nominal Imped. (ohms)	Nominal Capacitance (pf/ft)	Max. Oper. Temp. Range (°C)	Max. Oper. Voltage (Volts RMS)	Comments
17	0.1880" BC	PE	0.680	1: BC	PVC-II	0.870	0.460	52.0	29.5	−40 +80	11,000	Use up to 1000 MHz
17A	0.1880" BC	PE	0.680	1: BC	PVC-IIA	0.870	0.460	52.0	29.5	−40 +80	11,000	Use RG218
17B	CANCELLED. REASSIGNED NEW NOMENCLATURE, RG177											
18	0.1880" BC	PE	0.680	1: BC	PVC-II w. Armor	0.945 max.	0.585	52.0	29.5	−40 +80	11,000	Use RG219
18A	0.1880" BC	PE	0.680	1: BC	PVC-IIA w. Armor	0.945 max.	0.585	52.0	29.5	−40 +80	11,000	Use RG219
19	0.2500" BC	PE	0.910	1: BC	PVC-II	1.120	0.740	52.0	29.5	−40 +80	14,000	Use RG220
19A	0.2500" BC	PE	0.910	1: BC	PVC-IIA	1.120	0.740	52.0	29.5	−40 +80	14,000	Use RG220
20	0.2500" BC	PE	0.910	1: BC	PVC-II w. Armor	1.195 max.	0.925	52.0	29.5	−40 +80	14,000	Use RG221
20A	0.2500" BC	PE	0.910	1: BC	PVC-IIA w. Armor	1.195 max.	0.925	52.0	29.5	−40 +80	14,000	Use RG221
21	0.0508" high resistance wire	PE	0.185	2: SC	PVC-II	0.332	0.087	53.0	29.0	−40 +80	2,700	Use RG222
21A	0.0508" high resistance wire	PE	0.185	2: SC	PVC-IIA	0.332	0.087	53.0	29.0	−40 +80	2,700	Use RG222
22	2 cond. 0.0456" 7/0.0152" BC	PE	0.285	1: TC	PVC-I	0.405	0.105	95.0	16.0	−40 +80	1,000	Balanced line w twisted cond
22A	2 cond. 0.0456" 7/0.0152" BC	PE	0.285	2: TC	PVC-II	0.420	0.151	95.0	16.0	−40 +80	1,000	Balanced line w twisted cond
22B	2 cond. 0.0456" 7/0.0152" BC	PE	0.285	2: TC	PVC-IIA	0.420	0.151	95.0	16.0	−40 +80	1,000	Balanced line w twisted cond
23	2 cond. 0.0855" 7/0.0285" BC	PE, 2 cores	0.380	2: Individual inner Common outer BC	PVC-I	0.650 × 0.945	0.490	125.0	12.0	−40 +80	3,000	Use RG23A
23A	2 cond. 0.0855" 7/0.0285" BC	PE, 2 cores	0.380	2: Individual inner Common outer BC	PVC-IIA	0.650 × 0.945	0.490	125.0	12.0	−40 +80	3,000	Dual coaxial balanced line
24	2 cond. 0.0855" 7/0.0285" BC	PE, 2 cores	0.380	2: Individual inner Common outer BC	PVC-I w. Armor	1.034 × 0.735"	0.670	125.0	12.0	−40 +80	3,000	Use RG24A (*max.)
24A	2 cond. 0.0855" 7/0.0285" BC	PE, 2 cores	0.380	2: Individual inner Common outer BC	PVC-IIA w. Armor	1.034 × 0.735"	0.670	125.0	12.0	−40 +80	3,000	Armored RG23A (*max.)
25A	0.0585" 19/0.0117" TC	Rubber-E	0.288	2: TC	Rubber-IV	0.505	0.205	48.0	50.0	−40 +80	10,000	
26A	0.0585" 19/0.0117" TC	Rubber-E	0.288	1: TC	Rubber-IV w. Armor	0.505 max.	0.189	48.0	50.0	−40 +80	10,000	
27A	0.0925" 19/0.0185" TC	Rubber-D	0.455	1: TC	Rubber-IV w. Armor	0.670 max	0.304	48.0	50.0	−40 +80	15,000	
28B	0.0925" 19/0.0185" TC	Rubber-D	0.455	2: TC, GS	Rubber-IV	0.750	0.370	48.0	50.0	−40 +80	15,000	
29	0.0320" BC	PE	0.116	1: TC	PE-III	0.184 max.	0.021	53.5	28.5	−55 +80	1,900	Use RG58
30	0.0477" 7/0.0159" BC	PIB	0.185	1: BC	PVC-I	0.250	0.044	50.0	27.0	−40 +80	1,500	Use RG58
31	0.0855" 7/0.0285" BC	PIB	0.285	1: BC	PVC-I	0.405	0.106	51.0	31.0	−40 +80	2,000	Use RG213

Cables recommended for use per supplement 1A Mil-C-17D.

Cable Specifications (cont.)

RG/U Type	Inner Conductor	Dielectric Material	DOD (Inch)	Number & Type of Shielding Braids	Jacket Material	O.D. (Inch)	Weight (lbs/ft)	Nominal Imped. (ohms)	Nominal Capacitance (pf/ft)	Max. Oper. Temp. Range (°C)	Max. Oper. Voltage (Volts RMS)	Comments
32	0.0855" 7/0.0285" BC	PIB	0.285	1: BC	PVC-I w. Armor	0.465	0.141	51.0	29.0	−40 +80	2,000	Use RG215
33	0.1019" BC	PE	0.370	None	Lead	0.470	0.390	51.0	30.0	−55 +80	6,000	
34	0.0855" 7/0.0285" BC	PE	0.455	1: BC	PVC-I	0.625	0.224	71.0	21.5	−40 +80	5,200	Use RG31B
34A	0.0747" 7/0.0249" BC	PE	0.460	1: BC	PVC-IIA	0.630	0.224	75.0	20.6	−40 +80	6,500	Use RG34B
34B	0.0747" 7/0.0249" BC	PE	0.460	1: BC	PVC-IIA	0.630	0.224	75.0	20.6	−40 +80	6,500	Use up to 1000 MHz
35	0.1144" BC	PE	0.680	1: BC	PVC-II w. Armor	0.945 max.	0.525	71.0	21.5	−40 +80	10,000	Use RG35B
35A	0.1045" BC	PE	0.680	1: BC	PVC-IIA w. Armor	0.945 max.	0.525	75.0	20.6	−40 +80	10,000	Use RG35B
35B	0.1045" BC	PE	0.680	1: BC	PVC-IIA w. Armor	0.945 max.	0.525	75.0	20.6	−40 +80	10,000	Unarmored: see RG164
36	0.1620" PC	PE	0.910	1: BC	PVC-I	1.180	0.805	69.0	22.0	−40 +80	13,000	Use up to 1000 MHz
37	0.0320" TC	Rubber-C	0.140	1: TC	PE-III	0.210	0.040	52.5	38.0	−55 +80	750	
38	0.0453" TC	Rubber-C	0.196	2: TC	PE-III	0.312	0.110	52.5	38.0	−55 +80	1,000	
39	0.0253" CCS	Rubber-C	0.196	2: TC	PE-III	0.312	0.100	72.5	28.6	−55 +80	1,000	
40	0.0253" CCS	Rubber-C	0.196	2: TC	Rubber-IV	0.420	0.150	72.5	28.0	−40 +80	1,000	
41	0.0490" 16/0.0100" TC	Rubber-C	0.250	1: TC	Rubber-IV	0.425	0.150	67.5	27.6	−40 +80	3,000	
42	0.0285" Resistance wire	PE	0.196	2: SC	PVC-II	0.342	0.050	78.0	20.0	−40 +80	2,700	Use RG222
43	2 cond. 0.0855" 7/0.0285" BC	Rubber-B	0.472	1: BC	PVC-I	0.617		95.0	17.6	−40 +80	1,500	Use RG57
44-47	STUD SUPPORTED RIGID LINES See Mil-HDBK 216, Para. 5.5											
48-53	RECTANGULAR WAVEGUIDES COVERED BY MIL-W-85 See Mil HDBK 216, Para. 6.23											
54	0.0477" 7/0.0159" BC	PE	0.185	1: BC	PVC-I	0.275	0.045	58.0	27.0	−40 +80	2,500	Use RG54A
54A	0.0456" 7/0.0152" BC	PE	0.178	1: TC	PE-III	0.250 (max.)	0.041	58.0	26.5	−55 +80	3,000	Use up to 1000 MHz
55	0.0320" BC	PE	0.116	2: TC	PE-III	0.206 max.	0.032	53.5	28.5	−55 +80	1,900	Use RG55B
55A	0.0350" SC	PE	0.116	2: SC	PVC-IIA	0.216 max.	0.034	50.0	30.8	−40 +80	1,900	Use RG223
55B	0.0320" SC	PE	0.116	2: TC	PE-IIIA	0.206 max.	0.033	53.5	28.5	−55 +80	1,900	Use up to 1000 MHz
56	0.0585" 19/0.0117" BC	Rubber-D	0.308	2: BC	PVC-I	0.535	0.243	48.0	50.0	−40 +80	8,000	
57	2 cond. 0.0855" 7/0.0285" BC	PE	0.472	1: TC	PVC-I	0.625	0.225	95.0	17.0	−40 +80	3,000	Balanced line Parallel

Cables recommended for use per supplement 1A Mil-C-17D.

Cable Specifications (cont.)

RG/U Type	Inner Conductor	Dielectric Material	DOD (Inch)	Number & Type of Shielding Braids	Jacket Material	O.D. (Inch)	Weight (lbs/ft)	Nominal Imped. (ohms)	Nominal Capac-itance (pf/ft)	Max. Oper. Temp. Range (°C)	Max. Oper. Voltage (Volts RMS)	Comments
57A	2 cond. 0.0855" 7/0.0285" BC	PE	0.472	1: TC	PVC-IIA	0.625	0.225	95.0	17.0	−40 +80	3,000	Balanced line Parallel
58	0.0320" BC	PE	0.116	1: TC	PVC-I	0.195	0.029	53.5	28.5	−40 +80	1,900	Use RG58B
58A	0.0355" 19/0.0071" TC	PE	0.116	1:TC	PVC-I	0.195	0.029	52.0	28.5	−40 +80	1,900	Use RG58C
58B	0.0320" BC	PE	0.116	1: TC	PVC-IIA	0.195	0.029	53.5	28.5	−40 +80	1,900	Use up to 1000 MHz
58C	0.0355" 19/0.0071" TC	PE	0.116	1: TC	PVC-IIA	0.195	0.029	50.0	30.8	−40 +80	1,900	Extra flexible version RG58B
59	0.0253" CCS	PE	0.146	1: BC	PVC-I	0.242	0.032	73.0	21.0	−40 +80	2,300	Use RG59B
59A	0.0253" CCS	PE	0.146	1: BC	PVC-IIA	0.242	0.032	73.0	21.0	−40 +80	2,300	Use RG59B
59B	0.0230" CCS	PE	0.146	1: BC	PVC-IIA	0.242	0.032	75.0	20.6	−40 +80	2,300	Use up to 1000 MHz
60	0.0508" Str C	Rubber-C	0.250	1: BC	Rubber-IV	0.425	0.150	50.0	39.0	−40 +80	1,100	
61	SPECIAL 500 OHM LINE											
62	0.0253" CCS	Air-space PE	0.146	1: BC	PVC-I	0.242	0.038	93.0	13.5	−40 +80	750	Use RG62A
62A	0.0253" CCS	Air-space PE	0.146	1: BC	PVC-IIA	0.242	0.038	93.0	13.5	−40 +80	750	Low capacitance
62B	0.0240" 7/0.0080" CCS	Air-space PE	0.146	1: BC	PVC-IIA	0.242	0.038	93.0	13.5	−40 +80	750	Extra flexible RG62A
63	0.0253" CCS	Air-space PE	0.285	1: BC	PVC-I	0.405	0.083	125.0	10.0	−40 +80	1,000	Use RG63B
63A	0.0253" BC	Air-space PE	0.285	1: BC	PVC-I	0.405	0.083	125.0	10.0	−40 +80	1,000	Use RG63B
63B	0.0253" CCS	Air-space PE	0.285	1: BC	PVC-IIA	0.405	0.083	125.0	10.0	−40 +80	1,000	Low capacitance
64	0.0585" 19/0.0117 TC	Rubber-D	0.308	2: TC	Rubber-IV	0.495	0.225	48.0	60.0	−40 +80	10,000	
64A	0.0585" 19/0.0117" TC	Rubber-E	0.288	2: TC	Rubber-IV	0.475 max.	0.205	48.0	50.0	−40 +80	10,000	
65	0.0080" Formex-F 0.1280" dia Helix	PE	0.285	1: BC	PVC-I	0.405	0.096	950.0	44.0	−40 +80	1,000	High impedance video delay line
65A	0.0080" Formex-F 0.1280" dia Helix	PE	0.285	1: BC	PVC-IIA	0.405	0.096	950.0	44.0	−40 +80	1,000	High impedance video delay line
66-69	RECTANGULAR WAVEGUIDE COVERED BY MIL-W-25. See Mil HDBK 216. Para 6.17-6.23											
71	0.0253" CCS	Air-space PE	0.146	2: TC	PVC-I	0.250 max.	0.046	93.0	13.5	−40 +80	750	Use RG71B
71A	0.0253" CCS	Air-space PE	0.146	2: TC	PE-III	0.250 max.	0.046	93.0	13.5	−55 +80	750	Use RG71B
71B	0.0253" CCS	Air-space PE	0.146	2: TC	PE-IIIA	0.250 max.	0.046	93.0	13.5	−55 +80	750	Low capacitance
72	0.0253" CCS	Air-space PE	0.460	1: BC	PVC-I	0.630	0.169	150.0	7.8	−40 +80	750	Low capacitance

Cables recommended for use per supplement 1A Mil-C-17D.

Cable Specifications (cont.)

RG/U Type	Inner Conductor	Dielectric Material	DOD (Inch)	Number & Type of Shielding Braids	Jacket Material	O.D. (Inch)	Weight (lbs/ft)	Nominal Imped. (ohms)	Nominal Capacitance (pf/ft)	Max. Oper. Temp. Range (°C)	Max. Oper. Voltage (Volts RMS)	Comments
73	0.0320" BC	PE	0.116	2: BC	Copper braid	0.175	0.031	25.0	61.8	−55 +80	1,000	Low impedance
74	0.1020" BC	PE	0.370	2: BC	PVC-II w. Armor	0.615 max.	0.310	52.0	29.5	−40 +80	5,500	Use RG224
74A	0.1020" BC	PE	0.370	2: BC	PVC-IIA w. Armor	0.615 max.	0.310	52.0	29.5	−40 +80	7,000	Use RG224
75	RECTANGULAR WAVEGUIDE COVERED BY MIL-W-85. See Mil HDBK 216, Para. 6.17–6.21											
76	STUD SUPPORTED RIGID LINE. See Mil HDBK 216, Para. 5.5											
77A	0.0585" 19/0.117" TC	Rubber-E	0.288	2: TC	PVC-IIA	0.450	0.195	48.0	50.0	−40 +80	8,000 Peak	
78A	0.0585" 19/0.0117" TC	Rubber-E	0.288	1: TC	PVC-IIA	0.420	0.149	48.0	50.0	−40 +80	8,000 Peak	
79	0.0253" CCS	Air-space PE	0.285	1: BC	PVC-I w. Armor	0.475 max.	0.136		10.0	−40 +80	1,000	Low capacitance
79A	0.0253" CCS	Air-space PE	0.285	1: BC	PVC-I w. Armor	0.475 max.	0.130	125.0	10.0	−40 +80	1,000	Low capacitance
79B	0.0253" CCS	Air-space PE	0.285	1: BC	PVC-IIA w. Armor	0.475 max.	0.136	125.0	10.0	−40 +80	1,000	Low capacitance
80	RIGID LINE. See Mil HDBK 216, Para. 5.2											
81	0.0625" BC	Magnesium Oxide G	0.321	None	Copper tube	0.325	0.172	50.0	37.0	>250	3,000	
82	0.1250" BC	Magnesium Oxide-G	0.650	None	Copper tube	0.750	0.698	50.0	36.0	>250	5,000	
83	0.1020" BC	PE	0.240	1: BC	PVC-I	0.405	0.120	35.0	44.0	−40 +80	2,000	Low impedance
84A	0.1045" BC	PE	0.680	1: BC	PVC-IIA w. lead sheath	1.000	1.325	75.0	20.6	−40 +80	10,000	RG35B with
85A	0.1045" BC	PE	0.680	1: BC	PVC-IIA w lead Armor	1.565 max.	2.910	75.0	20.6	−40 +80	10,000	RG84A with special Armor
86	2 cond. 0.0855" 7/0.0285" BC	PE	0.300× 0.650	None	None	0.300× 0.650	0.100	200.0	7.8	−55 +80	10,000	Twin lead
87A	0.0960" 7/0.0320" SC	PTFE	0.280	2: SC	FG Braid-V	0.425	0.180	50.0	29.4	−55 +250	5,000	Use RG225
88	0.0585" 19/0.0117" TC	Rubber-E	0.288	4: TC	PVC-I	0.515 max.	0.211	48.0	50.0	−40 +80	10,000	
88A	0.0585" 19/0.0117" TC	Rubber-E	0.288	4: TC	PVC-IIA	0.515 max	0.211	48.0	50.0	−40 +80	10,000	
88B	0.0585" 19/0.0117" TC	Rubber-E	0.288	4: TC	Rubber-IV	0.565	0.238	48.0	50.0	−40 +80	10,000	
89	0.0253" CCS	Air-space PE	0.285	1: BC	PVC-I	0.632	0.195	125.0	10.0	−40 +80	1,000	Low capacitance
90	0.0603" 7/0.0201" SC	PE	0.195	3: SC, GS, GC	PVC-IIA	0.425		50.0	30.8	−40 +80	3,000	Excellent shielding
91	RECTANGULAR WAVEGUIDE COVERED BY MIL-W-85. See Mil HDBK 216, Para. 6.17–6.23											
92	RIGID COAXIAL LINE. See Mil HDBK 216, Para. 5.2											

Cables recommended for use per supplement 1A Mil-C-17D.

461

Cable Specifications (cont.)

RG/U Type	Inner Conductor	Dielectric Material	DOD (Inch)	Number & Type of Shielding Braids	Jacket Material	O.D. (Inch)	Weight (lbs/ft)	Nominal Imped. (ohms)	Nominal Capac- itance (pf/ft)	Max. Oper. Temp. Range (°C)	Max. Oper. Voltage (Volts RMS)	Comments
93	0.2000" 19/0.0400" BC	Taped PTFE	0.573	1: BC	FG Braid-V	0.710	0.475	50.0	29.4	−55 +250	10,000	Use RG211A
94	0.1125" 19/0.0225" SC	Taped PTFE	0.292	2: BC	FG Braid-V	0.445	0.270	50.0	29.4	−55 +250	7,000	Use RG226
94A	0.1270" 19/0.0254" SC	Taped PTFE	0.370	2: BC	FG Braid-V	0.500	0.445	50.0	29.4	−55 +250	7,000	Use RG226
95- 99	RECTANGULAR WAVEGUIDES COVERED BY MIL-W-85. See Mil HDBK 216, Para. 6.17–6.23											
100	0.0735" 19/0.0147" BC	PE	0.146	1: BC	PVC-I	0.242	0.046	35.0	44.0	−40 +80	2,000	Use up to 1000 MHz
101	0.0641" BC	Rubber		1: TC		0.588		75.0				
102	2 cond. 0.0808" BC	Rubber		1: TC		1.088		140.0				
103- 107	RECTANGULAR WAVEGUIDES COVERED BY MIL-W-85. See Mil HDBK 216, Para. 6.17–6.23											
108	2 cond. 0.0378" 7/0.0126" TC	PE	0.079 each	1: TC	PVC-II	0.235	0.032	78.0	19.6	−40 +80	1,000	Use RG108A
108A	2 cond. 0.0378" 7/0.0126" TC	PE	0.079 each	1: TC	PVC-IIA	0.235	0.032	78.0	19.6	−40 +80	1,000	Balanced line
109- 110	RECTANGULAR WAVEGUIDE COVERED BY MIL-W-85. See Mil HDBK 216, Para. 6.17–6.23											
111	2 cond. 0.0456" 7/0.0152" BC	PE	0.285	2: TC	PVC-II w. Armor	0.490 max.	0.146	95.0	16.0	−40 +80	1,000	Use RG111A
111A	2 cond. 0.0456" 7/0.0152" BC	PE	0.285	2: TC	PVC-IIA w. Armor	0.490 max.	0.146	95.0	16.0	−40 +80	1,000	Use RG22B w. Armor
112- 113	RECTANGULAR WAVEGUIDE COVERED BY MIL-W-85. See Mil HDBK 216, Para. 6.17–6.23											
114	0.0070" CCS	Air-space PE	0.285	1: BC	PVC-I	0.405	0.087	185.0	6.5	−40 +80	1,000	Use RG114A
114A	0.0070" CCS	Air-space PE	0.285	1: BC	PVC-IIA	0.405	0.087	185.0	6.5	−40 +80	1,000	Low capacitance
115	0.0840" 7/0.0280" SC	Taped PTFE	0.250	2: SC	FG Braid-V	0.375	0.148	50.0	29.4	−55 +250	5,000	Use RG115A
115A	0.0840" 7/0.0280" SC	Taped PTFE	0.255	2: SC	FG Braid-V	0.415	0.180	50.0	29.4	−55 +250	5,000	Extra flexible RG225
116	0.0960" 7/0.0320" SC	PTFE	0.280	2: SC	FG Braid-V w. Armor	0.475	0.198	50.0	29.4	−55 +250	5,000	Use RG227
117	0.1880" BC	PTFE	0.620	1: BC	FG Braid V	0.730	0.641	50.0	29.4	−55 +250	7,000	Use RG211A
117A	0.1880" BC	PTFE	0.620	1: BC	FG Braid-V	0.730	0.641	50.0	29.4	−55 +250	7,000	Use RG211A
118	0.1880" BC	PTFE	0.620	1: BC	FG Braid-V w. Armor	0.780	0.682	50.0	29.4	−55 +250	7,000	Use RG228A
118A	0.1880" BC	PTFE	0.620	1: BC	FG Braid-V w. Armor	0.780	0.682	50.0	29.4	−55 +250	7,000	Use RG228A
119	0.1020" BC	PTFE	0.332	2: BC	FG Braid-V	0.465	0.225	50.0	29.4	−55 +250	6,000	Use up to 1000 MHz
120	0.1020" BC	PTFE	0.332	2: BC	FG Braid-V w. Armor	0.525 max.	0.282	50.0	29.4	−55 +250	6,000	RG119 w Armor

Cables recommended for use per supplement 1A Mil-C-17D.

Cable Specifications (cont.)

RG/U Type	Inner Conductor	Dielectric Material	DOD (Inch)	Number & Type of Shielding Braids	Jacket Material	O.D. (Inch)	Weight (lbs/ft)	Nominal Imped. (ohms)	Nominal Capacitance (pf/ft)	Max. Oper. Temp. Range (°C)	Max. Oper. Voltage (Volts RMS)	Comments
121	RECTANGULAR WAVEGUIDE COVERED BY MIL-W-85. See Mil HDBK 216, Para. 17–6.23											
122	0.0300" 7/0.0050" TC	PE	0.096	1: TC	PVC-IIA	0.160	0.016	50.0	29.4	−40 +80	1,900	Use up to 1000 MHz
124	0.0253" TCCS	Taped PTFE	0.135	1: TC	FG Braid-V	0.240	0.210	73.0	20.3	−55 +250	2,300	Use RG140
125	0.0159" CCS	Air-space PE	0.460	1: BC	PVC-IIA	0.600	0.180	150.0	7.8	−40 +80	2,000	Low capacitance
126	0.0609" 7/0.0203" HR	PTFE	0.185	1: HR	FG Braid-V	0.280	0.070	50.0	29.4	−55 +250	3,000	High loss cable
127	RECTANGULAR WAVEGUIDE COVERED BY MIL-W-85. See Mil HDBK 216, Para. 6.17–6.23											
128	RIGID LINE. See Mil HDBK 216, Para. 5.2											
129	RECTANGULAR WAVEGUIDE COVERED BY MIL-W-85. See Mil HDBK, Para. 6.17–6.23											
130	2 cond. 0.0855" 7/0.0285" BC	PE	0.472	1: TC	PVC-I	0.625	0.220	95.0	17.0	−40 +80	3,000	RG57 with twisted cond.
131	2 cond. 0.0855" 7/0.0285" BC	PE	0.472	1: TC	PVC-I w. Armor	0.710 max.	0.290	95.0	17.0	−40 +80	3,000	Armored RG130
132	RECTANGULAR WAVEGUIDE COVERED BY MIL-W-85. See Mil HDBK 216, Para. 6.17–6.23											
133	0.0285" BC	PE	0.285	1: BC	PVC-I	0.405	0.094	95.0	16.2	−40 +80	4,000	Use RG133A
133A	0.0253" BC	PE	0.285	1: BC	PVC-IIA	0.405	0.094	95.0	16.2	−40 +80	4,000	95 ohm version RG8
134	RIGID LINE. See Mil HDBK 216, Para. 5.2											
135 139	RECTANGULAR WAVEGUIDE COVERED BY MIL-W-85. See Mil HDBK 216, Para. 6.17–6.23											
140	0.0250" SCCS	PTFE	0.146	1: SC	FG Braid-V	0.233	0.056	75.0	19.5	−55 +250	2,300	See RG302 for FEP jacket
141	0.0359" SCCS	PTFE	0.116	1: SC	FG Braid-V	0.190	0.036	50.0	29.4	−55 +250	1,900	Use RG141A
141A	0.0390" SCCS	PTFE	0.116	1: SC	FG Braid-V	0.190	0.036	50.0	29.4	−55 +250	1,900	See RG303 for FEP jacket
142	0.0359" SCCS	PTFE	0.116	2: SC	FG Braid-V	0.206 max.	0.047	50.0	29.4	−55 +250	1,900	Use RG142A
142A	0.0390" SCCS	PTFE	0.116	2: SC	FG Braid-V	0.206 max.	0.047	50.0	29.4	−55 +250	1,900	See RG142B for FEP jacket
142B	0.0390" SCCS	PTFE	0.116	2: SC	FEP	0.195	0.050	50.0	29.4	−55 +200	1,900	Stranded center cond. available
143	0.0570" SCCS	PTFE	0.185	2: SC	FG Braid-V	0.325	0.114	50.0	29.4	−55 +250	3,000	Use RG143A
143A	0.0590" SCCS	PTFE	0.185	2: SC	FG Braid-V	0.325	0.109	50.0	29.4	−55 +250	3,000	See RG304 for FEP jacket
144	0.0537" 7/0.0179" SCCS	PTFE	0.285	1: SC	FG Braid-V	0.410	0.137	75.0	19.5	−55 +250	5,000	High temp RG11A
145	2 cond. 0.0720" BC	Air-space PE		Copper tube	Lead/tar			75.0	14.6			
146	0.0070" CCS	Air-space PTFE	0.285	1: BC	FG Braid-V	0.375	0.108	190.0	6.0	−55 +250	1,000	Low capacitance

Cables recommended for use per supplement 1A Mil-C-17D.

463

Cable Specifications (cont.)

RG/U Type	Inner Conductor	Dielectric Material	DOD (Inch)	Number & Type of Shielding Braids	Jacket Material	O.D. (Inch)	Weight (lbs/ft)	Nominal Imped. (ohms)	Nominal Capacitance (pf/ft)	Max. Oper. Temp. Range (°C)	Max. Oper. Voltage (Volts RMS)	Comments
147	0.2500" BC	PE	0.910	1: BC	PVC-I w. Armor	1.937		52.0	29.5	−40 +80	14,000	RG19U with Armor
148	0.0855" 7/0.0285" BC	PE	0.285	1: BC	PVC-I w. Armor	0.800		52.0	29.5	−40 +80	4,000	
149	0.048" 7/0.0159" TC	PE	0.285	1: BC	PVC-IIA	0.405	0.105	75.0	20.6	−40 +80	5,000	Use RG391
150	0.048" 7/0.0159" TC	PE	0.285	1: BC	PVC-IIA w. Armor	0.475 max.	0.112	75.0	20.6	−40 +80	5,000	Use RG392
151 155	RIGID LINES COVERED BY MIL-L-3890. See Mil HDBK 216, Para. 5.4											
156	0.0855" 7/0.0285" TC	PE and cond. PE	0.285	3: TC, GS, TC	PVC-IIA	0.540	0.211	50.0	32.0	−40 +80	10,000	Triaxial pulse cable
157	0.1005" 19/0.0201" TC	PE and cond. PE	0.455	3: TC, GS, TC	PVC-IIA	0.725	0.317	50.0	38.0	−40 +80	15,000	Triaxial pulse cable
158	0.1988" 37/0.0284" TC	PE and cond. PE	0.455	3: TC, GS, TC	PVC-IIA	0.725	0.380	25.0	78.0	−40 +80	15,000	Triaxial pulse cable
159	0.0320" SC	Taped PTFE	0.116	1: SC	FG Braid-V	0.195	0.035	50.0	29.4	−55 +250	1,900	Use RG141
160	4 cond. 0.071" 19/.0142" 2TC, 2BC	PE	0.322	1: BC	PVC-I	1.055		125.0	12.0	−40 +80	3,000	4 cond. Balanced line
161	0.012" 7/0.004" S Cad BR	PTFE	0.057	1: SC	Nylon	0.082	0.015	70.0	20.0	−60 +120	1,000	Miniature
162	RIGID LINE. See Mil HDBK 216, Para. 5.2											
163	RECTANGULAR WAVEGUIDE COVERED BY MIL-W-85. See Mil HDBK 216, Para. 6.17–6.23											
164	0.1045" BC	PE	0.680	1: BC	PVC-IIA	0.870	0.490	75.0	20.6	−40 +80	10,000	RG35B without Armor
165	0.0960" 7/0.0320" SC	PTFE	0.285	1: SC	FG Braid-V	0.410	0.121	50.0	29.4	−55 +250	5,000	RG225 with one braid
166	0.096" 7/0.0320" SC	PTFE	0.285	1: SC	FG Braid-V w. Armor	0.460	0.144	50.0	29.4	−55 +250	5,000	RG165 with Armor
167 173	RECTANGULAR WAVEGUIDE COVERED BY MIL-W-85. See Mil HDBK 216, Para. 6.17–6.23											
174	0.0189" 7/0.0063" CCS	PE	0.060	1: TC	PVC-I	0.100	0.008	50.0	30.8	−40 +80	1,500	Miniature data transmission
175	RIGID LINE											
176	0.135" Helix over Magnetic core	PE	0.285	1: Magnet Wire	PVC-I	0.405	0.120	2,240	49.0	−40 +80	5,000	
177	0.195" BC	PE	0.680	2: SC	PVC-IIA	0.895	0.470	50.0	30.8	−40 +80	11,000	High frequency RG218
178	0.0120" 7/0.0040" SCCS	PTFE	0.036	1: SC	KEL-F	0.079 max.	0.0054	50.0	29.4	−40 +150	1,000	Use RG178B
178A	0.0120" 7/0.0040" SCCS	PTFE	0.034	1: SC	KEL-F	0.075 max.	0.005	50.0	29.4	−40 +150	1,000	Use RG178B
178B	0.01020" 7/0.0040" SCCS	PTFE	0.034	1 SC	FEP-IX	0.075 max.	0.0054	50.0	29.4	−55 +200	1,000	High strength conductors avail.
179	0.0120" 7/0.0040" SCCS	PTFE	0.057	1: SC	KEL-F	0.094 max.	0.010	70.0	20.4	−55 +150	1,200	Use RG179B

Cables recommended for use per supplement 1A Mil-C-17D.

Cable Specifications (cont.)

RG/U Type	Inner Conductor	Dielectric Material	DOD (Inch)	Number & Type of Shielding Braids	Jacket Material	O.D. (Inch)	Weight (lbs/ft)	Nominal Imped. (ohms)	Nominal Capac- itance (pf/ft)	Max. Oper. Temp. Range (°C)	Max. Oper. Voltage (Volts RMS)	Comments
179A	0.0120" 7/0.0040" SCCS	PTFE	0.063	1: SC	KEL-F	0.105 max.	0.010	75.0	19.5	−40 +150	1,200	Use RG179B
179B	0.0120" 7/0.0040" SCCS	PTFE	0.063	1: SC	FEP-IX	0.100	0.010	75.0	19.5	−55 +200	1,200	High strength conductors avail
180	0.0120" 7/0.0040" SCCS	PTFE	0.103	1: SC	KEL-F	0.141 max.	0.019	93.0	15.4	−40 +150	1,500	Use RG180B
180A	0.0120" 7/0.0040" SCCS	PTFE	0.102	1: SC	KEL-F	0.145 max.	0.019	95.0	15.4	−40 +150	1,500	Use RG180B
180B	0.0120" 7/0.0040" SCCS	PTFE	0.102	1: SC	FEP-IX	0.145 max.	0.019	95.0	15.4	−55 +200	1,500	High strength conductors avail.
181	2 cond. 0.0477" 7/0.0159" BC	PE	0.210	2: Individual inner Common Outer BC	PVC-IIA	0.640	0.198	125.0	12.0	−40 +80	3,500	Balanced line
182	2 of 19/0.0142" BC 2 of 19/0.0066" TC	4 cores, PE	2/.332 2/.146	Each core, 1: BC overall shield, 1: BC	PVC-IIA each PVC-I overall	1.055		125.0 Each	12.0 Each	−40 +80	2,300 3,000	Special 4 coax.
183	0.2510" BC	PS Helix	0.632	Al tube	None	0.750	0.380	50.0	23.0	−40 +80	1,800	
184	RECTANGULAR WAVEGUIDE											
185	0.0031" Mag wire Helix on PE core	Air-space PE	0.188	1: Magnet wire	PVC-IIA	0.282		2,000		−40 +80		Delay line cable
186	0.008" TFE Helix over core	Air-space PE	0.292	1: Magnet wire	PVC-IIA	0.405		1,000		−40 +80		Delay line cable
187	0.0120" 7/0.0040" SCCS	PTFE	.060	1: SC	PTFE	0.110 max.	0.010	75.0	19.5	−55 +250	1,200	Use RG179B
187A	0.0120" 7/0.0040" SCCS	PTFE	.060	1: SC	PTFE	0.110 max.	0.010	75.0	19.5	−55 +250	1,200	Use RG179B
188	0201 7/.0067" SCCS	PTFE	0.060	1: SC	PTFE	0.110 max.	0.011	50.0	29.4	−55 +250	1,200	Use RG316
188A	0201" 7/.0067" SCCS	PTFE	0.060	1: SC	PTFE	0.110 max.	0.011	50.0	29.4	−55 +250	1,200	Use RG316
189	0.2510" BC	PS Helix	0.632	2: SC	PE-IIIA	0.875	0.570	50.0	23.0	−55 +80	3,500	Use RG389
190	0.0585" 19/0.0117" TC	Rubber H, J	0.380	3: TC, GS, TC	Neoprene VIII	0.700	0.353	50.0	50.0	−55 +80	15,000	
191	0.485" TC Braid	Rubber H, J, H	1.065	3: TC, GS, GS	Neoprene VIII	1.460	1.469	25.0	85.0	−55 +80	15,000	
192	1.055" GS Tube TC Braid	Butyl Rubber		3: TC, GS, GS	Rubber	2.200		12.5	175.0	−55 +80	15,000 Peak	
193	1.055" GS Tube TC Braid	Silicone Rubber		3: TC, GS, GS	Rubber	2.100		12.5	159.0	−55 +80	30,000 Peak	
194	1.055" GS Tube TC Braid	Silicone Rubber		3: TC, GS, GS	Rubber w. Al, Armor	1.945		12.5	159.0	−55 +80	30,000 Peak	
195	0.0120" 7/0.004" SCCS	PTFE	0.102	1: SC	PTFE	0.155 max.	0.020	95.0	15.4	−55 +250	1,500	Use RG180B
195A	0.0120" 7/0.004" SCCS	PTFE	0.102	1: SC	PTFE	0.155 max.	0.020	95.0	15.4	−55 +250	1,500	Use RG180B
196	0.0120" 7/0.004" SCCS	PTFE	0.034	1: SC	PTFE	0.080 max.	0.006	50.0	29.4	−55 +250	1,000	Use RG178B
196A	0.0120" 7/0.004" SCCS	PTFE	0.034	1: SC	PTFE	0.080 max.	0.006	50.0	29.4	−55 +250	1,000	Use RG178B

Cables recommended for use per supplement 1A Mil-C-17D.

Cable Specifications (cont.)

RG/U Type	Inner Conductor	Dielectric Material	DOD (Inch)	Number & Type of Shielding Braids	Jacket Material	O.D. (Inch)	Weight (lbs/ft)	Nominal Imped. (ohms)	Nominal Capac- itance (pf/ft)	Max. Oper. Temp. Range (°C)	Max. Oper. Voltage (Volts RMS)	Comments
197	0.300" BC	PS Helix	0.758	.875" OD Al. Tube	None	0.875	0.500	50.0	22.0	−55 +80	2,400 Peak	
198	0.1140" BC	PS Helix	0.421	500" OD Al. Tube	PE	0.600	0.155	70.0	16.0	−55 +80	1,300 Peak	
199	0.209" BC	PS Helix	0.758	.875" OD Al. Tube	PE	1.015	0.435	70.0	16.0	−55 +80	2,400 Peak	
200	OD- 405" ID-0.301" BC Tube	PS Helix	1.472	1.625" OD Al. Tube	PE	1.765	0.900	70.0	16.0	−55 +80	4,600 Peak	
201- 208	RECTANGULAR WAVEGUIDE COVERED BY MIL-W-85. See Mil HDBK 216, Para. 6.17–6.23											
209	0.189" 19/0.0378" SC	Air-space PTFE	0.500	2: SC	SR and Polyester-VI	0.750 max.	0.432	50.0	25.0	−55 +150	3,200	Low loss RG211A
210	0.0253" SCCS	Air-space PTFE	0.146	1: SC	FG Braid-V	0.242	0.040	93.0	13.5	−55 +250	750	High temp Low capacitance
211	0.1900" BC	PTFE	0.620	1: BC	FG Braid-V	0.730	0.641	50.0	29.4	−55 +250	7,000	High temp High power
211A	0.1900" BC	PTFE	0.620	1: BC	FG Braid-V	0.730	0.641	50.0	29.4	−55 +250	7,000	High temp High power
212	0.0556" SC	PE	0.185	2: SC	PVC-IIA	0.332	0.063	50.0	29.4	−40 +80	3,000	Use to 10,000 MHz
213	0.0888" 7/0.0296" BC	PE	0.285	1: BC	PVC-IIA	0.405	0.099	50.0	30.8	−40 +80	5,000	Use up to 1000 MHz
214	0.0888" 7/0.0296" SC	PE	0.285	2: SC	PVC-IIA	0.425	0.126	50.0	30.8	−40 +80	5,000	Use up to 10,000 MHz
215	0.0888" 7/0.0296 BC	PE	0.285	1: BC	PVC-IIA w. Armor	0.475 max.	0.121	50.0	30.8	−40 +80	5,000	Armored RG213
216	0.0477" 7/0.0159" BC	PE	0.285	2: BC	PVC-IIA	0.425	0.114	75.0	20.6	−40 +80	5,000	Use up to 1000 MHz
217	0.106" BC	PE	0.370	2: BC	PVC-IIA	0.545	0.201	50.0	30.8	−40 +80	7,000	Use up to 1000 MHz
218	0.195" BC	PE	0.680	1: BC	PVC-IIA	0.870	0.460	50.0	30.8	−40 +80	11,000	Use up to 1000 MHz
219	0.195" BC	PE	0.680	1: BC	PVC-IIA w. Armor	0.945 max.	0.585	50.0	30.8	−40 +80	11,000	Armored RG218
220	0.260" BC	PE	0.910	1: BC	PVC-IIA	1.120	0.740	50.0	30.8	−40 +80	14,000	Use up to 1000 MHz
221	0.260" BC	PE	0.910	1: BC	PVC-IIA w. ARmor	1.195 max.	0.925	50.0	30.8	−40 +80	14,000	Armored RG220
222	0.0556" high Resistance wire	PE	0.185	2: SC	PVC-IIA	0.332	0.087	50.0	30.8	−40 +80	3,000	High attenuation
223	0.035" SC	PE	0.116	2: SC	PVC-IIA	0.216 max.	0.034	50.0	30.8	−40 +80	1,900	Usable to 10,000 MHz
224	0.106" BC	PE	0.370	2: BC	PVC-IIA w. Armor	0.615 max.	0.310	50.0	30.8	−40 +80	7,000	Armored RG217
225	0.0936" 7/0.0312" SC	PTFE	0.285	2: SC	FG Braid-V	0.430	0.180	50.0	29.4	−55 +250	5,000	SEe RG393 for FEP jacket
226	0.1270" 19/0.0254" SC	Taped PTFE	0.370	2: SC	FG Braid-V	0.500	0.445	50.0	29.4	−55 +250	7,000	

Cables recommended for use per supplement 1A Mil-C-17D.

Cable Specifications (cont.)

RG/U Type	Inner Conductor	Dielectric Material	DOD (Inch)	Number & Type of Shielding Braids	Jacket Material	O.D. (Inch)	Weight (lbs/ft)	Nominal Imped. (ohms)	Nominal Capacitance (pf/ft)	Max. Oper. Temp. Range (°C)	Max. Oper. Voltage (Volts RMS)	Comments
227	0.0936" 7/0.0312" SC	PTFE	0.285	2: SC	FG Braid-V w. Armor	0.490 max.	0.198	50.0	29.4	−55 +250	5,000	Armored RG225
228	0.1900" BC	PTFE	0.620	1: BC	FG Braid-V w. Armor	0.795 max.	0.682	50.0	29.4	−55 +250	,000	Armored RG211
228A	0.1900" BC	PTFE	0.620	1: BC	FG Braid-V w. Armor	0.795 max.	0.682	50.0	29.4	−55 +250	7,000	Armored RG211A
229	0.0960" 7/0.032" SC	PTFE	0.285	1: SC	FG Braid-V w. Armor	0.480 max.	0.144	50.0	29.4	−55 +250	5,000	Use RG166
230	0.1988" 37/0.0284" TC	Rubber-D	0.455	3: TC, GS, GS	Rubber-IV	0.740 max.		25.0	100.0	−40 +80	15,000	
231	OD-0.162 ID-0.112" BC	Foam PE	0.450	0.500" OD Al. Tube	None	0.500	0.118	50.0	25.0	−55 +80	5,000 Peak	RG331 for jacketed cable
232	0.300" BC	PE Helix	0.758	0.875" OD Al. Tube	PE-IIIA	1.015	0.570	50.0	22.0	−55 +80	2,400 Peak	
233	OD 0.591" BC ID 0.481"	PS Helix	1.472	1.625" OD Al. Tube	PE-IIIA	1.765	1.050	50.0	22.0	−55 +80	4,700 Peak	
234	OD-1.1570" BC ID-1.015"	PS Helix	2.775	3. 125" OD Al. Tube	PE-IIIA	3.295	3.110	50.0	22.0	−55 +80	8,700 Peak	
235	0.0852" 7/0.0284" SC	Taped PTFE	0.255	2: SC	SIL/DAC VI	0.470 Max.	0.160	50.0	29.5	−55 +250	5,000	RG115A w. VI jacket
236	0.1620" BC	PS Helix	0.421	0.500" OD Al. Tube	None	0.500	0.165	50.0	24.0	−55 +80	1,300 Peak	
237	0.1620" BC	PS Helix	0.421	0.500" Al. Tube	PE-IIIA	0.600	0.195	50.0	24.0	−55 +80	1,300 Peak	
238	CANCELLED, REPLACE WITH RG197/U											
239	CANCELLED, REPLACE WITH RG232/U											
240	OD 0.591" BC ID 0.481" BC	PS Helix	1.420	1.625 OD Al. Tube	None	1.625	0.930	50.0	22.0	−55 +80	4,700 Peak	
241	CANCELLED, REPLACE WITH RG233											
242	OD-1.157" ID-1.0150" BC	PS Helix	2.850	3.125" OD Al. Tube	None	3.125	2,700	50.0	22.0	−55 +80	8,700 Peak	
243	CANCELLED, REPLACE WITH RG234											
244	0.102" BC	PS Helix	0.421	0.500" OD Al. Tube	None	0.500	0.118	75.0	15.5	−55 −80	1,200 Peak	
245	0.102" BC	PS Helix	0.421	0.500" OD Al. Tube	PE-IIIA	0.600	0.148	75.0	15.5	−55 +80	1,200 Peak	
246	0.1880" BC	PS Helix	0.758	0.875" OD Al. Tube	None	0.875	0.348	75.0	15.2	−55 +80	2200 Peak	
247	0.1880" BC	PS Helix	0.758	0.875" OD Al. Tube	PE-IIIA	1.015	0.418	75.0	15.2	−55 +80	2200 Peak	
248	OD-0.3740" BC ID-0.2740"	PS Helix	1.472	1.625" OD Al. Tube	None	1.625	0.948	75.0	15.0	−55 +80	4300 Peak	
249	OD-0.374" BC ID-0.2740"	PS Helix	1.472	1.625" OD Al. Tube	PE-IIIA	1.765	1.068	75.0	15.0	−55 +80	4300 Peak	
250	0.732" BC 0.632"	PS Helix	2.850	3.125" OD Al. Tube	None	3.125	2.395	75.0	15.0	−55 +80	8,500 Peak	

Cables recommended for use per supplement 1A Mil-C-17D.

Cable Specifications (cont.)

RG/U Type	Inner Conductor	Dielectric Material	DOD (Inch)	Number & Type of Shielding Braids	Jacket Material	O.D. (Inch)	Weight (lbs/ft)	Nominal Imped. (ohms)	Nominal Capacitance (pf/ft)	Max. Oper. Temp. Range (°C)	Max. Oper. Voltage (Volts RMS)	Comments
251	0.732" BC 0.632"	PS Helix	2.850	3.125" OD Al. Tube	PE-IIIA	3.295	2.805	75.0	15.0	−55 +80	8,500 Peak	
252	0.1670" BC	PE Tubes	0.456	0.530" OD Al. Tube	None	0.530	0.175	50.0	24.0	−55 +80	1,000	
253	0.167" BC	PE Tubes	0.456	0.530" OD Al. Tube	PE	0.655	0.225	50.0	24.0	−55 +80	1,000	
254	0.3100" BC	PE Tubes	0.833	0.953" OD Al. Tube	PE	1.100	0.655	50.0	24.0	−55 +80	1,860	
255	0.3110" BC	PE Tubes	0.833	0.953" OD Al. Tube	None	0.953	0.555	50.0	24.0	−55 +80	1,860	
256	OD-0.3110" SC ID-0.2550"	PTFE Tubes	0.833	0.953" OD Al. Tube	None	0.953	0.550	50.0	24.0	−55 +80	1,860	
257	OD-0.6060" ID-0.04860" BC	PS Tubes	1.622	1.786" OD Al. Tube	None	1.786	1.200	50.0	24.0	−55 +80	3,640	
258	OD-0.606" ID-0.4860" BC Tube	PE Tubes	1.622	1.786" OD Al. Tube	PE	1.926	1.380	50.0	24.0	−55 +80	3,640	
259	0.1150" BC Tube	PTFE Tubes	0.318	0.390" OD Al. Tube	None	0.390	0.100	50.0	24.0	−55 +80	697	
260	0.1150" BC Tube	PE Tubes	0.318	0.390" OD Al. Tube	PE-IIIA	0.450	0.140	50.0	24.0	−55 +80	697	
261-262												Cancelled
263	0.1720" BC	Air-space PTFE	0.421	Al. Tube	None	0.500	0.170	50.0	21.5	−40 +250	1,300 Peak	
264	4 cond. 19/0.0142" 2 TC 2 BC	PE	0.176 e/core	2 TC, 2 BC BC over all	PVC-IIA	0.750	0.336	36.8	41.0	−40 +80	2,000	Use RG264C
264A	4 cond. 19/0.0142" 2 TC 2 BC	PE	0.176 e/core	2 TC, 2 BC, BC over all	PUR	0.750	0.327	36.8	41.0	−40 +80	2,000	Use RG264C
264C	4 cond. 0.068" 2 BC. 2 TC	PE	0.186 e/core	2 TC, 2 BC BC overall	PUR	0.765	0.327	40.0	38.4	−40 +80	2,000	Watertight 4 coax.
265	0.6770" BC Tube	PE Helix	1.578	Copper clad Mild Steel Tube	PE-IIIA	2.070		50.0	22.3	−40 +80	145KW Peak	
266	0.0113" cond over 0.144" Mag core	PE	0.285	75 spiral wound cond 68 BC & 7 are insulated	PVC-I	0.400	0.120	1530.0	53.0	−40 +80	5,000 DC	Delay line cable 50 ns/ft
267	0.3550" BC Tube	PS Helix		Copper clad Mild Steel Corrugated Tubing	PE-IIIA	1.190		50.0	22.2	−40 +80	44 KW Peak	
268	0.1610" BC	PE Helix	0.350	Corrugated BC Tube	None	0.498	0.234	50.0	23.0	−55 +80	10KW Peak	
269	ID-0.2870" OD-0.3580" BC Tube	PE Helix	0.795	Corrugated BC Tube	None	1.005	0.430	50.0	22.2	−55 +80	44KW Peak	
269A	ID-0.2870" OD-0.3580" BC Tube	PE Helix	0.795	Corrugated BC Tube	None	1.005	0.430	50.0	22.2	−55 +80	44KW Peak	
270	ID-0.5880" OD-0.6880" BC Tube	PE Helix	1.578	Corrugated BC Tube	None	1.830	0.875	50.0	22.3	−55 +80	145KW Peak	
270A	ID-0.5880" OD-0.6770" BC Tube	PE Helix	1.578	Corrugated BC Tube	None	1.830	0.875	50.0	22.3	−55 +80	145KW Peak	
271-278	RECTANGULAR WAVEGUIDE COVERED BY MIL-W-85. See Mil HDBK 216. Para. 6.17–6.23											
279	0.0250" 19/0.0050" SCCS	Air-space PTFE	0.110	1: SC	FG Braid-V	0.145	0.125	75.0	19.5	−55 +250	1,000	Extra flexible high temp. cable
280	0.1144" BC	Taped PTFE	0.327	2: SC	FEP-IX	0.468	0.200	50.0	25.4	−55 +200	3,000	Low loss High frequency

Cables recommended for use per supplement 1A Mil-C-17D.

Cable Specifications (cont.)

RG/U Type	Inner Conductor	Dielectric Material	DOD (Inch)	Number & Type of Shielding Braids	Jacket Material	O.D. (Inch)	Weight (lbs/ft)	Nominal Imped. (ohms)	Nominal Capac- itance (pf/ft)	Max. Oper. Temp. Range (°C)	Max. Oper. Voltage (Volts RMS)	Comments
281	0.1890" 19/0.0378" SC	Taped PTFE	0.500	2: SC	Sil/DAC-VI	0.750 max.	0.400	50.0	25.4	−55 +150	4,000	Low loss High power
282	0.0253" SC	Irradiated PE	0.099	2: SC	FEP	0.200	0.031	54.5	28.2	−40 +150	4,500	
283	0.0585" 19/0.0117" SC	Rubber-D	0.288	2: SC	Sil.	0.475	0.145	46.0	50.0	−55 +150	8,000	
284A	0.220" BC	PE Helix	0.795	Corrugated BC Tube	None	1.005	0.410	75.0	15.0	−55 +80	29KW Peak	
285A	0.1140" BC	PTFE Helix	0.795	Corrugated BC Tube	None	1.005	0.430	100.0	13.0	−55 +200	22KW Peak	
286	OD-0.4300" ID-0.3600" BC Tube	PE Helix	1.570	Corrugated BC Tube	None	1.830	0.720	75.0	15.1	−55 +80	100KW Peak	
287	0.1970" BC	PE Helix	1.570	Corrugated BC Tube	None	1.830	0.750	100.0	13.5	−55 +80	73KW Peak	
288	OD-1.3330" ID-1.2221" BC Tube	PE Helix	2.960	3.750" CCS	None	3.750	3.000	50.0	21.6	−40 +80	440KW Peak	
289	OD-0.8200" ID-0.7400" CCS Tube	PE Helix	2.960	3.750" CCS	None	3.750	3.000	75.0	14.7	−40 +80	290KW Peak	
290- 291	RECTANGULAR WAVEGUIDE COVERED BY MIL-W-85. See Mil HDBK 216, Para. 6.17–6.23											
292	0.4300" BC Tube	PE Helix	1.570	1.8300" Corr. BC Tube	PE and flooding comp	2.000	1.040	75.0	15.1	−55 +80	100KW Peak	
293	0.106" BC	PE	0.375	1: SC	PE-IIIA	0.545	0.160	50.0	30.8	−55 +80	7,000	Use RG293A
293A	0.1060" BC	PE	0.370	1: SC	PE-IIIA	0.545	0.160	50.0	30.8	−55 +80	7,000	Watertight RG217
294	2 cond. 0.0808" 1 BC, 1 TC	PE	0.472	1: TC	PE-IIIA	0.630	0.205	95.0	16.3	−55 +80	3,000	Use RG294A
294A	2 cond. 0.0808" 1 BC, 1 TC	PE	0.472	1: SC	PE-IIIA	0.630	0.205	95.0	16.3	−55 +80	3,000	Watertight RG130
295	0.1950" BC	PE	0.680	1: SC	PE-IIIA	0.895	0.420	50.0	30.8	−55 +80	11,000	Watertight RG218
296	0.2352" 37/0.0336" SC	Silicone Rubber	.906	1: SC	Neoprene	1.190		50.0	36.4	−55 +80	13,800	
297	OD-0.3550" ID-0.2870" BC Tube	PTFE Helix	0.795	Corr. BC Tube	None	1.005		50.0	21.4	−55 +200	44 KW Peak	
298	0.6030" 7/0.0201" CCS	PE	0.115	None	Foam PE	0.650	0.090			−55 +80		Buoyant Per Mil-C-22667
299- 300	RECTANGULAR WAVEGUIDE											
301	0.0609" 7/0.0203" HR	PTFE	0.185	1: HR	FEP-IX	0.245	0.056	50.0	29.4	−55 +200	3,000	FEP Jacketed RG126
302	0.0250" SCCS	PTFE	0.146	1: SC	FEP-IX	0.206 max.	0.031	75.0	19.5	−55 +200	2,300	FEP Jacketed RG140
303	0.0390" SCCS	PTFE	0.116	1: SC	FEP-IX	0.170	0.030	50.0	29.4	−55 +200	1,900	FEP Jacketed RG141A
304	0.0590" SCCS	PTFE	0.185	2: SC	FEP-IX	0.280	0.088	50.0	29.4	−55 +200	3,000	FEP Jacketed RG143A

Cables recommended for use per supplement 1A Mil-C-17D.

Cable Specifications (cont.)

RG/U Type	Inner Conductor	Dielectric Material	DOD (Inch)	Number & Type of Shielding Braids	Jacket Material	O.D. (Inch)	Weight (lbs/ft)	Nominal Imped. (ohms)	Nominal Capacitance (pf/ft)	Max. Oper. Temp. Range (°C)	Max. Oper. Voltage (Volts RMS)	Comments
305	OD-0.4300" ID-0.3600" BC Tube	FEP	1.570	1.830" BC Tube	PE-IIIA	1.990		75.0	14.4	−55 +80	2,720	
306A	0.1730" BC	Foam PE	0.801	0.8750" Al. Tube	PE-IIIA	1.015	0.545	75.0	16.5	−55 +80	5,700	Per Mil-C-23806
307	0.0290" 19/0.0058" SC	Foam PE	0.146	2: SC PUR Interlayer	PE-IIIA	0.270	0.070	75.0	16.7	−55 +80	1,000	Triax Use to 100 MHz
307A	0.0290" 19/0.0058" SC	Foam PE	0.146	2: SC PUR Interlayer	PE-IIIA	0.270	0.070	75.0	16.7	−55 +80	1,000	Triax Use to 100 MHz
308-315	BEAD SUPPORTED RIGID LINES. See Mil-R-9671.											
316	0.0201" 7/0.0067" SCCS	PTFE	0.060	1: SC	FEP-IX	0.102	0.012	50.0	29.4	−55 +200	1,200	FEP jacketed RG188A
317	2 cond. 0.0870" 7/0.0290" BC	FEP	0.446	1: TC	Neoprene	0.710		95.0	15.4	−55 +80	10,000	Water blocked
318	OD-0.3580" BC Tube ID-0.2870"	PE Helix	0.795	1.005" Corr. BC Tube	PE-IIIA	1.125	0.530	50.0	22.0	−55 +80	44KW Peak	
319	OD-0.6880" BC Tube ID-0.5880"	PE Helix	1.570	1.830" Corr. BC Tube	PE-IIIA	2.000	1.040	50.0	22.0	−55 +80	145KW Peak	
320	WAVEGUIDE											
321	OD-1.1400" corr. BC Tube	PE helix		2.850" corr. BC Tube	None	2.850	1.210	50.0	21.7	−55 +80	320KW Peak	
322	OD-1.1400" corr. BC Tube	PE Helix		2.850" corr. BC Tube	PE & flood comp	3.040	1.780	50.0	21.7	−55 +80	320KW Peak	
323	0.312" BC Tube	Foam PE		0.980" corr. BC Tube	PE & flood comp	1.060	0.420	50.0	25.6	−55 +80	1,480	
324	0.312" BC Tube	Foam PE		0.980" corr. BC Tube	None	0.980	0.320	50.0	25.6	−55 +80	1,480	
325	0.1000" 19/0.0200" SC Al.	PE Spline	0.260	2: SC Strip Braids	PUR	0.350	0.10	50.0	26.3	−55 +80	750	Low loss
326	0.2000" 19/0.0400" SC Al.	PE Spline	0.550	2: SC Strip Braids	PUR	0.697	0.24	50.0	26.3	−55 +80	1,700	Low loss
327	0.3200" 19" 0.0640" SC Al.	PE Spline	0.840	2: SC Strip Braids	PUR	1.010	0.55	50.0	26.3	−55 +80	2,500	Low loss
328	0.4850" TC Braid	Rubber H, J, H	1.065	3: TC, GS, TC	Neoprene	1.460	1.469	25.0	85.0	−55 +80	20,000	
329	0.0585" 19/0.117" TC	Rubber H, J, H	0.380	3: TC, GS, TC	Neoprene	0.700	0.353	50.0	50.0	−55 +80	15,000	
330	SC	Foam PE		1: SC		0.242		50.0	25.0			
331	0.1620" BC	Foam PE	0.450	0.500" Al. Tube	PE-IIIA	0.600	0.187	50.0	25.0	−55 +80	2,500	Jacketed 231 Solid conductor
332	0.2800" BC	Foam PE	0.801	0.8750" Al. Tube	None	0.875	0.466	50.0	25.0	−55 +80	4,500	Per Mil-C-23806
333	0.2800" BC	Foam PE	0.801	0.8750" Al. Tube	PE-IIIA	1.015	0.548	50.0	25.0	−55 +80	4,500	Jacketed RG332
334	0.0980" BC	Foam PE	0.450	0.500" Al. Tube	None	0.500	0.109	75.0	17.0	−55 +80	2,500	Per Mil-C-23806
335	0.0980" BC	Foam PE	0.450	0.500" Al. Tube	PE-IIIA	0.625	0.143	75.0	17.0	−55 +80	2,500	Jacketed RG334
336	0.1730" BC	Foam PE	0.801	0.8750" Al. Tube	None	0.875	0.315	75.0	17.0	−55 +80	4,000	Per Mil-C-23806

Cables recommended for use per supplement 1A Mil-C-17D.

Cable Specifications (cont.)

RG/U Type	Inner Conductor	Dielectric Material	DOD (Inch)	Number & Type of Shielding Braids	Jacket Material	O.D. (Inch)	Weight (lbs/ft)	Nominal Imped. (ohms)	Nominal Capac-itance (pf/ft)	Max. Oper. Temp. Range (°C)	Max. Oper. Voltage (Volts RMS)	Comments
337-359	RECTANGULAR WAVEGUIDE COVERED BY MIL-W-85. See Mil HDBK 216, Para. 6.17–6.23											
360	0.2430" BC	Foam PE	0.676	0.7500" Al. Tube	PE-IIIA	0.825	0.397	50.0	25.0	−55 +80	4,000	Per Mil-C-23806
361-365	DATA NOT AVAILABLE											
366	0.1600" BC	Foam PE		0.540" Corr. BC Tube	PE-IIIA	0.620		50.0	26.6	−55 +80	4,000	
367	Corr. BC Tube	PE Helix		5.000" Corr. BC Tube	PE-IIIA	5.2000	4.590	50.0	21.7	−55 +80	830KW Peak	
369	0.1170" BC	PE Tubes	0.318	0.390" Al. Tube	PE-IIIA	0.470	0.140	50.0	24.0	−55 +80	700	
370	0.1170" BC	PE Tubes	0.318	0.390" Al. Tube	None	0.390	0.100	50.0	24.0	−40 +80	700	
372-373	EXPERIMENTAL BUOYANT COAXIAL TRANSMISSION LINE											
374	0.0285" BC	PE	0.160	None	Foam PE	0.650	0.097			−55 +80		Buoyant antenna
375	RECTANGULAR WAVEGUIDE											
376	0.3120" BC Tube	Foam PE		Corr. Al. Tube	PE-IIIA	1.060	0.390	50.0	26.0	−55 +80	6,000	
377	0.1650" SC Tube	PTFE Tubes		0.5300" Al. Tube	None	0.530	0.170	50.0	24.0	−55 +250	1,000	
378	0.7130" BC Tube	PE Helix		1.830" Corr. Al. Tube	PE-IIIA	2.000	0.620	50.0	22.1	−55 +80	145KW Peak	
379-381	ELLIPTICAL WAVEGUIDES											
382	RIGID LINE			Al. Tube		1.625		50.0				
383	2 cond. 0.0403" 2,000 lb. Break	PE		None	Foam PE	0.650		100.0		−55 +80		Exper buoyant Twisted pair
384	0.0508" BC	PE		1: Flat BC Braid Waterproofed	Foam PE	0.650		50.0	30.8	−55 +80		Buoy. Ant. 600 psig.
385	0.1530" SC	Semi-solid PTFE	0.425	0.500" Corr. Al.	Optional	0.660	0.178	50.0	25.0	−55 +250	1,500	Low loss No pres. req.
386	0.0508" CCS	PE		None Non-hosing	Foam PE	0.650				−55 +80		Buoy Ant 400 lb break
387	DATA NOT AVAILABLE											
388	SC	PE		0.444" max. SC	PE-IIIA	0.545		50.0	30.8	−55 +80		Watertight See RG14A
389	0.2500" BC Al.	PE Spline	0.635	2: SC	PE-IIIA	0.875	0.366	50.0	22.8	−55 +80	2,000	Low Loss Replaces RG189
390	DATA NOT AVAILABLE											
391	0.048" 7/0.0159" TC	Cond. PE and PE	0.285	1: TC	PVC-IIA	0.405	0.092	72.0		−55 +80	5,000	Low noise cable
392	0.048" 7/0.0159" TC	Cond. PE and PE	0.285	1: TC	PVC-IIA w. Armor	0.475	0.114	72.0		−55 +80	5,000	Armored RG391
393	0.0936" 7/0.0312" SC	PTFE	0.285	2: SC	FEP-IX	0.390	0.165	50.0	29.4	−55 +200	5,000	Moisture proof RG225

Cables recommended for use per supplement 1A Mil-C-17D.

Cable Specifications (cont.)

RG/U Type	Inner Conductor	Dielectric Material	DOD (Inch)	Number & Type of Shielding Braids	Jacket Material	O.D. (Inch)	Weight (lbs/ft)	Nominal Imped. (ohms)	Nominal Capacitance (pf/ft)	Max. Oper. Temp. Range (°C)	Max. Oper. Voltage (Volts RMS)	Comments
397	0.096" 7/0.032" SC	Air-space PTFE	0.270	2: SC	FEP-IX	0.360 Max	0.125	50.0	28.0	−55 +200	2,000	Low loss RG393
400	0.0385" 19/0.0077" SPC	PTFE	0.116	2: SC	FEP-IX	0.195	0.050	50.0	29.3	−55 +200	1,900	
401	0.0645" SPC	PTFE	0.215	250" OD Copper Tube	None	0.250	0.081	50.0	29.3	−55 +200	3,000	Semiflex RG304
402	0.0360" SCCS	PTFE	0.119	141" OD Copper Tube	None	0.141	0.032	50.0	29.3	−55 +200	2,500	Semiflex RG142B
403	0.012" SCCS 7/0.004"	PTFE	0.034	2: SC FEP Interlayer	FEP-IX	0.116	0.0075	50.0	29.3	−55 +200	2,500	Triaxial RG178B
404	0.012" SCCS 7/0.004"	PTFE and Cond PTFE	0.034	1: SC	FEP-IX	0.075 Max	0.0054	50.0	31.5 Max	−55 +200	2,000	Low noise RG178B
174A	0.0189" 7/0.0063" CCS	PE	0.060	1: TC	PVC-IIA	0.100	0.008	50.0	30.8	−40 +80	1,500	Miniature data transmission

Cables recommended for use per supplement 1A Mil-C-17D.

Appendix F

Characteristics of Waveguides

Characteristics of Standard Waveguides

EIA Waveguide Designation (Standard RS-261-A)	JAN Waveguide Designation (MIL-HDBK-216, 4 January 1962)	Outer Dimensions and Wall Thickness (in inches)	Frequency Range in Gigahertz for Dominant ($TE_{1,0}$) Mode	Cutoff Wavelength λ_c in Centimeters for $TE_{1,0}$ Mode	Cutoff Frequency f_c in Gigahertz for $TE_{1,0}$ Mode	Theoretical Attenuation, Lowest to Highest Frequency in dB/100 ft	Theoretical Power Rating in Megawatts for Lowest to Highest Frequency*
WR-2300	RG-290/U†	23.250 × 11.750 × 0.125	0.32–0.49	116.8	0.256	0.051–0.031	153.0–212.0
WR-2100	RG-291/U†	21.250 × 10.750 × 0.125	0.35–0.53	106.7	0.281	0.054–0.034	120.0–173.0
WR-1800	RG-201/U†	18.250 × 9.250 × 0.125	0.425–0.620	91.4	0.328	0.056–0.038	93.4–131.9
WR-1500	RG-202/U†	15.250 × 7.750 × 0.125	0.49–0.740	76.3	0.393	0.069–0.050	67.6–93.3
WR-1150	RG-203/U†	11.750 × 6.000 × 0.125	0.64–0.96	58.4	0.514	0.128–0.075	35.0–53.8
WR-975	RG-204/U†	10.000 × 5.125 × 0.125	0.75–1.12	49.6	0.605	0.137–0.095	27.0–38.5
WR-770	RG-205/U†	7.950 × 4.100 × 0.125	0.96–1.45	39.1	0.767	0.201–0.136	17.2–24.1
WR-650	RG-69/U	6.660 × 3.410 × 0.080	1.12–1.70	33.0	0.908	0.317–0.212	11.9–17.2
WR-510	—	5.260 × 2.710 × 0.080	1.45–2.20	25.9	1.16	—	—
WR-430	RG-104/U	4.460 × 2.310 × 0.080	1.70–2.60	21.8	1.375	0.588–0.385	5.2–7.5
WR-340	RG-112/U	3.560 × 1.860 × 0.080	2.20–3.30	17.3	1.735	0.877–0.572	—
WR-284	RG-48/U	3.000 × 1.500 × 0.080	2.60–3.95	14.2	2.08	1.102–0.752	2.2–3.2
WR-229	RG-49/U	2.418 × 1.273 × 0.064	3.30–4.90	11.6	2.59		
WR-187	—	2.000 × 1.000 × 0.064	3.95–5.85	9.50	3.16	2.08–1.44	1.4–2.0
WR-159	—	1.718 × 0.923 × 0.064	4.90–7.05	8.09	3.71		
WR-137	RG-50/U	1.500 × 0.750 × 0.064	5.85–8.20	6.98	4.29	2.87–2.30	0.56–0.71
WR-112	RG-51/U	1.250 × 0.625 × 0.064	7.05–10.00	5.70	5.26	4.12–3.21	0.35–0.46
WR-90	RG-52/U	1.000 × 0.500 × 0.050	8.20–12.40	4.57	6.56	6.45–4.48	0.20–0.29
WR-75	—	0.850 × 0.475 × 0.050	10.00–15.00	3.81	7.88		
WR-62	RG-91/U	0.702 × 0.391 × 0.040	12.40–18.00	3.16	9.49	9.51–8.31	0.12–0.16
WR-51	—	0.590 × 0.335 × 0.040	15.00–22.00	2.59	11.6		
WR-42	RG-53/U	0.500 × 0.250 × 0.040	18.00–26.50	2.13	14.1	20.7–14.8	0.043–0.058
WR-34	—	0.420 × 0.250 × 0.040	22.00–33.00	1.73	17.3	—	—
WR-28	RG-96/U‡	0.360 × 0.220 × 0.040	26.50–40.00	1.42	21.1	21.9–15.0	0.022–0.031
WR-22	RG-97/U†	0.304 × 0.192 × 0.040	33.00–50.00	1.14	26.35	31.0–20.9	0.014–0.020
WR-19	—	0.268 × 0.174 × 0.040	40.00–60.00	0.955	31.4	—	—
WR-15	RG-98/U‡	0.228 × 0.154 × 0.040	50.00–75.00	0.753	39.9	52.9–39.1	0.0063–0.0090
WR-12	RG-99/U‡	0.202 × 0.141 × 0.040	60.00–90.00	0.620	48.4	93.3–52.2	0.0042–0.0060
WR-10	—	0.180 × 0.130 × 0.040	75.00–110.00	0.509	59.0	—	—
WR-8	RG-138/U§	0.140 × 0.100 × 0.030	90.00–140.00	0.406	73.84	152–99	0.0018–0.0026

Characteristics of Standard Waveguides

EIA Waveguide Designation (Standard RS-261-A)	JAN Waveguide Designation (MIL-HDBK-216, 4 January 1962)	Outer Dimensions and Wall Thickness (in inches)	Frequency Range in Gigahertz for Dominant ($TE_{1,0}$) Mode	Cutoff Wavelength λ_c in Centimeters for $TE_{1,0}$ Mode	Cutoff Frequency f_c in Gigahertz for $TE_{1,0}$ Mode	Theoretical Attenuation, Lowest to Highest Frequency in dB/100 ft	Theoretical Power Rating in Megawatts for Lowest to Highest Frequency*
WR-7	RG-136/U§	0.125 × 0.0925 × 0.030	110.00–170.00	0.330	90.84	163–137	0.0012–0.0017
WR-5	RG-135/U§	0.111 × 0.0855 × 0.030	140.00–220.00	0.259	115.75	308–193	0.00071–0.00107
WR-4	RG-137/U§	0.103 × 0.0815 × 0.030	170.00–260.00	0.218	137.52	384–254	0.00052–0.00075
WR-3	RG-139/U§	0.094 × 0.0770 × 0.030	220.00–325.00	0.173	173.28	512–348	0.00035–0.00047

* For these computations, the breakdown strength of air was taken as 15,000 volts per centimeter. A safety factor of approximately 2 at sea level has been allowed.
† Aluminum, 2.83×10^{-6} ohm-cm resistivity. ‡ Silver, 1.62×10^{-6} ohm-cm resistivity. § JAN types are silver, with a circular outer diameter of 0.156 inch and a rectangular bore matching EIA types. All other types are of a Cu–Zn alloy, 3.9×10^{-6} ohm-cm resistivity.
Note: Equivalent designations of waveguides follow.

EIA	British	IEC
WR-2300	00	−R3
WR-2100	0	−R4
WR-1800	1	−R5
WR-1500	2	−R6
WR-1150	3	−R8
WR-975	4	−R9
WR-770	5	−R12
WR-650	6	−R14
WR-510	7	−R18
WR-430	8	−R22
WR-340	9A	−R26
WR-284	10	−R32
WR-229	11A	−R40
WR-187	12	−R48
WR-159	13	−R58
WR-137	14	−R70
WR-112	15	−R84
WR-90	16	−R100
WR-75	17	−R120
WR-62	18	−R140
WR-51	19	−R180
WR-42	20	−R220
WR-34	21	−R260
WR-28	22	−R320
WR-22	23	−R400
WR-19	24	−R500
WR-15	25	−R620
WR-12	26	−R740
WR-10	27	−R900
WR-8	28	−R1200

Characteristics of Single-Ridged Waveguides, From *MIL-HDBK-216, "RF Transmission Lines and Fittings," 4 January 1962*

Frequency Range (GHz)	$f_{c1,0}$ (GHz)	$\lambda_{c1,0}$ (in.)	$f_{c2,0}$ (GHz)	Dimensions in Inches							$AT f = (3)^{1/2} f_{c1,0}$	
				a	b	d	s	t	R_1 (max)	R_2	Atten* (dB/ft)	Power Rating† (kW)
Bandwidth 2.4:1												
0.175–0.42	0.148	79.803	0.431	28.129	12.658	5.278	4.360	—	—	1.056	0.00024	32 870.
0.267–0.64	0.226	52.260	0.658	18.421	8.289	3.457	2.855	—	—	0.691	0.00045	14 100.
0.42–1.0	0.356	33.177	1.036	11.695	5.263	2.195	1.813	0.125	0.047	0.439	0.00087	5 682.
0.64–1.53	0.542	21.792	1.577	7.682	3.457	1.442	1.191	0.125	0.047	0.288	0.00164	2 451.
0.84–2.0	0.712	16.588	2.072	5.847	2.631	1.097	0.906	0.080	0.047	0.219	0.00248	1 421.
1.5–3.6	1.271	9.293	3.699	3.276	1.474	0.615	0.508	0.080	0.047	0.123	0.00591	445.8
2.0–4.8	1.695	6.968	4.933	2.456	1.105	0.461	0.381	0.080	0.047	0.092	0.00908	250.6
3.5–8.2	2.966	3.982	8.632	1.404	0.632	0.264	0.218	0.064	0.031	0.053	0.0212	81.87
4.75–11.0	4.025	2.934	11.714	1.034	0.465	0.194	0.160	0.050	0.031	0.039	0.0333	44.43
7.5–18.0‡	6.356	1.858	18.498	0.655	0.295	0.123	0.1015	0.050	0.015	0.025	0.0661	17.82
11.0–26.5‡	9.322	1.267	27.130	0.4466	0.2010	0.0838	0.0692	0.040	0.015	0.017	0.117	8.285
18.0–40.0‡	15.254	0.7743	44.393	0.2729	0.1228	0.0512	0.0423	0.040	0.015	0.010	0.246	3.035
Bandwidth 3.6:1												
0.108–0.39	0.092	128.37	0.404	31.218	14.048	2.402	5.307	—	—	0.480	0.0016	14 550.
0.27–0.97	0.229	51.572	1.006	12.542	5.644	0.965	2.132	—	—	0.193	0.0065	2 348.
0.39–1.4	0.331	35.680	1.454	8.677	3.905	0.668	1.475	0.125	0.047	0.134	0.0112	1 124.
0.97–3.5	0.822	14.367	3.611	3.494	1.572	0.269	0.594	0.080	0.047	0.054	0.0438	182.2
1.4–5.0	1.186	9.958	5.210	2.422	1.090	0.186	0.412	0.080	0.047	0.037	0.0758	87.56
3.5–12.4	2.966	3.982	13.030	0.968	0.436	0.075	0.165	0.050	0.031	0.015	0.300	13.99
5.0–18.0‡	4.237	2.787	18.613	0.678	0.305	0.052	0.115	0.050	0.015	0.010	0.513	6.857
12.4–40.0‡	10.508	1.124	46.162	0.273	0.123	0.021	0.046	0.040	0.015	0.004	2.008	1.115

*Copper.

†Based on breakdown of air—15,000 volts per cm (safety factor of approx 2 at sea level). Corner radii considered.

‡Figure 13B in these frequency ranges only.

Characteristics of Single-Ridged Waveguides. *From MIL-HDBK-216, "RF Transmission Lines and Fittings," 4 January 1962*

Frequency Range (GHz)	$f_{c1,0}$ (GHz)	$\lambda c_{1,0}$ (in.)	$f_{c2,0}$ (GHz)	Dimensions in Inches							AT $f = (3)^{1/2} f_{c1,0}$	
				a	b	d	s	t	R_1 (max)	R_2	Atten* (dB/ft)	Power Rating† (kW)
Bandwidth 2.4:1												
0.175–0.42				29.667	13.795	5.863	7.417	—	—	1.173		
0.267–0.64				19.428	9.034	3.839	4.857	—	—	0.768		
0.42–1.0				12.333	5.737	2.437	3.083	0.125	0.050	0.487		
0.64–1.53				8.100	3.767	1.601	2.025	0.125	0.050	0.320		
0.84–2.0				6.167	2.868	1.219	1.542	0.125	0.050	0.244		
1.5–3.6				3.455	1.607	0.683	0.864	0.080	0.050	0.137		
2.0–4.8				2.590	1.205	0.512	0.648	0.080	0.050	0.102		
3.5–8.2				1.480	0.688	0.292	0.370	0.064	0.030	0.058		
4.75–11.0				1.090	0.506	0.215	0.272	0.050	0.030	0.043		
7.5–18.0				0.691	0.321	0.136	0.173	0.050	0.020	0.027		
11.0–26.5‡				0.471	0.219	0.093	0.118	0.040	0.015	0.019		
18.0–40.0‡				0.288	0.134	0.057	0.072	0.040	0.015	0.011		
Bandwidth 3.6:1												
0.108–0.39	0.092	128.37	0.401	34.638	14.894	2.904	8.660	—	—	0.581	0.0014	28 830.
0.27–0.97	0.229	51.572	0.999	13.916	5.984	1.167	3.479	—	—	0.233	0.0055	4 653.
0.39–1.4	0.331	35.680	1.444	9.628	4.140	0.807	2.407	0.125	0.050	0.161	0.0097	2 227.
0.97–3.5	0.822	14.367	3.587	3.877	1.667	0.325	0.969	0.080	0.050	0.065	0.0378	361.2
1.4–5.0	1.186	9.958	5.176	2.687	1.155	0.225	0.672	0.080	0.050	0.045	0.0656	173.5
3.5–12.4	2.966	3.982	12.944	1.074	0.462	0.090	0.269	0.050	0.030	0.018	0.259	27.74
5.0–18.0	4.237	2.787	18.490	0.752	0.323	0.063	0.188	0.050	0.020	0.013	0.443	13.59
12.4–40.0‡	10.508	1.124	45.857	0.303	0.130	0.025	0.076	0.040	0.015	0.005	1.730	2.210

*Copper.

†Based on breakdown of air—15,000 volts per cm (safety factor of approx 2 at sea level). Corner radii considered.

‡Figure 13B in these frequency ranges only.

Appendix G

Filter Specification Sheets

LOW-PASS AND HIGH-PASS FILTER SPECIFICATION SHEET

SPEC SUMMARY ☐ High

☐ Low   - PASS FILTER

Passband

    From: _____ MHz

      To: _____ MHz

  3 dB at _____ MHz

Passband Ripple:

_____ dB p-p

Max. Insertion Loss in Passband:

_____ dB

Rejection Data:

_____ dB   at _____ MHz

_____ dB   at _____ MHz

_____ dB   at _____ MHz

VSWR:

_____ to 1 (1.5 to 1 preferred)

Temp. Range:

_____ °C to + _____ °C (Without degraded

performance)

Input Power:

_____

Housing:

  ☐ Radial Pins (PCB)   ☐ Axial Pins (stripline)

  ☐ SMA-F   ☐ SMA-M   ☐ SMA-M/F

  Plating: ☐ Gold   ☐ Nickel   ☐ Other

  Max. Size (L x W x H) _____ x _____ x _____ in

Mechanical, Environmental, and Special Requirements

(Please List) _____

_____

_____

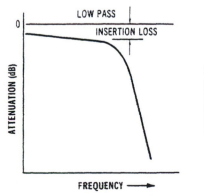

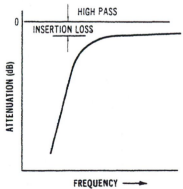

BANDPASS FILTER SPECIFICATION SHEET

SPEC SUMMARY — BANDPASS FILTER

Center Frequency:

_____ MHz nominal

Passband Data:

_____ dB Pts _____ MHz & _____ MHz

_____ dB Pts _____ MHz & _____ MHz

Passband Ripple:

_____ dB Max

Rejection Data:

_____ dB at _____ MHz & _____ MHz

_____ dB at _____ MHz & _____ MHz

_____ dB at _____ MHz & _____ MHz

Insertion Loss:

_____ dB Max at Center Frequency

VSWR:

_____ to 1 (1.5 to 1 if not specified)

Temp. Range:

_____ °C to + _____ °C (without degraded
                                        performance)

Input Power:

_____

Housing:

☐ Radial Pins (PCB)   ☐ Axial Pins (stripline)

☐ SMA-F   ☐ SMA-M   ☐ SMA-M/F

　　Plating: ☐ Gold   ☐ Nickel   ☐ Other

　　Max. Size (L x W x H) ____ x ____ x ____ In

Mechanical, Environmental, and Special Requirements

(Please List) _____

_____

_____

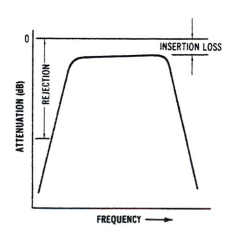

480

Bibliography

Adams, S. F. *Microwave Theory and Applications*. Englewood Cliffs, NJ: Prentice-Hall, 1969.

Altman, J. L. *Microwave Circuits*. New York: D. Van Nostrand, 1964.

Collins, R. C. *Foundations for Microwave Engineering*. New York: McGraw-Hill, 1966.

Edwards, T. C. *Foundations for Microstrip Circuit Design*. New York: John Wiley, 1981.

Frey, J. *Microwave Integrated Circuits*. Dedham, MA: Artech House, 1975.

Gunston, M. A. R. *Microwave Transmission Line Impedance Data*. New York: Van Nostrand Reinhold, 1972.

Gupta, K. C., Garg, Ramesh, and Bahl, I. J. *Microstrip Lines and Slotlines*. Dedham, MA: Artech House, 1979.

Gupta, K. C., Garg, Ramesh, and Chadra, R. *Computer Aided Design of Microwave Circuits*. Dedham, MA: Artech House, 1981.

Gupta, K. C. and Singh, A. (eds.). *Microwave Integrated Circuits*. New York: Halsted Press (Wiley), 1974.

Hardy, J. *High-Frequency Circuit Design*. Englewood Cliffs, NJ: Prentice-Hall, 1980.

Harvey, A. F. *Microwave Engineering*. New York: Academic Press, 1963.

Hilberg, W. *Electrical Characteristics of Transmission Lines*. Dedham, MA: Artech House, 1979.

Howe, H., Jr. *Stripline Circuit Design*. Dedham, MA: Artech House, 1974.

Lance, A. L. *Introduction to Microwave Theory and Measurements*. New York: McGraw-Hill, 1964.

Laverghetta, Thomas S. *Microwave Materials and Fabrication Techniques,* 2d ed. Norwood, MA: Artech House, 1991.

Marcuvitz, N. *Waveguide Handbook.* New York: McGraw-Hill, 1951.

Matthaei, G. L., Young, L., and Jones, E. M. T. *Microwave Filters, Impedance-Matching Networks, and Coupling Structures.* Dedham, MA: Artech House, 1980.

Montgomery, C. G., et al. *Principles of Microwave Circuits.* New York: McGraw-Hill, 1948.

Phillips, A. B. *Transistor Engineering.* New York: McGraw-Hill, 1962.

Ragan, G. L. *Microwave Transmission Circuits.* New York: Dover Publications, 1965.

Rizzi, Peter A. *Microwave Engineering.* Englewood Cliffs, NJ: Prentice-Hall, 1988.

Saad, T. (ed.). *Microwave Engineers' Handbook.* Dedham, MA: Artech House, 1971.

Slater, J. C. *Microwave Electronics.* Princeton, NJ: Van Nostrand, 1950.

Smith, P. H. *Electronic Applications of the Smith Chart.* New York: McGraw-Hill, 1969.

Thomas, R. *A Practical Introduction to Impedance Matching.* Dedham, MA: Artech House, 1976.

Vendelin, G. D. *Design of Amplifiers and Oscillators by the S-Parameter Method.* New York: John Wiley, 1980.

Watson, H. A. *Microwave Semiconductor Devices and Their Circuit Applications.* New York: McGraw-Hill, 1968.

White, J. F. *Microwave Semiconductor Engineering.* New York: Van Nostrand Reinhold, 1982.

Young, V. J. *Understanding Microwaves.* New York: John F. Rider Co., 1965.

Index